QUARKS, BARYONS AND CHIRAL SYMMETRY

QUARKS, BARYONS AND CHIRAL SYMMETRY

Atsushi Hosaka
Research Center for Nuclear Physics
Osaka University, Japan

Hiroshi Toki
Research Center for Nuclear Physics
Osaka University, Japan

World Scientific
Singapore • New Jersey • London • Hong Kong

Published by

World Scientific Publishing Co. Pte. Ltd.
5 Toh Tuck Link, Singapore 596224
USA office: 27 Warren Street, Suite 401-402, Hackensack, NJ 07601
UK office: 57 Shelton Street, Covent Garden, London WC2H 9HE

British Library Cataloguing-in-Publication Data
A catalogue record for this book is available from the British Library.

QUARKS, BARYONS AND CHIRAL SYMMETRY

ISBN-13 978-981-02-4629-7
ISBN-10 981-02-4629-3

Preface

When the editor of the World Scientific offered us an opportunity to expand our previous review article on the chiral bag model [Hosaka and Toki 1996] into the form of a book, we thought that the topic was rather specific and wondered if it worth doing. However, while we were collecting series of works, we realized that the chiral bag model seems to contain many interesting physics which are quite common in various physical systems. Furthermore, the physics of baryons is now regarded as one of exciting subjects, where many new developments are being made both in theories and experiments.

We know that a number of excellent articles have already appeared dealing with topics related to baryon physics. They are the ones, for instance, by Kokkedee [1969], by Close [1979], by Thomas [1984], by Bhaduri [1988], and by Donoghue, Golowich and Holstine [1992]. While thinking about the structure of our book we have noticed that many of our colleagues, including ourselves often ask naive but very important questions related mostly to chiral symmetry. For instance; Why the axial-vector coupling constant of the nucleon is so close to unity? Why the chiral field can be expressed in the matrix form of U? What is the chiral partner of the nucleon? It is not difficult to see that these questions confuse us.

Therefore, we thought that we would like to produce a volume which will help answer such questions. Also we would like to write one useful, not only to graduate course students, but also to ourselves. In order to accommodate these requirements, we have decided to discuss materials exclusively related to baryon physics from quarks and chiral symmetry. But, for selected topics, we have tried to describe all the details starting from very fundamental issues.

Readers should have knowledge of quantum mechanics, and some basics of the field theory and the many-body problem. We do not assume that the readers have achieved complete understanding of the latter advanced subjects. But they should have some experiences with standard textbook materials.

For the discussion of baryon structure, the group theoretical method is particularly important. The reason is that many features of the strong interaction physics result from symmetries such as isospin, flavor and chiral symmetries. Another issue is the semiclassical nature of strongly and non-linearly interacting systems of quarks and pions. These two issues form the main body of this volume. The first four chapters describe introductory and elementary topics, while the latter chapters are rather devoted to applications. But, whenever a new topic is introduced the discussion starts with necessary fundamentals. We have also included problems in the end of the chapters, where readers can learn the derivations, details and related topics not sufficiently discussed in the text.

In this book we decided not to discuss the dynamics of color confinement explicitly. This is also a very exciting subject in hadron physics and a lot of new progress has been achieved lately. Still the concept of confinement is not so simple. Also, we did not discuss related subjects such as dynamical symmetry breaking, chiral anomaly, chiral perturbation theories and etc.

In editing this book, we were helped by Miho Takayama in many ways. Especially, we thank her for creating tables for quark model wave functions, baryon mass plots and baryon tables in chapters 3 and 8. We would like also to express our thanks to Veljko Dmitrasinovic who read through the manuscript and gave us useful suggestions. AH would like to thank the colleagues from Numazu College of Technology, where he worked out most of his part. Finally, we can not forget acknowledging all of our colleagues whom we list in the bibliography. Without their help this book would not have appeared.

January 2001

Numazu — Atsushi Hosaka
Osaka — Hiroshi Toki

Contents

Chapter 1

INGREDIENTS OF THE STANDARD MODEL

The main subject treated in this textbook, the physics of baryon dynamics, is rather specific as it is a small part of strong interaction physics. For instance, by now we know that there are six kinds of quarks with different flavors, whereas in the most part of this textbook we consider only three of them with light flavors. Yet it would be useful to briefly summarize some basic facts about the standard model for electroweak and strong interactions.

We do not attempt a pedagogic and thorough description in this chapter; some terminologies are introduced without enough explanations. Rather, we try to present here a flavor as how the standard model accommodates building blocks of matter and forces: quarks, leptons and gauge fields. Introduction of gauge fields has been accomplished based on the principle of local gauge symmetry, whereas flavors for quarks and leptons were discovered after phenomenological observation. For electroweak interaction the gauge principle with its spontaneous breakdown is essential, while for the strong interaction the non-abelian nature of the color gauge symmetry is essential. These aspects are briefly sketched. A concise but complete description is found in the latest volume of Review of Particle Physics from the Particle Data Group [PDG 2000].

1.1 Strong Interaction – QCD

The elementary fields for the physics of the strong interaction are quarks and gluons. Quarks q^a $(a = 1, 2, 3)$ are fermions in the fundamental representation of the color $SU(3)$ gauge group. Gluons A^a_μ $(a = 1, \ldots, 8)$

are the gauge fields in the octet (adjoint) representation. Furthermore, the quarks carry flavor degrees of freedom which are independent of the color. The flavors form generations each of which has two components. Until today, six flavors have been discovered and hence three generations, (u, d), (c, s) and (t, b). Some properties of these quarks are given in Table 2.1 in the next chapter.

The lagrangian of QCD is given by

$$L_{\rm QCD} = -\frac{1}{2}\,{\rm tr}\, G^{\mu\nu}G_{\mu\nu} + \sum_{f=u,\ldots,b} \bar{q}_f\,(i\not{D} - m)\, q_f\,. \tag{1.1}$$

Here

$$\begin{aligned} G_{\mu\nu} &= \partial_\mu A_\nu - \partial_\nu A_\mu - ig[A_\mu, A_\nu]\,, \\ D_\mu &= \partial_\mu - igA_\mu\,, \\ A_\mu &= \sum_{a=1}^{8} \frac{\lambda^a}{2} A_\mu^a \end{aligned} \tag{1.2}$$

and λ^a are Gell-Mann matrices. The color $SU(3)$ gauge theory requires the gauge invariance under the following gauge transformation:

$$\begin{aligned} q_f &\to U(\theta)q_f\,, \\ A_\mu &\to U(\theta)A_\mu U^\dagger(\theta) + \frac{i}{g}U(\theta)\partial_\mu U^\dagger(\theta) \\ &= U(\theta)(A_\mu + \frac{i}{g}\partial_\mu)U^\dagger(\theta)\,, \end{aligned} \tag{1.3}$$

where $U(\theta) = \exp(i\lambda^a\theta_a)$. It is straightforward to verify that the covariant derivative $\not{D}$ and the field tensor $G_{\mu\nu}$ are transformed in the same way;

$$\not{D} \to U\not{D}U^\dagger\,, \quad G \to UGU^\dagger\,. \tag{1.4}$$

If we consider only the strong interaction, the mass matrix is diagonal in flavor space:

$$m = {\rm diag}(m_f) = (m_u, m_d, \ldots m_b)\,. \tag{1.5}$$

When the weak interaction is switched on, however, the eigenstates of the weak and strong interactions are different. As a consequence, the mass term involves the Cabbibo-Kobayashi-Maskawa (CKM) mixing matrix [Cabbibo 1963; Kobayashi and Maskawa 1973].

A distinguished feature of QCD may be expressed by the following two words: asymptotic freedom and infrared slavery [Gross and Wilczek 1973; Politzer 1973; Politzer 1974]. They can be characterized by the momentum (scale) dependent coupling constant g when the theory is renormalized. By calculating the one-loop contributions to gluon self energy and vertex as shown in Fig. 1.1, one can extract the β function

$$\beta(g) = -\frac{1}{16\pi^2}(11 - (2/3)N_f)g^3 \equiv bg^3 . \tag{1.6}$$

for the renormalization group equation at the one-loop level:

$$\frac{dg}{dt} = -bg^3 . \tag{1.7}$$

Here the scale parameter t is defined by the momentum scale parameter

$$p \to \lambda p , \quad t = \ln \lambda . \tag{1.8}$$

Eq. (1.7) is an ordinary differential equation and can be solved easily:

$$g^2(t) = \frac{g(0)^2}{1 + 2bg(0)^2 t} . \tag{1.9}$$

Defining $\lambda^2 = Q^2/\mu^2$ and so $t = (1/2)\ln Q^2/\mu^2$, we obtain

$$\alpha(Q^2) = \frac{\alpha(\mu^2)}{1 + 4\pi b\alpha(\mu^2)\ln(Q^2/\mu^2)} \tag{1.10}$$

or, defining $\ln \Lambda^2 = \ln \mu^2 - 1/(\alpha(\mu^2)4\pi b)$, we can write

$$\alpha(Q^2) = \frac{4\pi}{(11 - (2/3)N_f)\ln(Q^2/\Lambda^2)} . \tag{1.11}$$

In Fig. 1.2, we plot the momentum dependent QCD coupling constant as a function of Q for $\Lambda \sim$ a few hundred MeV. Actual value of the scale parameter depends on both theoretical (order of perturbation, renormalization scheme, etc) and experimental analysis [PDG 2000]. For instance, typical value determined by the scaling violation in deep inelastic scatterings [Muta 1997] is about five hundred MeV. Other analyses predict usually smaller value around two hundred MeV. Obviously the coupling constant becomes small as the momentum Q is increased, while it increases as Q is decreased. In particular the coupling constant diverges at $Q = \Lambda$. For $\alpha(Q^2) \gtrsim 1$ the perturbation method breaks down and the non-perturbative dynamics becomes dominant. We do not discuss the details of this very interesting

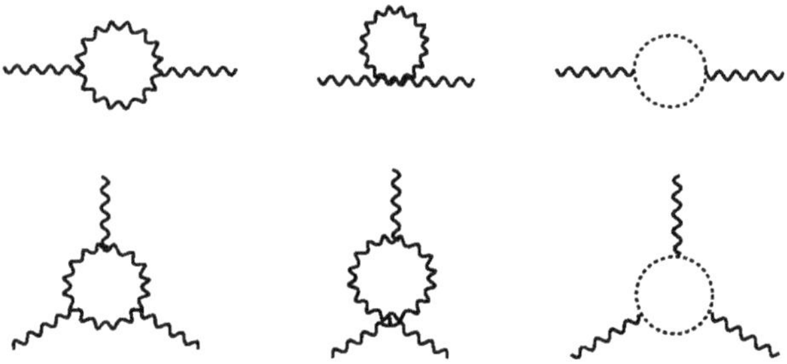

Fig. 1.1 One loop contributions to the gluon self energy and vertex functions. Both of them contribute to the β function. Here the wavy lines represent gluons while the dashed lines ghost field.

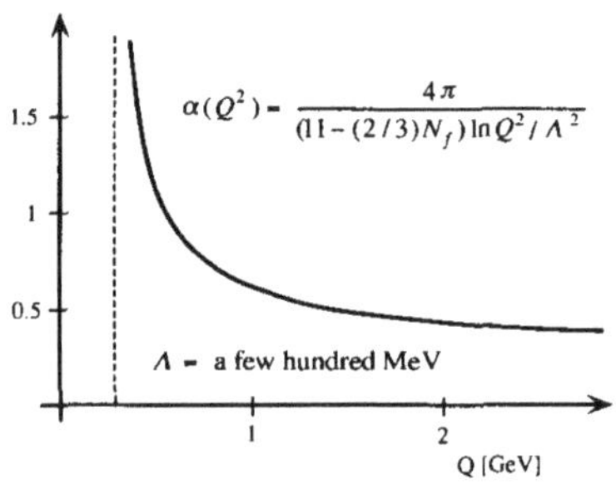

Fig. 1.2 Effective strong coupling constant as a function of Q.

aspect now, except for mentioning that chiral symmetry breaking and color confinement are the two dominant features of the non-perturbative QCD.

1.2 Electroweak Theory

The theory of the electroweak interaction is based on the gauge theory of $SU(2)_L \times U(1)_Y$ [Weinberg 1967a; Salam 1968; Glashow et al. 1970], which is spontaneously broken down to the electromagnetic $U(1)$ subgroup. Here we outline essences of the theory. The lagrangian of the electroweak theory consists of four terms:

$$L_{EW} = L_{gauge} + L_{fermion} + L_{Higgs} + L_{Yukawa}\,. \tag{1.12}$$

The gauge part contains two terms for the $SU(2)_L$ and $U(1)_Y$ gauge fields:

$$L_{gauge} = -\frac{1}{2}\,\mathrm{tr}\, F^{\mu\nu}F_{\mu\nu} - \frac{1}{4}B^{\mu\nu}B_{\mu\nu}\,, \tag{1.13}$$

where

$$\begin{aligned} F_{\mu\nu} &= \sum_{a=1,2,3} \frac{\tau_a}{2} F^a_{\mu\nu}\,, \\ F^a_{\mu\nu} &= \partial_\mu A^a_\nu - \partial_\nu A^a_\mu - ig_2\epsilon_{abc}A^b_\mu A^c_\nu\,, \\ B_{\mu\nu} &= \partial_\mu B_\nu - \partial_\nu B_\mu\,, \end{aligned}$$

where ϵ_{abc} the totally antisymmetric tensor which is the structure constants of the $SU(2)$ group, and g_2 is the gauge coupling constant of $SU(2)_L$.

The fermion term is given by

$$\begin{aligned} L_{fermion} &= \sum_{generation}^{3} (\bar{l}_L i\not{D} l_L + \bar{q}_L i\not{D} q_L) \\ &+ \sum_{e_R, q_R} (\bar{e}_R i\not{D} e_R + \bar{q}_R i\not{D} q_R)\,. \end{aligned} \tag{1.14}$$

The left handed fermions form doublet (fundamental) representations of $SU(2)_L$ for each three generations, while the right handed fermions are $U(1)_Y$ singlet *:

$$\begin{aligned} l_L &= \begin{pmatrix} \nu_e \\ e_L \end{pmatrix}, \begin{pmatrix} \nu_\mu \\ \mu_L \end{pmatrix}, \begin{pmatrix} \nu_\tau \\ \tau_L \end{pmatrix} \qquad (1.15) \\ q_L &= \begin{pmatrix} u_L \\ d_L \end{pmatrix}, \begin{pmatrix} c_L \\ s_L \end{pmatrix}, \begin{pmatrix} t_L \\ b_L \end{pmatrix}, \\ q_R &= u_R,\ d_R,\ c_R,\ s_R,\ t_R,\ b_R\,, \\ e_R &= e_R,\ \mu_R,\ \tau_R\,. \qquad (1.16) \end{aligned}$$

The covariant derivatives are given by

$$\begin{aligned} D_\mu \psi_L &= \left(\partial_\mu + i\frac{g_2}{2}\boldsymbol{\tau}\cdot\boldsymbol{A}_\mu + i\frac{g_1}{2}Y_W B_\mu\right)\psi_L\,, \\ D_\mu \psi_R &= \left(\partial_\mu + i\frac{g_1}{2}Y_W B_\mu\right)\psi_R\,, \end{aligned}$$

where g_1 is the gauge coupling constant for $U(1)_Y$, and Y_W is the weak hyper charge of the fermions.

*For the definition of the left and right handed fermions, see section 3.1.

The Higgs field is a bosonic doublet of $SU(2)_L$ and the lagrangian contains the kinetic and the potential terms:

$$L_{Higgs} = (D_\mu \Phi^\dagger)(D^\mu \Phi) - V(|\Phi|^2)\,, \tag{1.17}$$

where

$$\begin{aligned} \Phi &= \begin{pmatrix} \phi^\dagger \\ \phi^0 \end{pmatrix}, \\ D_\mu \Phi &= \left(\partial_\mu - \frac{i}{2} g_2 \boldsymbol{\tau}\cdot \boldsymbol{A}_\mu - \frac{i}{2} g_1 B_\mu\right)\Phi \\ V(\Phi) &= -\mu^2 \Phi^\dagger \Phi + (\Phi^\dagger \Phi)^2\,. \end{aligned} \tag{1.18}$$

Finally the Yukawa term provides the coupling between the fermion and Higgs fields:

$$L_{Yukawa} = \sum_{u,d,s} \left(-f_u \bar{q}_L \tilde{\Phi} u_R - f_d \bar{q}_L \Phi d_R - f_e \bar{l}_L \Phi e_R\right)\,, \tag{1.19}$$

where the isospin transformed Higgs field $\tilde{\Phi} = i\tau_2 \Phi$ is needed to provide the right form of the quark mass term. Furthermore, the sum over u, d, e is understood to run over $u \equiv (u, c, t)$, $d \equiv (d, s, b)$ and $e \equiv (e, \mu, \tau)$, respectively. q_L is understood to be a doublet whose upper component has the same flavor as the right quark $u_R, d_R, \cdots$. In Eq. (1.19), the left handed quarks $\bar{q}_L$ and the Higgs field $\tilde{\Phi}$ (or Φ) form an isospin singlet while $\bar{q}_L$ and u (or d) form a Dirac scalar. The similar rule is also adopted for leptons.

The electroweak theory induces spontaneous braking due to the self interaction of the Higgs field. When $\mu^2 > 0$, the Higgs potential takes the mexican hat shape and the vacuum can be chosen at an arbitrary point on the bottom of the potential. By a gauge transformation, the point can be always put on the ϕ^0 axis, to be consistent with the chargeless vacuum. In other words, we can take the vacuum at

$$\Phi = (0, v/\sqrt{2})\,. \tag{1.20}$$

This is the starting point of the Higgs mechanism [Higgs 1964; Higgs 1966]. In the spontaneous breaking of the gauge theory, the would-be Nambu-Goldstone mode corresponding to the zero energy mode along the potential minima is absorbed into the longitudinal component of the massive gauge fields. As a consequence, out of the four gauge fields A^a_μ and B_μ, three of them become massive and one massless gauge field remains, which is

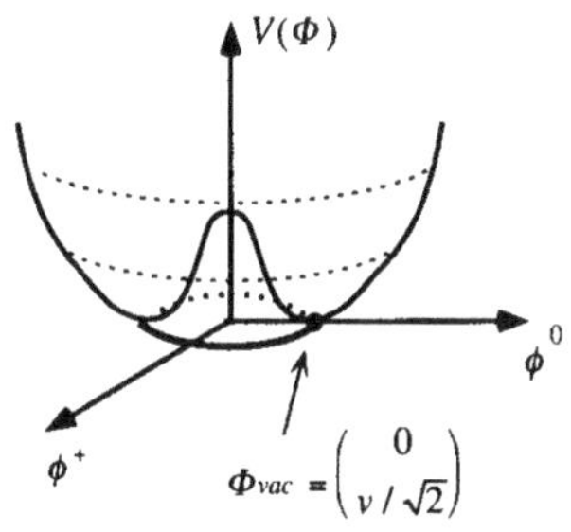

Fig. 1.3 Potential of the Higgs field and spontaneous symmetry breaking.

identified with the photon. After the spontaneous breaking of Eq. (1.20), the Yukawa term Eq. (1.19) yields the quark mass term as

$$L_{Yukawa} = -\sum_{u,d,e} \left(m_u \bar{u}_L u_R + m_d \bar{d}_L d_R + m_e \bar{e}_L e_R + h.c. \right) . \tag{1.21}$$

The masses of quarks are then defined by the vacuum expectation value of the Higgs field by

$$m = \frac{fv}{\sqrt{2}} . \tag{1.22}$$

1.3 CKM Mass Matrix

The interaction part of the lagrangian $L_{fermion}$ of Eq. (1.14) is diagonal in generation space and therefore it can not describe the actual processes of for instance strangeness changing weak decays. This can be accomplished by introducing a unitary matrix which mixes lower (or equivalently upper) components of different generations. Such a mixing matrix is now called the Cabbibo-Kobayashi-Maskawa (CKM) matrix [Kobayashi and Maskawa 1973]. There were three major progresses in establishing the present form of the CKM matrix. Originally, Cabbibo introduced the single mixing angle for $\Delta S = 1$ transitions of $s \to u$. This was the story within three flavors of u, d, s. Then the fourth quark c was introduced by Glashow, Iliopoulos and Maiani in order to explain the absence of the flavor changing neutral current (GIM mechanism [Glashow et al. 1970]). Finally three generations were completed by Kobayashi and Maskawa for CP violation processes.

The 3×3 matrix V mixes d, s and b quarks and modifies the interaction term of the quark charged current as:

$$\begin{aligned} L_{fermion} &= \sum_{generation}^{3} (\bar{l}_L i\not{D} l_L + \bar{q}_L i\not{D} q_L) \\ &\quad + \sum_{e_R, q_R} (\bar{e}_R i\not{D} e_R + \bar{q}_R i\not{D} q_R) + \cdots \\ &\to \bar{\nu} i\not{\partial} \nu + \bar{e}_L i\not{\partial} e_L - \frac{g_2}{2} \bar{\nu} \gamma^\mu e_L A_\mu^+ \\ &\quad + \bar{u}_L i\not{\partial} u_L + \bar{d}_L i\not{\partial} d_L - \frac{g_2}{2} \bar{u}_L \gamma^\mu V d_L A_\mu^+ . \end{aligned} \tag{1.23}$$

Here on the last line

$$V d_L \sim \begin{pmatrix} V_{ud} & V_{us} & V_{ub} \\ V_{cd} & V_{cs} & V_{cb} \\ V_{td} & V_{ts} & V_{tb} \end{pmatrix} \begin{pmatrix} d_L \\ s_L \\ b_L \end{pmatrix} . \tag{1.24}$$

Let us briefly discuss the reason that the matrix V enters only the quark part but not lepton part. In general one can introduce the unitary transformed field for all fermions including leptons and quarks,

$$\begin{aligned} u_{iL} &= \sum_j S_{ij}^{(u)} u'_{jL} , \quad u_{iR} = \sum_j T_{ij}^{(u)} u'_{jR} , \\ d_{iL} &= \sum_j S_{ij}^{(d)} d'_{jL} , \quad d_{iR} = \sum_j T_{ij}^{(d)} d'_{jR} , \\ \nu_{iL} &= \sum_j S_{ij}^{(\nu)} \nu'_{jL} , \\ e_{iL} &= \sum_j S_{ij}^{(e)} e'_{jL} , \quad e_{iR} = \sum_j T_{ij}^{(e)} e'_{jR} . \end{aligned}$$

Here all matrices S and T are unitary. Using these primed fields, one can rewrite the fermion term as

$$\begin{aligned} L_{fermion} &\sim \bar{\nu}' i\not{\partial} \nu' + \bar{e}'_L i\not{\partial} e'_L - \frac{g_2}{2} \bar{\nu}' \gamma^\mu S^{(\nu)} S^{(e)\dagger} e'_L A_\mu^+ \\ &+ \bar{u}'_L i\not{\partial} u'_L + \bar{d}'_L i\not{\partial} d'_L - \frac{g_2}{2} \bar{u}'_L \gamma^\mu S^{(u)} S^{(d)\dagger} d'_L A_\mu^+ . \end{aligned} \tag{1.25}$$

Now if the masses of the neutrinos were zero (this actually is not likely the case from i) the discrepancy between the observed solar neutrino and that of the standard solar model [Fukuda et al. 1999], and ii) the asymmetry in the atmospheric neutrinos [Fukuda et al. 1998]. Both neutrinos are detected by the underground Super-Kamiokande detector [PDG 2000], see the web page of Super-Kamiokande), one is able to redefine the neutrino fields by $\bar{\nu}' S^{(\nu)} S^{(e)\dagger} \to \bar{\nu}'$ and remove the off diagonal components from the lepton part (the third term of Eq. (1.25)). On the other hand, the quark mixing $S^{(u)} S^{(d)\dagger}$ can not be discarded due to the non-zero mass term. Identifying the primed fields with mass eigenstates of the strong interaction, writing $S^{(d)\dagger} S^{(u)}$ as V and removing the primes from Eq. (1.25), we find the desired lagrangian of Eq. (1.23).

Finally it is convenient to show the parameterization of the CKM matrix. In general an $n \times n$ unitary matrix can be parameterized in terms of n^2 real numbers. Out of them, $n(n-1)/2$ are angles and $n(n+1)$ are phases. Among the phases, $2n$ quark fields can absorb $2n-1$ phases (one overall phase can not be independent), leaving $\frac{1}{2}(n-1)(n-2)$ phases. Therefore, when $n = 3$, there are three angles and one phase, which parameterize the CKM matrix. One realization is due to the original work by Kobayashi and Maskawa who represented

$$
\begin{aligned}
V \;=\; & \begin{pmatrix} 1 & 0 & 0 \\ 0 & c_2 & -s_2 \\ 0 & s_2 & c_2 \end{pmatrix} \begin{pmatrix} c_1 & -s_1 & 0 \\ s_1 & c_1 & 0 \\ 0 & 0 & 1 \end{pmatrix} \\
& \times \begin{pmatrix} 1 & 0 & 0 \\ 0 & 1 & 0 \\ 0 & 0 & -e^{-i\delta} \end{pmatrix} \begin{pmatrix} 1 & 0 & 0 \\ 0 & c_3 & s_3 \\ 0 & -s_3 & c_3 \end{pmatrix} . \qquad (1.26)
\end{aligned}
$$

The complex phase in the matrix elements induces CP (or T) violation in the standard theory. One way to determine the parameters in the mass matrix is the context of the chiral perturbation theory, which is one of current topics of the strong interaction physics [Gasser and Leutwyler 1982; Gasser and Leutwyler 1985; Gasser and Leutwyler 1984; Ecker 1995; Pich 1995].

Chapter 2

SYMMETRIES AND WAVE FUNCTIONS

Symmetry plays a central role in modern physics. From condensed matter to particle physics, many discoveries were related to symmetry. Baryon physics is no exception. Here and in subsequent chapters, we consider mostly a class of what are called global internal symmetries.

The primary use of the internal symmetry was recognized in particle classification, where mesons and baryons were grouped into approximately degenerate multiplets of a symmetry group. The fact that observed states seem to belong to higher dimensional representations led to the introduction of quarks as their constituents. Of particular importance are $SU(2)$ and $SU(3)$ for light flavors.

Our ultimate goal here is to write down the baryon wave functions in the quark model. We introduce ideas and tools that are necessary for that purpose. After a brief illustration of how symmetry has been recognized to be important, we discuss basic properties of $SU(2)$ and $SU(3)$. The method of highest weight is adopted, since it leads naturally to polygonal diagrams for group representations. Then we discuss permutation symmetry for two and three-particle systems. Performing inner-products of spin, isospin and orbital wave functions, we finally write down the baryon wave functions.

Our description here is not very general, but leads straightforwardly to the baryon wave functions. Readers can learn not only techniques but also symmetry ideas by explicitly working out quark model wave functions.

2.1 Why is Symmetry Important?

Particle degeneracy

Let us first take a brief look at particle degeneracy associated with symmetry. We denote a group by a capital letter G and its element by a small letter g, e.g., $g \in G$. When a system has a symmetry, the hamiltonian H of the system commutes with every element g of the group which is associated with the symmetry:

$$[H, g] = 0\,. \tag{2.1}$$

Consider an eigenstate $|i\rangle$ of the hamiltonian:

$$H|i\rangle = E_i|i\rangle\,. \tag{2.2}$$

Then Eq. (2.1) implies that

$$gH|i\rangle = H(g|i\rangle) = E_i(g|i\rangle) \tag{2.3}$$

and that the g-transformed state $g|i\rangle$ is also an eigenstate of H with the same energy E_i. Of particular importance are the irreducible representations of the group G, since they are generally observed as degenerate states in the physical world *.

Eight-fold way and introduction of quarks

If observed states belong to a higher dimensional representation, it is likely that they are composites of lower dimensional states. Noticing this, Gell-Mann and Zweig introduced quarks as constituents of mesons and baryons [Gell-Mann and Ne'eman 1964; Zweig 1963]. In baryon physics many of the dynamical properties are dictated by symmetry properties among quarks.

By the end of 1950's, eight baryons of spin one-half were discovered. Their masses are similar; all of them are within 20 % of their mean value. Similarly, eight pseudo-scalar mesons were known, although their masses are not as similar as baryon masses. The number eight was considered too large for these mesons or baryons to be fundamental. It was then natural to develop an idea in which they emerged as composite states of fewer fundamental elements.

*If the symmetry is broken spontaneously, such degeneracy is not observed. This will be considered in Chapter 3.

Historically, two possibilities were proposed. The first one was Sakata's idea in which he considered p, n and Λ as the fundamental particles and other mesons and baryons were their composites. Although this was on the right track from the group theoretical point of view, this idea did not meet with much success. The subsequent idea which was proposed independently by Gell-Mann and by Zweig has eventually turned out to be correct. They introduced three fundamental constituents for the octet mesons and baryons. We call them now up (u), down (d) and strange (s) quarks. For more history, see, for instance, Refs. [Nambu 1985; Han 1999].

Now the Gell-Mann-Nishijima-Nakano relation holds for hadronic processes [Gell-Mann 1953; Nishijima and Nakano 1953]:

$$Q = I_3 + \frac{Y}{2}, \quad Y = B + S, \tag{2.4}$$

where Q, I_3, Y, B and S are the charge, the third component of isospin, hyper-charge, baryon number and strangeness, respectively. Under the strong interaction, each of these quantities is conserved. However, only the charge Q and the baryon number B are conserved in weak interaction processes.

Having conservation of I_3 and Y separately, together with isospin symmetry, one can guess the underlying symmetry

$$SU(2)_{\text{isospin}} \times U(1)_{\text{hypercharge}} \,. \tag{2.5}$$

This group, however, has many low dimensional representations which are not observed in nature.

It is natural to extend the group of Eq. (2.5) to $SU(3)$ which contains it as a subgroup:

$$SU(3) \supset SU(2) \times U(1) \,. \tag{2.6}$$

The particles associated with the three dimensional fundamental representation of flavor $SU(3)$ are identified with (constituent) quarks and are denoted as u, d and s. Although we have not seen quarks in their isolated form, in groups of two (quark-antiquark) or three (quark-quark-quark) they form higher dimensional representations eight and ten dimensional representations corresponding to observed mesons and baryons. As will be shown in section 2.4, the higher dimensional representations for mesons

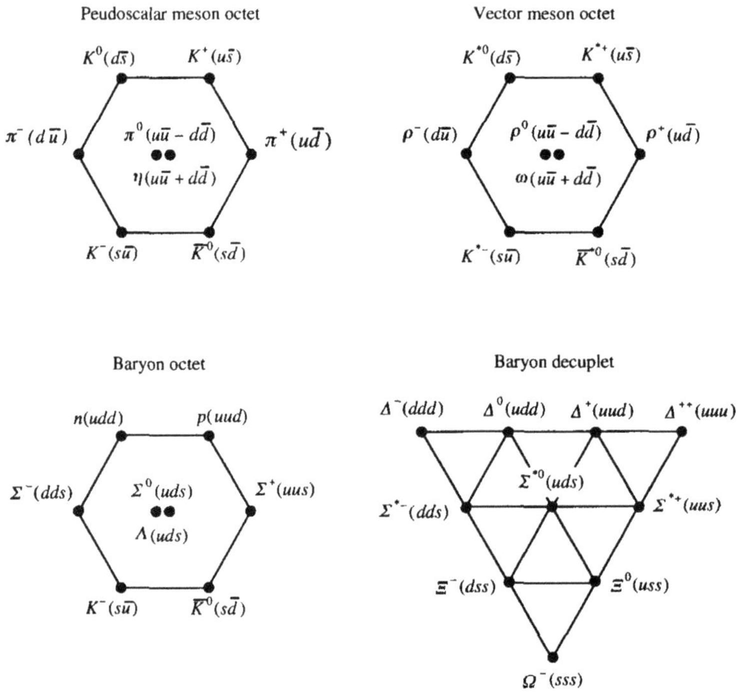

Fig. 2.1 Meson and baryon octets and baryon decuplet. Their quark contents are also shown.

and baryons are:

$$\begin{aligned} 3^* \times 3 &= 1 + 8\,, \\ 3 \times 3 \times 3 &= 1 + 8 + 8 + 10\,. \end{aligned}$$

The identification of octet (8) and decuplet (10) states with observed states is shown in Fig. 2.1.

Now, $U(1)$ charge is additive, and so it is assumed that all quarks carry the same baryon number one-third. On the other hand, strangeness S is carried only by the s-quark, $S^{u,d} = 0, S^s = -1$. The isospin of the quarks are assigned as $I^{u,d} = 1/2, I_z^u = +1/2, I_z^d = -1/2$ and $I^s = 0$. From these assignments and by using the Gell-Mann-Nishijima-Nakano relation (2.4)

Table 2.1 Properties of current quarks.

	mass	flavor	charge
u	$1.5 - 5$ MeV	$I_3 = 1/2$	$+2/3$
d	$3 - 9$ MeV	$I_3 = -1/2$	$-1/3$
	$m_u/m_d = 0.2 - 0.7$		
s	$60 - 170$ MeV	$S = -1$	$-1/3$
	$m_s/m_d = 17 - 25$		
c	$1.15 - 1.35$ GeV	$c = -1$	$+2/3$
b	$4.0 - 4.4$ GeV	$b = -1$	$-1/3$
t	174.3 ± 5.1 GeV[1)]	$t = -1$	$+2/3$
	$168.2^{+9.6}_{-7.4}$ GeV[2)]		

1) Direct observation of top events
2) Standard model electroweak fit

we find fractional charges of quarks:

$$\begin{aligned} Q^u &= \frac{1}{2} + \frac{1}{2} \times \frac{1}{3} = \frac{2}{3}, \\ Q^d &= -\frac{1}{2} + \frac{1}{2} \times \frac{1}{3} = -\frac{1}{3}, \\ Q^s &= 0 + \frac{1}{2} \times \left(\frac{1}{3} - 1\right) = -\frac{1}{3}. \end{aligned}$$

Although it has been confirmed that these fractional charges are carried by quarks inside hadrons in deep inelastic scatterings, we have not observed them in an isolated form. At low energies, quarks (or generally colors) are believed to be permanently confined in hadrons, and what we observe are integer charges.

Up to the present date, we have found six quarks of different flavors. They are summarized in Table 2.1. All quarks carry the baryon number one-third and spin one-half. In this text our interest is focused on the light flavor sector of u, d and s quarks.

2.2 Symmetry Current

Symmetry entails the presence of conserved currents [Noether 1918]. To see this we study briefly the Noether's theorem. Let us consider a Lie group where group elements are labeled by continuous parameters. (For more details of the group theory, see, for instance, the textbook by Georgi [Georgi 1982].) Elements g of the group G are generated by a set of generators $\{t_a, a = 1, \ldots M\}$ with real parameters $\{\alpha_a, a = 1, \ldots M\}$ as

$$g(\alpha) = \exp(i \sum \alpha_a t^a) . \tag{2.7}$$

Properties of the group are determined by the commutation relations among the generators (Lie algebra):

$$[t_a, t_b] = i f_{abc} t_c . \tag{2.8}$$

When the generators $\{t_i\}$ are hermitian and the parameter space α is bounded, the group is called compact. For instance, the unitary group $U(N)$ is compact.

Consider a set of (complex) field $\phi(x) \equiv (\phi^a(x); a = 1, \cdots N)$, which is assumed to form a linear representation of a continuous group G. For an element $g(\alpha) \in G$, the linear representation ϕ is transformed as

$$\phi(x) \to D(g(\alpha))\, \phi(x) , \tag{2.9}$$

where $D(g(\alpha))$ is an $N \times N$ representation matrix acting on the N-component ϕ vector. Here we consider the unitary group, where the matrix $D(g(\alpha))$ satisfies the unitary condition:

$$D^{-1} = D^\dagger , \quad (D^\dagger)_{ij} = (D_{ji})^* . \tag{2.10}$$

In what follows, we use g and t for representation matrices also, for notational simplicity:

$$\begin{aligned} D(t_a) &\to t_a , \\ D(g(\alpha)) &\to g(\alpha) . \end{aligned} \tag{2.11}$$

Let us write the lagrangian for $\phi(x)$ as

$$\mathcal{L}(\phi) \equiv \mathcal{L}(\phi(x), \partial_\mu \phi(x)) . \tag{2.12}$$

We say that the system has a symmetry when $\mathcal{L}(\phi)$ is invariant under the group transformation $g \in G$:

$$\mathcal{L}(\phi) = \mathcal{L}(\phi' = g(\alpha)\phi) . \tag{2.13}$$

For an infinitesimal transformation of Eq. (2.7), we have

$$\phi' \sim (1 + i\alpha_a t^a)\phi . \tag{2.14}$$

The invariance of Eq. (2.13) implies that the variation of order α_a vanishes:

$$\delta\mathcal{L} = \frac{\partial\mathcal{L}}{\partial\phi} i\alpha_a t^a\phi + \frac{\partial\mathcal{L}}{\partial(\partial_\mu\phi)} i\alpha_a t^a \partial_\mu\phi = 0 \tag{2.15}$$

Using the (Euler-Lagrange) equation of motion

$$\partial_\mu \frac{\partial\mathcal{L}}{\partial(\partial_\mu\phi)} = \frac{\partial\mathcal{L}}{\partial\phi}, \tag{2.16}$$

one finds that

$$\alpha_a\partial_\mu \left(i\frac{\partial\mathcal{L}}{\partial(\partial_\mu\phi)} t^a\phi \right) = 0. \tag{2.17}$$

Since α_a are arbitrary, this equation implies the conservation of each component of the current J^a_μ:

$$\partial^\mu J^a_\mu = 0\,, \quad J^a_\mu = -i\frac{\partial\mathcal{L}}{\partial(\partial_\mu\phi)} t^a\phi\,. \tag{2.18}$$

The spatial integral of the zeroth component of the current has a special meaning in the quantum theory. It defines the charge operator

$$\begin{aligned} Q^a &\equiv \int d^3x\, J^a_0 \\ &= -\int d^3x\, i\frac{\partial\mathcal{L}}{\partial(\partial_0\phi)} t^a\phi \\ &= -i\int d^3x\, \pi(x) t^a\phi(x)\,, \end{aligned} \tag{2.19}$$

where we have used the definition of the canonical momentum

$$\pi(x) = \frac{\partial\mathcal{L}}{\partial(\partial_0\phi)}\,. \tag{2.20}$$

Then Q^a can be regarded as a generator of the group transformation for the field operator ϕ:

$$\begin{aligned} & i[Q^a, \phi(x)] \\ =\; & \left[\int d^3y\, \pi(y) t^a\phi(y), \phi(x)\right] \\ =\; & -it^a\phi(x)\,, \end{aligned} \tag{2.21}$$

where we have used the equal-time commutation relations

$$[\phi^a(x_0, \boldsymbol{x}), \pi^b(y_0 = x_0, \boldsymbol{y})] = i\delta(\boldsymbol{x} - \boldsymbol{y})\delta^{ab} . \tag{2.22}$$

Example 1: U(1) current of a complex scalar field

This is associated with a phase transformation for a complex scalar field $\phi = \phi_1 + i\phi_2$:

$$\phi \to e^{i\alpha}\phi , \quad \phi^* \to e^{-i\alpha}\phi^* , \tag{2.23}$$

where the only one generator of $U(1)$ is 1. The lagrangian density for a free field is given by

$$\begin{aligned} \mathcal{L} &= \frac{1}{2}\partial_\mu\phi^*\partial^\mu\phi - \frac{1}{2}\mu^2\phi^*\phi \\ &= \frac{1}{2}\left((\partial_\mu\phi_1)^2 + (\partial_\mu\phi_2)^2\right) - \frac{\mu^2}{2}\left(\phi_1^2 + \phi_2^2\right) . \end{aligned} \tag{2.24}$$

Thus, by setting simply $t^a = 1$ (and -1 for ϕ^*) in Eq. (2.18), we find the $U(1)$ current

$$\begin{aligned} J_\mu^a &= -i\frac{\partial\mathcal{L}}{\partial(\partial_\mu\phi)}1\phi - i\frac{\partial\mathcal{L}}{\partial(\partial_\mu\phi^*)}(-1)\phi^* \\ &= -\frac{i}{2}(\partial_\mu\phi^*)\phi + \frac{i}{2}\phi^*\partial_\mu\phi \\ &= -(\partial_\mu\phi_2)\phi_1 + \phi_2(\partial_\mu\phi_1) . \end{aligned} \tag{2.25}$$

The associated charge is then given by

$$Q = \int d^3x \left(-\frac{i}{2}(\partial_t\phi^*)\phi + \frac{i}{2}\phi^*\partial_t\phi\right) . \tag{2.26}$$

As expected, $Q^* = -Q$ which means that the complex conjugate field (anti-particle) carries the charge of opposite sign.

Example 2: Isospin current of a fermion

Isospin is a part of flavor symmetry for two component spinors. An isodoublet fermion $\psi = (\psi_1, \psi_2)$ forms the fundamental representation of $SU(2)$. Thus ψ_1 and ψ_2 transform under an $SU(2)$ transformation as

$$\begin{pmatrix} \psi_1 \\ \psi_2 \end{pmatrix} \longrightarrow \exp\left(i\frac{1}{2}\boldsymbol{\tau}\cdot\boldsymbol{\alpha}\right)\begin{pmatrix} \psi_1 \\ \psi_2 \end{pmatrix}$$

$$\xrightarrow{\text{infinitesimal}} \left(1 + i\frac{1}{2}\boldsymbol{\tau}\cdot\boldsymbol{\alpha}\right)\begin{pmatrix}\psi_1\\ \psi_2\end{pmatrix}, \tag{2.27}$$

where the generators of $SU(2)$ are Pauli matrices (see Eq. (2.35))

$$t_i = \frac{\tau_i}{2}\,. \tag{2.28}$$

The lagrangian density for the free fermion field is

$$\mathcal{L} = \bar{\psi}(i\partial\!\!\!/ - m)\psi\,, \tag{2.29}$$

and therefore, the isospin current is given by

$$J^a_\mu = -i\frac{\partial\mathcal{L}}{\partial(\partial_\mu\psi)}\frac{\tau^a}{2}\psi = \bar{\psi}\gamma_\mu\frac{\tau^a}{2}\psi\,. \tag{2.30}$$

The associated charge is then given by

$$Q^a = \int d^3x\left(\psi^\dagger\frac{\tau^a}{2}\psi\right), \tag{2.31}$$

which is nothing but the familiar isospin charge.

2.3 SU(2)

This unimodular unitary group is the most familiar symmetry and frequently appears in hadron physics, as it describes not only spin but also isospin symmetry between the proton and neutron, and of three charged states of the pion.

Basics of SU(2)

The $SU(2)$ matrices are complex valued 2×2 matrices:

$$U = \begin{pmatrix} u^1_{\ 1} & u^1_{\ 2}\\ u^2_{\ 1} & u^2_{\ 2}\end{pmatrix} \tag{2.32}$$

with the constraints

$$U^\dagger = U^{-1}\,,\quad \det U = 1\,. \tag{2.33}$$

By these constraints the matrix U can be parameterized by three real parameters. As usual, we generate the $SU(2)$ matrix in terms of the traceless hermitian matrices, which are identified with the generators of the $SU(2)$ group:

$$U = \exp\left(i\sum_{i=1}^{3}\phi^i t_i\right), \tag{2.34}$$

where the generators are given by the Pauli matrices, $t_i = \tau_i/2$:

$$\tau_1 = \begin{pmatrix} 0 & 1 \\ 1 & 0 \end{pmatrix}, \quad \tau_2 = \begin{pmatrix} 0 & -i \\ i & 0 \end{pmatrix}, \quad \tau_3 = \begin{pmatrix} 1 & 0 \\ 0 & -1 \end{pmatrix}. \tag{2.35}$$

The generators t_i satisfy the commutation relations

$$[t_i, t_j] = i f_{ijk} t_k , \tag{2.36}$$

where the structure constants are $f_{ijk} = \epsilon_{ijk}$. They determine the properties of $SU(2)$. Another useful and equivalent expression for the $SU(2)$ matrix is

$$U = a_0 + i\boldsymbol{\tau}\cdot\boldsymbol{a} \quad \text{with} \quad a_0^2 + \boldsymbol{a}^2 = 1 . \tag{2.37}$$

This is often used in the literature and discussed in Appendix C.

The fundamental representation of $SU(2)$ is a complex valued two component spinor, $\psi = (\psi^1, \psi^2)$, which transforms under $SU(2)$ transformations according to

$$\psi' = U\psi \quad \text{or} \quad (\psi')^i = \sum_{j=1}^{2} u^i{}_j \psi^j . \tag{2.38}$$

The quantities with upper indices are called contravariant spinors. The metric tensor can be introduced as $\epsilon_{ij} = \epsilon^{ij}$, which is the two-dimensional Levi-Civita tensor. Using this metric, upper (contravariant) indices are lowered (covariant) and lower indices are raised:

$$\psi_i = \sum_j \epsilon_{ij}\psi^j , \quad \text{etc}. \tag{2.39}$$

Consider the inner product of the two spinors (ψ^1, ψ^2) and (ϕ^1, ϕ^2)

$$\phi^{1*}\psi^1 + \phi^{2*}\psi^2 \equiv \phi^{i*}\psi^i , \tag{2.40}$$

where the repeated indices are dummies for summation. Obviously this quantity is $SU(2)$ invariant:

$$\begin{aligned}\phi^{i*}\psi^{i} &\to u^{i*}_{l}\phi^{l*}\,u^{i}_{\ m}\psi^{m}\\ &= (u^{-1})^{l}_{\ i}u^{i}_{\ m}\phi^{l*}\psi^{m}\\ &= \delta^{l}_{\ m}\phi^{l*}\psi^{m} = \phi^{l*}\psi^{l}\,.\end{aligned} \tag{2.41}$$

Also, the combination $\phi_i\psi^i = \epsilon_{ij}\phi^j\psi^i$ is invariant, for †

$$\begin{aligned}\epsilon_{ij}\phi^{i}\psi^{j} &\to \epsilon_{ij}u^{i}_{\ l}u^{j}_{\ m}\phi^{l}\psi^{m}\\ &= \det U\epsilon_{ij}\phi^{i}\psi^{j}\,,\end{aligned} \tag{2.42}$$

and $\det U = 1$.

Now the two equations (2.41) and (2.42) imply that the complex conjugate ψ^* transforms in the same way as ψ which is transformed by ϵ_{ij},

$$\psi^{i*} \sim \epsilon_{ij}\psi^{j}\,. \tag{2.43}$$

This property is called the reality of $SU(2)$. By this it is meant that the complex conjugate ψ^{i*} does not introduce any new representations which are independent of ψ^i (compare with $SU(3)$). Explicitly, there exists a unitary matrix S which satisfies

$$St^{a}S^{-1} = -t^{a*}\,. \tag{2.44}$$

By inspection, we find

$$S = \begin{pmatrix} 0 & 1 \\ -1 & 0 \end{pmatrix}. \tag{2.45}$$

Taking a complex conjugate of Eq. (2.36), one can verify that $-t^*_a$ satisfies the desired commutation relation

$$[-t^*_a, -t^*_b] = if_{abc}(-t^*_c)\,.$$

Representations

Representations of the $SU(n)$ group are characterized by the eigenvalues of the generators which can be simultaneously diagonalized. There are $n-1$ of such operators, where the number $n-1$ is called the rank of the group. In the case of $SU(2)$, there is only one which is usually taken to be t_3. In

†This is nothing but a proof of $\det AB = \det A \det B$.

the fundamental representation of the (non-trivial) lowest dimension, it is nothing but the third Pauli matrix, $t_3 = \tau_3/2$. Any higher dimensional representation may be constructed from tensor products of fundamental representations. Here, instead of doing so, we illustrate the method of the highest weight to generate irreducible representations of higher dimensions.

Let a state in a certain representation be labeled by an eigenvalue of t_3:

$$t_3|m\rangle = m|m\rangle\,. \tag{2.46}$$

Then the other two generators t_1 and t_2 behave as raising or lowering operators,

$$t_\pm = \frac{1}{\sqrt{2}}(t_1 \pm it_2) \tag{2.47}$$

which satisfy, together with t_3, the commutation relations

$$\begin{aligned} [t_3, t_\pm] &= \pm t_\pm\,, \\ [t_+, t_-] &= t_3\,. \end{aligned} \tag{2.48}$$

From the first equation, it follows that

$$\begin{aligned} t_3 t_\pm|m\rangle &= (t_\pm t_3 \pm t_\pm)|m\rangle \\ &= (m \pm 1)t_\pm|m\rangle\,. \end{aligned} \tag{2.49}$$

Hence the states $t_\pm|m\rangle$ are eigenstates of $|m \pm 1\rangle$.

Suppose that a set of $|m\rangle$ forms an M-dimensional representation, with the maximum value of $m \equiv j$. The quantity m is called the weight of the representation. Then by definition,

$$t_+|j\rangle = 0\,. \tag{2.50}$$

Applying the lowering operator t_- to $|m\rangle$, one finds

$$t_-|m\rangle = N_m|m-1\rangle\,, \tag{2.51}$$

where N_m is a normalization constant which is determined as follows. From Eq. (2.51), we find

$$\langle m-1|t_-|m\rangle = N_m$$

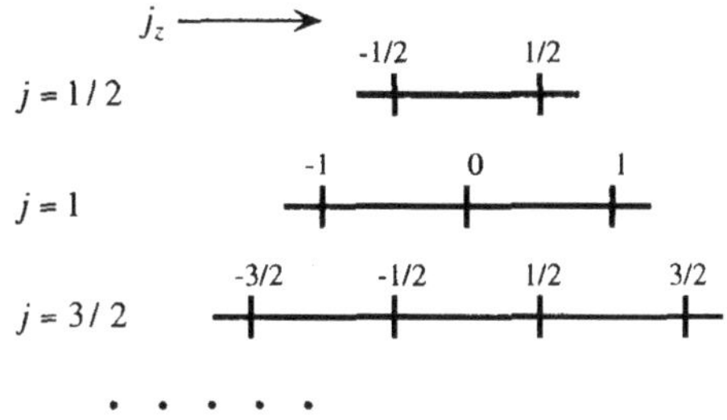

Fig. 2.2 One-dimensional (t_3) plot for $SU(2)$ representations.

and its complex conjugate

$$\langle m|t_+|m-1\rangle = N_m^* \,.$$

Therefore, choosing the phase of N_m appropriately, we have

$$\begin{aligned} t_-|m\rangle &= N_m|m-1\rangle \,, \\ t_+|m-1\rangle &= N_m|m\rangle \,, \end{aligned}$$

Taking the square of Eq. (2.51), we have

$$\begin{aligned} N_m^2 &= \langle m|t_+t_-|m\rangle \\ &= \langle m|t_-t_+|m\rangle + m \\ &= N_{m+1}^2 + m \,. \end{aligned} \tag{2.52}$$

This is a recursion formula for N_m and can easily be solved under the initial condition $N_j^2 = j$. The solution is

$$N_m = \sqrt{\frac{1}{2}(j-m+1)(j+m)} \,. \tag{2.53}$$

There are $2j$ coefficients N_m that are non-zero and real for $-(j-1) \leq m \leq j$. From Eq. (2.51), N_m appears when a state $|m-1\rangle$ is created from $|m\rangle$ by applying t_-. They are $|j-1\rangle, |j-1\rangle, \cdots, |-j\rangle$. Adding the initial state $|j\rangle$, the number (dimension) of total states is $M = 2j+1$. This completes the $M(=2j+1)$-dimensional representation of $SU(2)$. Of course, j corresponds to the total spin and m to the third component of the spin. The method demonstrated here, without using the properties of the Casimir operator $j^2 \equiv j_x^2 + j_y^2 + j_z^2$, can be extended to $SU(3)$.

It is convenient to show $SU(2)$ representations graphically, which are just one-dimensional plot as shown in Fig. 2.2. They are weight diagrams of $SU(2)$. The axis is labeled by the eigenvalues of the diagonal generator t_3.

2.4 SU(3)

There are $n^2 - 1 = 8$ generators of $SU(3)$, which are usually represented by the Gell-Mann matrices:

$$t_a = \frac{1}{2}\lambda_a \tag{2.54}$$

where

$$\begin{aligned}
\lambda_1 &= \begin{pmatrix} 0 & 1 & 0 \\ 1 & 0 & 0 \\ 0 & 0 & 0 \end{pmatrix} & \lambda_2 &= \begin{pmatrix} 0 & -i & 0 \\ i & 0 & 0 \\ 0 & 0 & 0 \end{pmatrix} \\
\lambda_3 &= \begin{pmatrix} 1 & 0 & 0 \\ 0 & -1 & 0 \\ 0 & 0 & 0 \end{pmatrix} & \lambda_4 &= \begin{pmatrix} 0 & 0 & 1 \\ 0 & 0 & 0 \\ 1 & 0 & 0 \end{pmatrix} \\
\lambda_5 &= \begin{pmatrix} 0 & 0 & -i \\ 0 & 0 & 0 \\ i & 0 & 0 \end{pmatrix} & \lambda_6 &= \begin{pmatrix} 0 & 0 & 0 \\ 0 & 0 & 1 \\ 0 & 1 & 0 \end{pmatrix} \\
\lambda_7 &= \begin{pmatrix} 0 & 0 & 0 \\ 0 & 0 & -i \\ 0 & i & 0 \end{pmatrix} & \lambda_8 &= \frac{1}{\sqrt{3}}\begin{pmatrix} 1 & 0 & 0 \\ 0 & 1 & 0 \\ 0 & 0 & -2 \end{pmatrix}
\end{aligned} \tag{2.55}$$

Then elements of the $SU(3)$ group are generated as

$$U(\alpha_1, \cdots, \alpha_8) = \exp(i\alpha_a t^a)\,, \quad a = 1, \cdots 8\,. \tag{2.56}$$

The generators are normalized as

$$\mathrm{tr}\,(t_a t_b) = \frac{1}{2}\delta_{ab} \tag{2.57}$$

and satisfy the following commutation and anti-commutation relations:

$$\begin{aligned}
[t_a, t_b] &= i f_{abc} t_c \\
\{t_a, t_b\} &= \delta_{ab} + d_{abc} t_c\,.
\end{aligned} \tag{2.58}$$

Table 2.2 Non-vanishing totally anti-symmetric and symmetric coefficients f and d.

f_{abc}		d_{abc}			
abc	f	abc	d	abc	d
123	1	118	$1/\sqrt{3}$	355	$1/2$
147	$1/2$	146	$1/2$	366	$-1/2$
156	$-1/2$	157	$1/2$	377	$-1/2$
246	$1/2$	228	$1/\sqrt{3}$	448	$-1/(2\sqrt{3})$
257	$1/2$	247	$-1/2$	558	$-1/(2\sqrt{3})$
345	$1/2$	256	$1/2$	668	$-1/(2\sqrt{3})$
367	$-1/2$	338	$1/\sqrt{3}$	778	$-1/(2\sqrt{3})$
458	$\sqrt{3}/2$	344	$1/2$	888	$-1/\sqrt{3}$
678	$\sqrt{3}/2$				

The totally anti-symmetric and symmetric coefficients f_{abc} and d_{abc} are tabulated in Table 2.2.

Just as in $SU(2)$, the fundamental representation of $SU(3)$ is subject to the transformation

$$\psi^{i\prime} = \sum_{j=1}^{3} u^i_{\ j}\, \psi^j \,, \tag{2.59}$$

where $u^i_{\ j}$ are components of a 3×3 special unitary matrix U of Eq. (2.56). Unlike the $SU(2)$ case, the $SU(3)$ representation (ψ^i) is no longer real, and the complex conjugate introduces another transformation:

$$\psi^{i*\prime} = \sum_{j=1}^{3} u^{i\ *}_{\ j}\, \psi^{j*} \tag{2.60}$$

which is independent of Eq. (2.59).

The inner product of two arbitrary fundamental representations ψ and φ is preserved:

$$(\varphi^\dagger \psi)' \to \varphi^\dagger U^\dagger U \psi = \varphi^\dagger \psi \,. \tag{2.61}$$

Another invariant is the determinant $\epsilon_{ijk}\phi^i \varphi^j \psi^k$, for

$$\begin{aligned} \epsilon_{ijk}\phi^i \varphi^j \psi^k &\to \epsilon_{ijk} u^i_{\ l} u^j_{\ m} u^k_{\ n} \phi^l \varphi^m \psi^n \\ &= \det U \epsilon_{ijk}\phi^i \varphi^j \psi^k \\ &= \epsilon_{ijk}\phi^i \varphi^j \psi^k \,. \end{aligned} \tag{2.62}$$

Here we have used $\det U = 1$. Thus the metric tensor is regarded as the totally anti-symmetric tensor ϵ_{ijk} $(i, j, k = 1, 2, 3)$. Comparing Eq. (2.61) and Eq. (2.62), we recognize that the complex conjugate ψ^{i*} behaves as

$$\psi^{i*} \sim \epsilon_{ijk}\varphi^j \phi^k \,. \tag{2.63}$$

Note that the right hand side is the anti-symmetric combination of the second rank tensor $\varphi^j \phi^k$. Irreducible decomposition of the second (and in general higher) rank tensors will be discussed in the subsequent sections.

Representations

Among eight generators of $SU(3)$, two of them can be diagonalized simultaneously. Therefore, the rank of $SU(3)$ is two. In the fundamental representation of Eq. (2.55) they are already given by t_3 and t_8. Therefore, $SU(3)$ representations are labeled by eigenvalues of t_3 and t_8. For a given representation, two eigenvalues define a point on a two-dimensional t_3-t_8 plane. Other generators $t_{1,2,4,5,6,7}$ behave as raising or lowering operators which shift one point to another.

Define

$$\begin{aligned} I_\pm &= \frac{1}{\sqrt{2}}(t_1 \pm it_2) \,, \\ V_\pm &= \frac{1}{\sqrt{2}}(t_4 \pm it_5) \,, \\ U_\pm &= \frac{1}{\sqrt{2}}(t_6 \pm it_7) \,. \end{aligned} \tag{2.64}$$

Then the following commutation relations follow:

$$\begin{array}{ll} [t_3, I_\pm] = \pm I_\pm \,, & [t_8, I_\pm] = 0 \,, \\ [t_3, V_\pm] = \pm\frac{1}{2}V_\pm \,, & [t_8, V_\pm] = \pm\frac{\sqrt{3}}{2}V_\pm \,, \\ [t_3, U_\pm] = \mp\frac{1}{2}U_\pm \,, & [t_8, U_\pm] = \pm\frac{\sqrt{3}}{2}U_\pm \,. \end{array} \tag{2.65}$$

These imply that $I_\pm$, $U_\pm$ and $V_\pm$ raise or lower the values of t_3 and t_8 by the coefficients on the right hand sides. Therefore, they are conveniently expressed by two-dimensional vectors which point from the origin to one of the vertices of an equilateral hexagon as shown in Fig. 2.3.

Using these raising and lowering operators, it is possible to construct $SU(3)$ representations. In the case of $SU(2)$, we have done so by starting

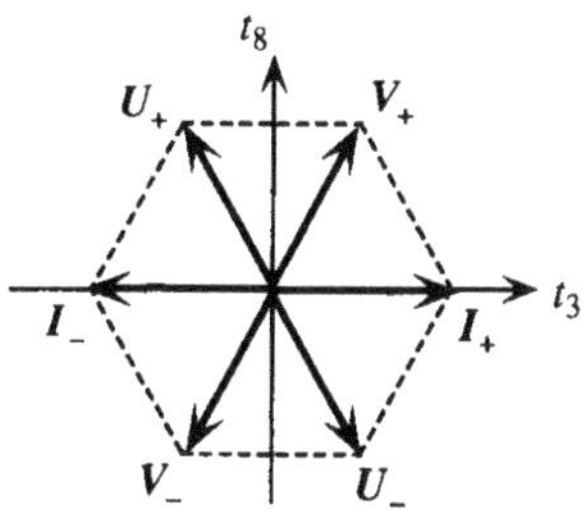

Fig. 2.3 Diagrammatic representation of six raising and lowering operators in the t_3-t_8 plane.

from the state of the highest weight located at the right end of a one-dimensional plot (Fig. 2.2) and by applying successively the lowering operators. Similarly, for $SU(3)$, we can start from the state of the highest weight and apply the lowering operators in Eq. (2.64).

Let us first establish a diagrammatic representation for the fundamental representation. Denoting them as

$$\psi^1 \sim |1\rangle = \begin{pmatrix} 1 \\ 0 \\ 0 \end{pmatrix}, \quad \psi^2 \sim |2\rangle = \begin{pmatrix} 0 \\ 1 \\ 0 \end{pmatrix}, \quad \psi^3 \sim |3\rangle = \begin{pmatrix} 0 \\ 0 \\ 1 \end{pmatrix}, \tag{2.66}$$

we see that

$$\begin{aligned} t_3|1\rangle &= \tfrac{1}{2}|1\rangle \\ t_8|1\rangle &= \tfrac{\sqrt{3}}{6}|1\rangle \end{aligned} \quad \Longrightarrow \quad \boldsymbol{\mu}_1^1 = \left(\frac{1}{2}, \frac{\sqrt{3}}{6}\right) \tag{2.67}$$

$$\begin{aligned} t_3|2\rangle &= -\tfrac{1}{2}|2\rangle \\ t_8|2\rangle &= \tfrac{\sqrt{3}}{6}|2\rangle \end{aligned} \quad \Longrightarrow \quad \boldsymbol{\mu}_2^1 = \left(-\frac{1}{2}, \frac{\sqrt{3}}{6}\right) \tag{2.68}$$

$$\begin{aligned} t_3|3\rangle &= 0 \\ t_8|3\rangle &= -\tfrac{\sqrt{3}}{3}|3\rangle \end{aligned} \quad \Longrightarrow \quad \boldsymbol{\mu}_3^1 = \left(0, -\frac{\sqrt{3}}{3}\right) \tag{2.69}$$

where we have introduced three two-dimensional vectors $\boldsymbol{\mu}_i^1, i = 1, 2, 3$ corresponding to the states $|i\rangle$. The vectors $\boldsymbol{\mu}_i^1$ can be used to specify the three states $|i\rangle$ and are called the weight vectors. The superscript 1 for $\boldsymbol{\mu}$ is to distinguish them from another weight $\boldsymbol{\mu}^2$ which will be introduced shortly. In Fig. 2.4 (a), the three vectors are shown; they form a unit equilateral triangle centered at the origin.

The fundamental representation ψ^i is denoted by

$$\text{either} \quad (1,0) \quad \text{or} \quad 3. \tag{2.70}$$

The former consists of two integer numbers (p_1, p_2) which specify an $SU(3)$ representation, while the latter denotes the dimension of the representation. Both of them are used frequently and several examples are shown in the following discussions.

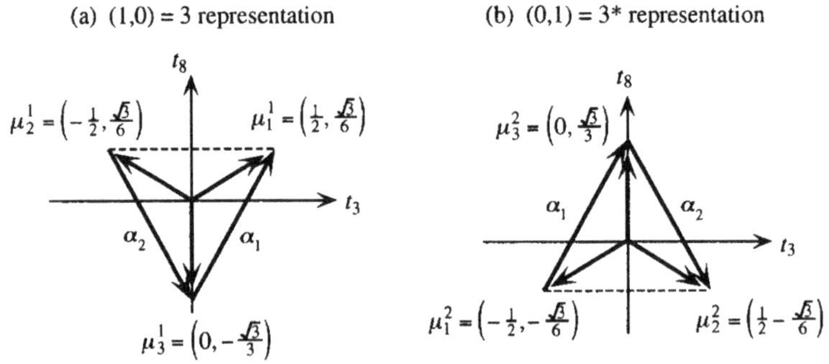

Fig. 2.4 Diagrammatic representation for the two fundamental representations of $SU(3)$.

The three states of the fundamental representation are related to each other through the raising or lowering operators. For instance, it is easy to check in the three component vector form that

$$|1\rangle = V_+|3\rangle\,. \tag{2.71}$$

In terms of weight vectors, this is expressed as

$$\boldsymbol{\mu}_1^1 = \boldsymbol{\alpha}_1 + \boldsymbol{\mu}_3^1\,, \tag{2.72}$$

where the "root" vector $\boldsymbol{\alpha}_1$

$$\boldsymbol{\alpha}_1 = \left(\frac{1}{2}, \frac{\sqrt{3}}{2}\right) \tag{2.73}$$

relates two weights additively and increases the weight by the "unit $\boldsymbol{\alpha}_1$", as shown in Fig. 2.4 (a). Similarly, one can consider another root $\boldsymbol{\alpha}_2$

$$\boldsymbol{\mu}_3^1 = \boldsymbol{\alpha}_2 + \boldsymbol{\mu}_2^1\,, \tag{2.74}$$

which raises a weight by another unit α_2. The root vectors α_1 and α_2 are independent. In general, all weight vectors are related by

$$\mu' = \mu + l\alpha_1 + m\alpha_2 \,, \tag{2.75}$$

where l and m are some integers.

Notice the correspondence between the root vectors and lowering and raising operators:

$$\alpha_1 \sim V_+ \,, \quad \alpha_2 \sim U_- \,. \tag{2.76}$$

In principle, one may choose any two independent operators out of six $I_\pm, U_\pm, V_\pm$. In the particular choice of Eq. (2.76), the two vectors are called "simple" roots.

In $SU(3)$ there is another fundamental representation which is the complex conjugate ψ^{i*} of Eq. (2.60). It is denoted as $(0,1)$ or 3^*. Complex conjugation of the commutation relations (2.58) reads

$$[(-t_a^*), (-t_b^*)] = if_{abc}(-t_c^*) \,, \tag{2.77}$$

implying that $-t_a^* = -\lambda_a^*/2$ can be another representation. The diagonal generators are replaced simply by the minus of the original one, $-t_{3,8}$, and therefore the weight vectors change their signs. In other words,

$$|1^*\rangle \to \mu_1^2 \;=\; \left(-\frac{1}{2}, -\frac{\sqrt{3}}{6}\right) \tag{2.78}$$

$$|2^*\rangle \to \mu_2^2 \;=\; \left(+\frac{1}{2}, -\frac{\sqrt{3}}{6}\right) \,, \tag{2.79}$$

$$|3^*\rangle \to \mu_3^2 \;=\; \left(0, \frac{\sqrt{3}}{3}\right) \,, \tag{2.80}$$

which form another triangle as shown in Fig. 2.4 (b).

Higher dimensional representations are constructed diagrammatically using the weight and root vectors. To do this, we need to order and define larger or smaller vectors under an appropriate rule. Compare the first components of two vectors μ and ν. If $\mu_1 > \nu_1$, we define that $\mu > \nu$. But if $\mu_1 = \nu_1$, we compare the second components. For instance,

$$(3,1) \;>\; (2,3)$$
$$(2,1) \;>\; (2,0)$$

Now we can introduce the "highest weight" for each representation. For the fundamental representations $(1,0) = 3$ and $(0,1) = 3^*$, they are

$$\boldsymbol{\mu}^1 \equiv \boldsymbol{\mu}_1^1 = \left(\frac{1}{2}, \frac{\sqrt{3}}{6}\right),$$
$$\boldsymbol{\mu}^2 \equiv \boldsymbol{\mu}_2^2 = \left(\frac{1}{2}, -\frac{\sqrt{3}}{6}\right). \tag{2.81}$$

For an arbitrary representation D, the highest weight is given by

$$\boldsymbol{\mu} = p_1 \boldsymbol{\mu}^1 + p_2 \boldsymbol{\mu}^2, \tag{2.82}$$

where p_1 and p_2 are non-negative integers. Thus the representation D is specified by (p_1, p_2). Starting from the highest weight $\boldsymbol{\mu}$, application $-\boldsymbol{\alpha}_1$ by p_1 times determines one side $-p_1\boldsymbol{\alpha}_1$. Similarly one has another side $-p_2\boldsymbol{\alpha}_2$. As shown in Fig. 2.5, the two sides can be reflected with respect to the t_8 axis, determining a hexagon. If p_1 (p_2) is zero a triangle is formed with a side of length p_2 (p_1).

Angles between neighboring sides of the hexagon are all 120 degrees (for a triangle, they are of course 60 degrees). Thus the hexagon is filled by equilateral triangles which are the units of the hexagon lattice (see Fig. 2.9). Then all states of the (p_1, p_2) representations correspond to one site inside or on the boundary of the hexagon. Weights are, of course, vectors pointing to sites from the origin.

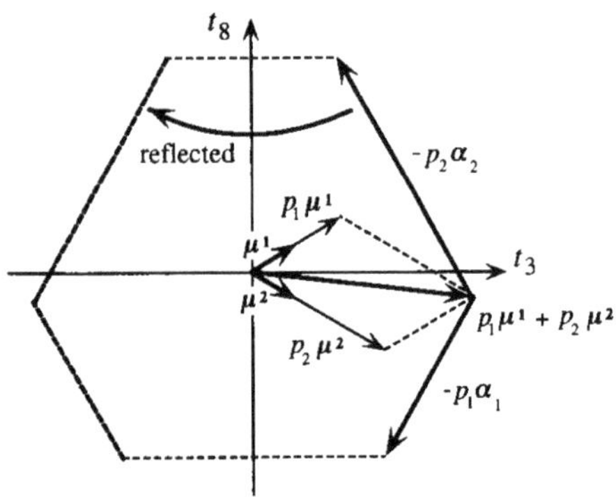

Fig. 2.5 Forming the hexagon for the (p_1, p_2) representation.

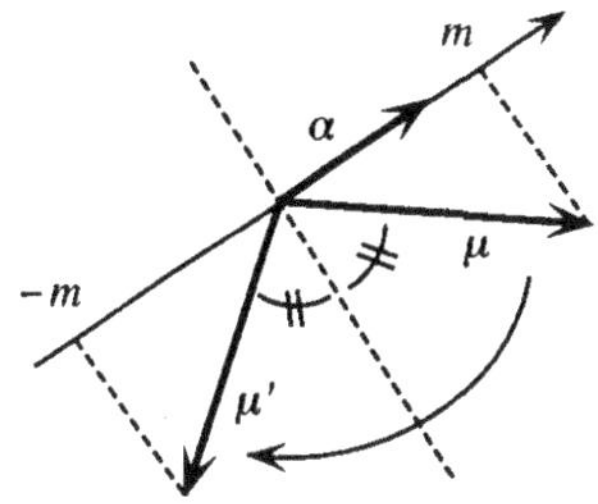

Fig. 2.6 Weyl transformation of a weight vector $\boldsymbol{\mu}$.

Weight vectors have the so called Weyl symmetry. It says that for an arbitrary weight $\boldsymbol{\mu}$ and an arbitrary root $\boldsymbol{\alpha}$, the vector

$$\boldsymbol{\mu}' = \boldsymbol{\mu} - \frac{2(\boldsymbol{\alpha}\cdot\boldsymbol{\mu})}{\boldsymbol{\alpha}^2}\boldsymbol{\alpha} \tag{2.83}$$

is also a weight. Geometrically, $\boldsymbol{\mu}'$ is a mirror image of $\boldsymbol{\mu}$ with respect to a hyperplane which is perpendicular to $\boldsymbol{\alpha}$ (Fig. 2.6). Group theoretically, it is a consequence of the fact that along the root $\boldsymbol{\alpha}$, an $SU(2)$ subgroup is considered as $\boldsymbol{\alpha}$ plays the role of the raising operator along its direction. Let us see some examples in the following.

***Example 1: Fundamental representation* (1, 0) = 3**

The highest weight of this representation is

$$\boldsymbol{\mu}_1 \equiv 1\cdot\boldsymbol{\mu}^1 + 0\cdot\boldsymbol{\mu}^2 = \left(\frac{1}{2}, \frac{\sqrt{3}}{6}\right). \tag{2.84}$$

Lowering $\boldsymbol{\mu}_1$ by applying $-\boldsymbol{\alpha}_1$, we obtain (see Fig. 2.4 (a))

$$\boldsymbol{\mu}_1 - \boldsymbol{\alpha}_1 = \left(0, -\frac{\sqrt{3}}{3}\right) = \boldsymbol{\mu}_3\,. \tag{2.85}$$

This is the lowest weight reached from the highest weight $\boldsymbol{\mu}_1$ by successively applying $\boldsymbol{\alpha}_1$. For other weights, we can use the Weyl symmetry in terms of the combinations $(\boldsymbol{\mu}_1, \boldsymbol{\alpha}_1)$, $(\boldsymbol{\mu}_1, \boldsymbol{\alpha}_2)$, $(\boldsymbol{\mu}_3, \boldsymbol{\alpha}_1)$, and $(\boldsymbol{\mu}_3, \boldsymbol{\alpha}_2)$. They yield,

$$(\boldsymbol{\mu}_1, \boldsymbol{\alpha}_1) \;\rightarrow\; \boldsymbol{\mu}' = \boldsymbol{\mu}_1 - \boldsymbol{\alpha}_1 = \left(0, -\frac{\sqrt{3}}{3}\right) = \boldsymbol{\mu}_3\,,$$

$$
\begin{aligned}
(\boldsymbol{\mu}_1, \boldsymbol{\alpha}_2) &\rightarrow \boldsymbol{\mu}' = \boldsymbol{\mu}_1\,,\\
(\boldsymbol{\mu}_3, \boldsymbol{\alpha}_1) &\rightarrow \boldsymbol{\mu}' = \boldsymbol{\mu}_3 + \boldsymbol{\alpha}_1 = \left(\frac{1}{2}, \frac{\sqrt{3}}{6}\right) = \boldsymbol{\mu}_1\,,\\
(\boldsymbol{\mu}_3, \boldsymbol{\alpha}_2) &\rightarrow \boldsymbol{\mu}' = \boldsymbol{\mu}_3 - \boldsymbol{\alpha}_2 = \left(-\frac{1}{2}, \frac{\sqrt{3}}{6}\right) = \boldsymbol{\mu}_2\,.
\end{aligned}
$$

These exhaust all the weights in the fundamental representation $(1,0) = 3$.

***Example 2: Adjoint representation* $(1,1) = 8$**

The weight diagram is given by an equilateral hexagon as shown in Fig. 2.7. The highest weight is

$$\boldsymbol{\mu}_1 = 1 \cdot \boldsymbol{\mu}^1 + 1 \cdot \boldsymbol{\mu}^2 = (1,0)\,.$$

There are two states which can be reached by applying $\boldsymbol{\alpha}_1$ and $\boldsymbol{\alpha}_2$:

$$
\begin{aligned}
\boldsymbol{\mu}_2 = \boldsymbol{\mu}_1 - \boldsymbol{\alpha}_1 &= \left(\frac{1}{2}, -\frac{\sqrt{3}}{2}\right),\\
\boldsymbol{\mu}_3 = \boldsymbol{\mu}_1 - \boldsymbol{\alpha}_2 &= \left(\frac{1}{2}, \frac{\sqrt{3}}{2}\right).
\end{aligned}
$$

All other weights are obtained by the Weyl symmetry. There are altogether seven weights which form a hexagon including its center.

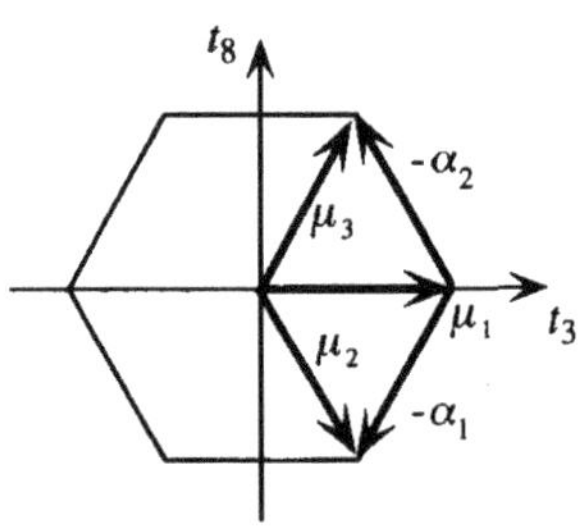

Fig. 2.7 Diagrammatic representation of the adjoint representation, $(1,1) = 8$, of $SU(3)$.

***Example 3: Decuplet representation* (3, 0) = 10**

We see this representation often in the quark model as it is supposed to correspond to the multiplet involving the delta resonance, $\Delta(1232)$. The highest weight is given by

$$\boldsymbol{\mu}_1 = 3 \cdot \boldsymbol{\mu}^1 + 0 \cdot \boldsymbol{\mu}^2 = \left(\frac{3}{2}, \frac{\sqrt{3}}{2}\right),$$

and is identified with the upper-right edge of an equilateral triangle as shown in Fig. 2.8. By applying the root vector $-\boldsymbol{\alpha}_1$ three times, the point reaches the bottom edge of the triangle on the t_8 axis. All points of the representation are obtained by the Weyl symmetry.

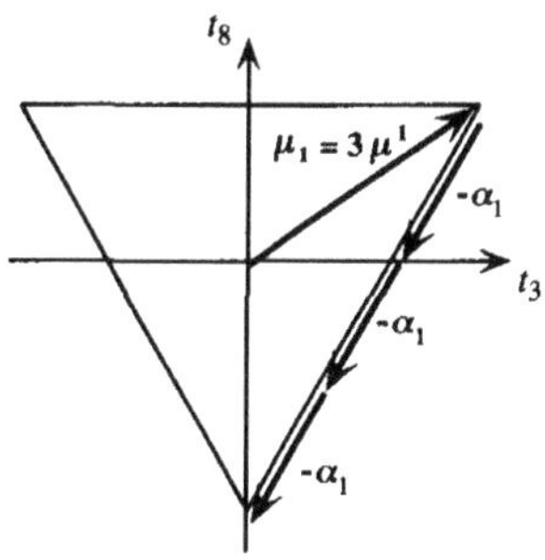

Fig. 2.8 Diagrammatic representation of the decuplet representation of $SU(3)$.

Degeneracy

There is a one to one correspondence between a (p_1, p_2) representation and a hexagon in the weight diagram. The hexagon is characterized by the two sides $-p_1\boldsymbol{\alpha}_1$ and $-p_2\boldsymbol{\alpha}_2$ starting from the point of the highest weight $\boldsymbol{\mu}$ and has reflection symmetry with respect to the t_8-axis. By decreasing p_1 and p_2 by one unit, we obtain a smaller hexagon inside the original one, corresponding to the representation $(p_1 - 1, p_2 - 1)$. One can repeat this process until one or both of the two parameters becomes zero, $(0, p_2')$, $(p_1', 0)$ or $(0, 0)$. The diagrams for $(0, p_2')$ and $(p_1', 0)$ are triangles and that for $(0, 0)$ is the point at the origin.

Starting from the point of highest weight, application of roots in an

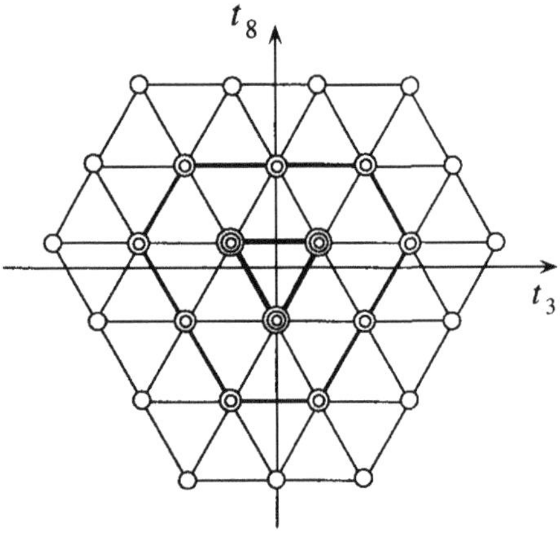

Fig. 2.9 Degeneracy of the (3, 2) representation.

arbitrary way generates sites on a triangle lattice whose boundary is the hexagon of (p_1, p_2). A physical state corresponds to one site of inside or the boundary of the hexagon. The number of states (degeneracy) which is attached to one site increases by one as we go from outside to inside layers of hexagon, starting from the singly occupied sites on the boundary. An example is depicted in Fig. 2.9 for the (3, 2) representation.

The multiplicity of a given representation is the number of sites of the hexagon including the degeneracy. One can verify that the dimension of the representation (p_1, p_2) is

$$\dim(p_1, p_2) = \frac{1}{2}(p_1 + 1)(p_2 + 1)(p_1 + p_2 + 2) . \tag{2.86}$$

The last remark is on how to distinguish degenerate states on the same site. This can be done by the isospin value. It is the horizontal width (passing through the site) of a hexagon which involves the site in its interior or the boundary. For example, in Fig. 2.9, the states on the second row from the top have either isospin $I = 2$ or $I = 1$ depending on whether they are seen to be on the outer or inner hexagon.

Tensor products

It often happens that we need to form direct products of two (or more) representations. It is sufficient to solve the Clebsch-Gordan decomposition

$$D_1 \times D_2 = \sum_i C_{12}^i D_i \,, \tag{2.87}$$

where D_i are representations labeled by the index i. Then our task is to find the Clebsch-Gordan coefficients C_{12}^i.

In order to work out this problem, it is rather convenient to follow the method of tensor product. Denoting the fundamental representations as

$$\begin{aligned} (1,0) &\sim \psi^i \,, \quad i = 1,2,3 \\ (0,1) &\sim \phi_k \,, \quad k = 1,2,3 \end{aligned} \tag{2.88}$$

higher dimensional representations (m, n) are formed by products of m ψ's and n ϕ's, carrying m upper indices and n lower indices:

$$\begin{aligned} (m,n) &\sim T^{i_1,i_2,\cdots,i_m}_{k_1,k_2,\cdots,k_n} \\ &\sim \psi^{i_1}\psi^{i_2}\cdots\psi^{i_m}\ \phi_{k_1}\phi_{k_2}\cdots\phi_{k_n} \,. \end{aligned} \tag{2.89}$$

A single term alone on the right hand side is not an irreducible representation and may contain lower dimensional representations. In order to remove reducible components, traceless and symmetric combination in contravariant indices { i's } and in covariant { k's } should be formed. The trace here is a contraction of one upper index and one lower index.

Now consider a product of two irreducible representations

$$(m,n) \times (m',n') \sim T^{i_1,i_2,\cdots,i_m}_{k_1,k_2,\cdots,k_n}\ U^{j_1,j_2,\cdots,j_{m'}}_{l_1,l_2,\cdots,l_{n'}} \,. \tag{2.90}$$

It is traceless in the indices (i, k) and in (j, l), but not in (i, l) and in (j, k). Also, it is symmetric in the indices i's, j's, k's and l's separately, but not among (i, j) and (k, l). Therefore, we perform irreducible decomposition first by forming traceless combinations and then by symmetrizing indices.

Forming traceless tensors

Consider a pair of contracted indices; one from i's and one from l's, or one from j's and one from k's. Starting from the tensor with $m + m'$ upper indices and $n + n'$ lower indices, we can form subsequently a tensor with

$(m-1)+m'$ upper and $n+(n'-1)$ lower indices, a tensor with $m+(m'-1)$ upper and $(n-1)+n'$ lower indices, a tensor with $(m-1)+(m'-1)$ upper and $(n-1)+(n'-1)$ lower indices, and so on. At every step, the contracted lower tensors must be subtracted from the original one to make it traceless. We can write this decomposition as

$$\begin{aligned} & (m,n)\times(m',n') \\ = \; & (m,m';n,n') \\ + \; & (m-1,m';n,n'-1)+(m,m'-1;n-1,n') \\ + \; & (m-1,m'-1;n-1,n'-1)+\cdots . \end{aligned} \tag{2.91}$$

The expression

$$(m,m';n,n') \tag{2.92}$$

is for a tensor with $m+m'$ upper and $n+n'$ lower indices; it is symmetric among the indices $i\in m$, among the indices $j\in m'$, among the indices $k\in n$ and among the indices $l\in n'$, and traceless. But it is not symmetrized between indices i and j, and between indices k and l.

Symmetrization

Each term of the above decomposition shown in Eq. (2.91) is not yet symmetrized under the exchange of indices $i\leftrightarrow j$ and $k\leftrightarrow l$. Let us introduce the notation

$$T^{i_1,\cdots,i_m,\,j_1,\cdots,j_{m'}}_{k_1,\cdots,k_n,\,l_1,\cdots,l_{n'}} \equiv (m,m';n,n')\,. \tag{2.93}$$

We can decompose it into symmetric and anti-symmetric parts. Without loss of generality, we can choose i_1 and j_1:

$$\begin{aligned} T^{i_1,\cdots,\,j_1,\cdots}_{k_1,\cdots,\,l_1,\cdots} &= \frac{1}{2}\left(T^{i_1,\cdots,\,j_1,\cdots}_{k_1,\cdots,\,l_1,\cdots}+T^{j_1,\cdots,\,i_1,\cdots}_{k_1,\cdots,\,l_1,\cdots}\right) \\ &+ \frac{1}{2}\left(T^{i_1,\cdots,\,j_1,\cdots}_{k_1,\cdots,\,l_1,\cdots}-T^{j_1,\cdots,\,i_1,\cdots}_{k_1,\cdots,\,l_1,\cdots}\right) \\ &\equiv S^{i_1,\cdots,\,j_1,\cdots}_{k_1,\cdots,\,l_1,\cdots}+A^{i_1,\cdots,\,j_1,\cdots}_{k_1,\cdots,\,l_1,\cdots}\,. \end{aligned} \tag{2.94}$$

But for the antisymmetric part $A^{i_1,\cdots,\,j_1,\cdots}_{k_1,\cdots,\,l_1,\cdots}$, the indices i_1 and j_1 may be lowered by the metric ϵ-tensor:

$$\epsilon_{pi_1j_1}A^{i_1,i_2\cdots,j_1,j_2\cdots}_{k_1,\cdots,l_1,\cdots} \equiv A^{i_2,\cdots,j_2,\cdots}_{p,k_1,\cdots,l_1,\cdots} \tag{2.95}$$

A somewhat nontrivial fact is that the tensor

$$A^{i_2,\cdots i_m,j_2,\cdots,j_{m'}}_{p,k_1,\cdots,k_n,l_1,\cdots,l_{n'}} \tag{2.96}$$

has already become symmetric under exchanges of $k \leftrightarrow l$ (see problem 2.3).

Hence the symmetry decomposition is simple. Remove one upper index from i's and one from j's, and at the same time add one lower index. Similarly, one can remove one lower index from k's and one from l's, and at the same time add one upper index. But once we remove upper indices, we can not remove lower indices any more, since the lower indices are already symmetrized in the former step. One can repeat this process until no pair of indices is left to be removed. We write the procedure as

$$\begin{aligned}(m,m';n,n') &= (m+m',n+n') \\ &+ (m+m'-2,n+n'+1)+(m+m'+1,n+n'-2) \\ &+ (m+m'-4,n+n'+2)+(m+m'+2,n+n'-4) \\ &+ \cdots . \end{aligned} \tag{2.97}$$

Examples:

(1) $(1,0)\times(0,1) = 3\times 3^*$

$$\begin{aligned}(1,0)\times(0,1) &= (1,0;0,1)+(0,0;0,0) \\ &= (1,1)+(0,0)\end{aligned} \tag{2.98}$$

In other words,

$$3\times 3^* = 8+1 \tag{2.99}$$

(2) $(1,0)\times(1,0) = 3\times 3$

$$\begin{aligned}(1,0)\times(1,0) &= (1,1;0,0) \\ &= (2,0)+(0,1)\end{aligned} \tag{2.100}$$

In other words,

$$3\times 3 = 6+3^* \tag{2.101}$$

(3) $(2,0)\times(1,0) = 6\times 3$

$$\begin{aligned}(2,0)\times(1,0) &= (2,1;0,0) \\ &= (3,0)+(1,1)\end{aligned} \tag{2.102}$$

In other words,

$$6 \times 3 = 10 + 8 \tag{2.103}$$

(4) From the above two examples we see

$$\begin{aligned} 3 \times 3 \times 3 &= (6 + 3^*) \times 3 \\ &= 1 + 8 + 8 + 10 \end{aligned} \tag{2.104}$$

(5) $(1,1) \times (1,1) = 8 \times 8$

$$\begin{aligned} (1,1) \times (1,1) &= (1,1;1,1) + (0,1;1,0) + (1,0;0,1) + (0,0;0,0) \\ (1,1;1,1) &= (2,2) + (0,3) + (3,0) \\ (1,0;0,1) &= (1,1) \\ (1,0;0,1) &= (1,1) \\ (0,0;0,0) &= (0,0) \end{aligned}$$

In other words,

$$8 \times 8 = 27 + 10 + 10^* + 8 + 8 + 1 \tag{2.105}$$

2.5 Multi-Particle States

In the quark model, we need baryon wave functions made of three valence quarks. As fermions, baryon wave functions are totally anti-symmetric; they change the overall sign under every single interchange of any quark pair. Now baryons are color singlet, for which the totally antisymmetric color wave function is responsible. Hence, the rest of the baryon wave functions must be totally symmetric.

The question is therefore how to construct totally symmetric wave functions in the combined space of orbital motion, spin and flavor. This problem can be systematically formulated using the theory of permutation symmetry [Bohr and Mottelson 1975]. The usefulness of the permutation symmetry lies also in the fact that the irreducible decomposition of $SU(n)$ representations is equivalent to that of permutation symmetry of multi-particle tensor products.

In this section we discuss this subject in some detail, but our descriptions are not very systematic. Starting from a simple example of two particle sys-

Table 2.3 Products of two permutation elements. Elements in the left column are multiplied by those of the top raw.

	$\mathcal{E}$	$\mathcal{O}$
$\mathcal{E}$	$\mathcal{E}$	$\mathcal{O}$
$\mathcal{O}$	$\mathcal{O}$	$\mathcal{E}$

tems, we introduce necessary concepts and tools as we proceed. Eventually, we write down quark model wave functions for baryons.

Permutations of two particles

The permutation group for two-particle systems, S_2, contains two elements $\mathcal{E} \equiv 1$ (identity) and $\mathcal{O}$; the latter is for an interchange of the two particles and is usually denoted by P_{12} in literatures,

$$\mathcal{O}\psi(1,2) = \psi(2,1)\,. \tag{2.106}$$

Apparently, two successive operations of $\mathcal{O}$ leaves the function invariant; $\mathcal{O}^2 = \mathcal{E}$. Products of the two elements are summarized in Table 2.3.

Any mathematical functions for two particles $\psi(1,2)$ can be decomposed into the symmetric (even) and anti-symmetric (odd) combination under the interchange $1 \leftrightarrow 2$:

$$\begin{aligned} \psi_S(1,2) &= \frac{1}{\sqrt{2}}\left(\psi(1,2) + \psi(2,1)\right) \\ \psi_A(1,2) &= \frac{1}{\sqrt{2}}\left(\psi(1,2) - \psi(2,1)\right)\,. \end{aligned} \tag{2.107}$$

which satisfy

$$\begin{aligned} \mathcal{O}\psi_S &= \psi_S(1,2)\,, \\ \mathcal{O}\psi_A &= -\psi_A(1,2)\,. \end{aligned} \tag{2.108}$$

Hence we see that ψ_S and ψ_A form one-dimensional (irreducible) representations of S_2. We can introduce the Young diagrams for these functions as

$\psi_S(1,2)$ = 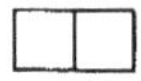

$\psi_A(1,2)$ = 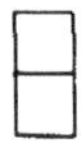

(2.109)

If we define the projection operators for S and A states by

$$\begin{aligned}\Pi_S &= \frac{1}{2}(1+\mathcal{O})\,,\\ \Pi_A &= \frac{1}{2}(1-\mathcal{O})\,,\end{aligned} \tag{2.110}$$

then the states $\psi_{S,A}$ can be projected out from an arbitrary function $\psi(1,2)$.

Example 1: Spin SU(2)

We consider irreducible decomposition of two-particle product states (tensor products) of spin up ($\uparrow$) and spin down ($\downarrow$). In order to distinguish quantum numbers of the two-particle states, we can allot a label $\uparrow$ or $\downarrow$ to each box of the Young diagram. The labeled diagrams are referred to as (state) tableaux [Bohr and Mottelson 1975]. We have the following four possibilities:

$\psi_S(1,2)$ = 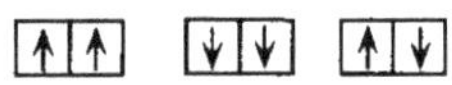

$\psi_A(1,2)$ = 

(2.111)

Apparently, it is not possible to give a single state (label) in two boxes in the same column.

Eqs. (2.111) provide dimensions of $SU(2)$ irreducible representations of two-particle states. They are the numbers of symmetric (S) and antisymmetric (A) combinations formed by two elements $\uparrow$ and $\downarrow$:

$$\begin{array}{llll}\text{Symmetric:} & 3 & \sim & \uparrow\uparrow,\ \downarrow\downarrow,\ \frac{1}{\sqrt{2}}(\uparrow\downarrow+\downarrow\uparrow)\\ \text{Antisymmetric:} & 1 & \sim & \frac{1}{\sqrt{2}}(\uparrow\downarrow-\downarrow\uparrow)\end{array}$$

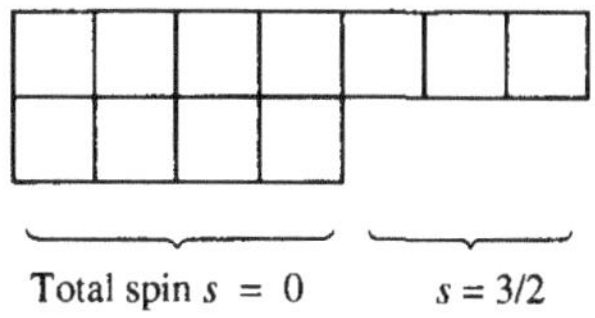

Fig. 2.10 A typical Young diagram of multi-particle state of $SU(2)$.

We write this as

$$2 \times 2 = \underbrace{3}_{S} + \underbrace{1}_{A}.$$

We note that the symmetry (or the associated dimension) determines the total spin s of the two-particle system also,

$$\begin{aligned} s\,(\text{symmetric state}) &= 1\,, \\ s\,(\text{anti}-\text{symmetric state}) &= 0\,. \end{aligned} \tag{2.112}$$

The equivalence between the symmetry and total spin applies to any-number states and is a special feature of $SU(2)$. It is easy to understand this. As shown in Fig. 2.10, a general Young diagram of $SU(2)$ has at most two rows, where each column carries spin $s = 0$. The net spin of the state is then given by the number of cells in the upper row which do not have a partner with a cell in the second row. The example in Fig. 2.10 carries the total spin $s = 3/2$.

Example 2: Flavor SU(3)

Examples described here have a large overlap with those in the previous section for the tensor products of $SU(3)$. We repeat them here in order to establish the use of the Young tableau.

For u, d and s states, we can form six symmetric and three antisymmetric combinations for the Young diagram in Eq. (2.109),

$$\psi_S \sim uu,\ dd, \ldots, ud,\ us \ldots \tag{2.113}$$

$$\psi_A \sim ud - du,\ us - su,\ ds - sd\,. \tag{2.114}$$

The antisymmetric combination behaves as a conjugate representation as shown in Eq. (2.63). Therefore we write the irreducible decomposition as

$$3 \times 3 = 6 + 3^* . \tag{2.115}$$

The corresponding Young diagrams are,

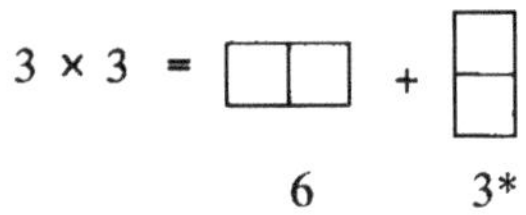

(2.116)

In terms of the notation of $SU(3)$ representations, (p_1, p_2) in section 2.4, we see that

$$3 = (1,0)\,, \quad 3^* = (0,1)\,, \quad 6 = (2,0)\,.$$

Another product we consider is

$$3^* \times 3\,.$$

At first sight, there are nine symmetric combinations

$$\psi_S \sim u^* u,\ d^* u,\ s^* u,\ u^* d,\ \ldots \tag{2.117}$$

But

$$u^* u + d^* d + s^* s \sim \psi_i^* \psi_i \tag{2.118}$$

is $SU(3)$ invariant which should be subtracted. Using the relation (2.63), this invariant is equivalent to the antisymmetric combination

$$\psi_i^* \psi_i \sim \epsilon_{ijk} \psi_i \psi_j \psi_k\,. \tag{2.119}$$

Note that the ordering here is important as it distinguishes particle numbers; otherwise the right-hand side of Eq. (2.119) vanishes identically. These explain the decomposition

$$3^* \times 3 = 1 + 8\,. \tag{2.120}$$

In terms of the Young diagram, it is written as

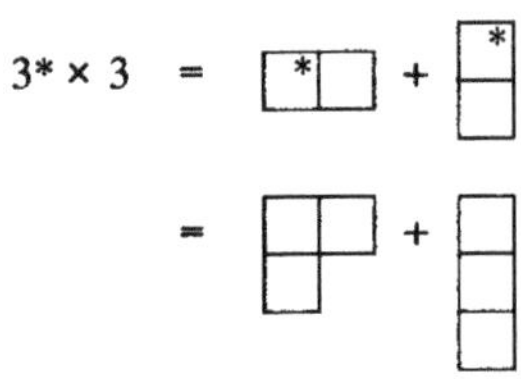

(2.121)

Table 2.4 Result of the operation of P_i's.

$\mathcal{E}_1$	$\mathcal{E}_2$	$\mathcal{E}_3$	$\mathcal{O}_1$	$\mathcal{O}_2$	$\mathcal{O}_3$
$\psi(1,2,3)$	$\psi(2,3,1)$	$\psi(3,1,2)$	$\psi(2,1,3)$	$\psi(1,3,2)$	$\psi(3,2,1)$

We have shown that two-particle states $3^* \times 3$ behave as three-particle states of either total antisymmetry or mixed symmetry.

Permutations of three particles

There are six elements of the permutation group of three particles (denoted as S_3). In this textbook we denote them as

$$\mathcal{E}_1(\equiv 1)\,,\ \mathcal{E}_2\,,\ \mathcal{E}_3\,,\ \mathcal{O}_1\,,\ \mathcal{O}_2\,,\ \mathcal{O}_3\,. \tag{2.122}$$

These notations imply that $\mathcal{E}_i$ (including the identity) corresponds to even permutations and $\mathcal{O}_i$ to odd permutations. They are denoted in the literature as P_{ij} and P_{ijk} [Bohr and Mottelson 1975]. Operation of these elements on a function $\psi(1,2,3)$ is defined in Table 2.4.

In order to represent the permutations we introduce three states α, β and γ. In general, there are six independent three-particle states as

$$\begin{array}{lll} \alpha(1)\beta(2)\gamma(3)\,, & \beta(1)\gamma(2)\alpha(3)\,, & \gamma(1)\alpha(2)\beta(3)\,, \\ \beta(1)\alpha(2)\gamma(3)\,, & \alpha(1)\gamma(2)\beta(3)\,, & \gamma(1)\beta(2)\alpha(3)\,. \end{array} \tag{2.123}$$

Note that if, for instance, $\beta = \gamma$, the number of independent states reduces to three; and if $\alpha = \beta = \gamma$, there is only one state. In what follows we omit arguments for particle numbers in parentheses; instead, the position of the state distinguishes particles. Now we attempt to decompose six elements $\alpha\beta\gamma$ etc. into irreducible representations of S_3.

Apparently, there are two singlet representations; the totally symmetric and anti-symmetric combinations of Eq. (2.123). They are

$$\begin{aligned} \psi_S &= \frac{1}{6}(\alpha\beta\gamma + \beta\gamma\alpha + \gamma\alpha\beta + \beta\alpha\gamma + \alpha\gamma\beta + \gamma\beta\alpha)\,, \\ \psi_A &= \frac{1}{6}(\alpha\beta\gamma + \beta\gamma\alpha + \gamma\alpha\beta - \beta\alpha\gamma - \alpha\gamma\beta - \gamma\beta\alpha)\,. \end{aligned} \tag{2.124}$$

The corresponding Young diagram is expressed either by the single row or by single column diagram as

$\psi_S(1,2,3) =$ 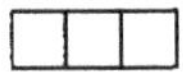

$\psi_A(1,2,3) =$ 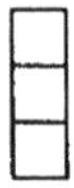

(2.125)

They can be projected out (if there is a component) from an arbitrary function by the projection operators,

$$\Pi_S = \frac{1}{6}(\mathcal{E}_1 + \mathcal{E}_2 + \mathcal{E}_3 + \mathcal{O}_1 + \mathcal{O}_2 + \mathcal{O}_3)\,, \tag{2.126}$$

$$\Pi_A = \frac{1}{6}(\mathcal{E}_1 + \mathcal{E}_2 + \mathcal{E}_3 - \mathcal{O}_1 - \mathcal{O}_2 - \mathcal{O}_3)\,. \tag{2.127}$$

The functions ψ_S and ψ_A satisfy

$$\begin{aligned} \mathcal{E}_i\psi_S &= \psi_S\,, \quad \mathcal{O}_i\psi_S = +\psi_S \\ \mathcal{E}_i\psi_A &= \psi_A\,, \quad \mathcal{O}_i\psi_A = -\psi_A\,, \end{aligned} \tag{2.128}$$

for all $i = 1, 2, 3$.

There are other types of functions of mixed symmetry $\psi_{MS}(1,2,3)$ which are neither symmetric nor antisymmetric, and hence satisfy

$$\Pi_S\psi_{MS}(1,2,3) = \Pi_A\psi_{MS}(1,2,3) = 0\,. \tag{2.129}$$

These functions are expressed by the diagram of mixed rows and columns:

$\psi_{MS}(1,2,3) =$ 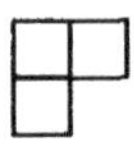

(2.130)

They are characterized by the symmetry between a pair of i-th and j-th particles; in the standard representation, $(i,j) = (1,2)$. There are two distinctive types depending on the even or odd nature under the interchange:

$$\begin{aligned} \mathcal{O}_1(\sim P_{12})\psi^{\lambda}_{MS} &= +\psi^{\lambda}_{MS}\,, \\ \mathcal{O}_1(\sim P_{12})\psi^{\rho}_{MS} &= -\psi^{\rho}_{MS}\,. \end{aligned} \tag{2.131}$$

In order to distinguish these states of different permutation symmetries, we can distribute particle numbers in the Young diagram:

$$\psi^{\rho}_{MS}(1,2,3) = \begin{array}{|c|c|}\hline 1 & 3 \\ \hline 2 & \\ \cline{1-1} \end{array} \qquad \psi^{\lambda}_{MS}(1,2,3) = \begin{array}{|c|c|}\hline 1 & 2 \\ \hline 3 & \\ \cline{1-1} \end{array} \tag{2.132}$$

They are referred to as particle number tableaux and should be distinguished from state tableaux in (2.111).

The numbering obeys the rule such that the numbers do not decrease when read from left to right and from top to bottom. Then a subset of the tableau in which the numbers increase successively carries the right symmetry determined by that sub-tableau itself. For instance, in ψ^{λ}_{MS}, the numbers 1 and 2 appear successively in the first row, and therefore, it must be symmetric under the interchange between 1 and 2. Similarly, ψ^{ρ}_{MS} is anti-symmetric under the same interchange.

At this point, we can establish the recursive rule of forming a Young tableau for a more-particles from a tableau of fewer particles. As the number of particles increases by one, one cell is attached to the original smaller tableau under the constraint that the number of cells in each row (or column) does not increase when read from left to right and from top to bottom. For instance, we can perform the following products:

$$\begin{array}{|c|c|}\hline 1 & 2 \\ \hline \end{array} \times \begin{array}{|c|}\hline 3 \\ \hline \end{array} \longrightarrow \begin{array}{|c|c|c|}\hline 1 & 2 & 3 \\ \hline \end{array} + \begin{array}{|c|c|}\hline 1 & 2 \\ \hline 3 & \\ \cline{1-1} \end{array}$$

$$\begin{array}{|c|}\hline 1 \\ \hline 2 \\ \hline \end{array} \times \begin{array}{|c|}\hline 3 \\ \hline \end{array} \longrightarrow \begin{array}{|c|c|}\hline 1 & 3 \\ \hline 2 & \\ \cline{1-1} \end{array} + \begin{array}{|c|}\hline 1 \\ \hline 2 \\ \hline 3 \\ \hline \end{array} \tag{2.133}$$

In this way we obtain one S, one A and two MS states as expected.

The projection operators for the λ and ρ mixed symmetries are given by

$$\begin{aligned} \Pi_{\lambda} &= \frac{1}{6}\left(2 - \mathcal{E}_2 - \mathcal{E}_3 + 2\mathcal{O}_1 - \mathcal{O}_2 - \mathcal{O}_3\right) , \\ \Pi_{\rho} &= \frac{1}{6}\left(2 - \mathcal{E}_2 - \mathcal{E}_3 - 2\mathcal{O}_1 + \mathcal{O}_2 + \mathcal{O}_3\right) . \end{aligned} \tag{2.134}$$

We note that the four projection operators Π_S, Π_A, Π_{λ} and Π_{ρ} satisfy the completeness:

$$\Pi_S + \Pi_A + \Pi_{\lambda} + \Pi_{\rho} = 1 . \tag{2.135}$$

Table 2.5 Products of permutation operators. An element in the left column must be multiplied by an element in the top row from the right.

	$\mathcal{E}_1$	$\mathcal{E}_2$	$\mathcal{E}_3$	$\mathcal{O}_1$	$\mathcal{O}_2$	$\mathcal{O}_3$
$\mathcal{E}_1$	$\mathcal{E}_1$	$\mathcal{E}_2$	$\mathcal{E}_3$	$\mathcal{O}_1$	$\mathcal{O}_2$	$\mathcal{O}_3$
$\mathcal{E}_2$	$\mathcal{E}_2$	$\mathcal{E}_3$	$\mathcal{E}_1$	$\mathcal{O}_2$	$\mathcal{O}_3$	$\mathcal{O}_1$
$\mathcal{E}_3$	$\mathcal{E}_3$	$\mathcal{E}_1$	$\mathcal{E}_2$	$\mathcal{O}_3$	$\mathcal{O}_1$	$\mathcal{O}_2$
$\mathcal{O}_1$	$\mathcal{O}_1$	$\mathcal{O}_3$	$\mathcal{O}_2$	$\mathcal{E}_1$	$\mathcal{E}_3$	$\mathcal{E}_2$
$\mathcal{O}_2$	$\mathcal{O}_2$	$\mathcal{O}_1$	$\mathcal{O}_3$	$\mathcal{E}_2$	$\mathcal{E}_1$	$\mathcal{E}_3$
$\mathcal{O}_3$	$\mathcal{O}_3$	$\mathcal{O}_2$	$\mathcal{O}_1$	$\mathcal{E}_3$	$\mathcal{E}_2$	$\mathcal{E}_1$

Also they satisfy the properties of the projection operators

$$\Pi_\alpha \Pi_\beta = \delta_{\alpha\beta}\Pi_\alpha \,. \tag{2.136}$$

To check these formula, the table for the products in Table 2.5 is helpful.

Example 1: SU(2)

The irreducible decomposition of three-particle states for $SU(2)$ is expressed as

$$2 \times 2 \times 2 = \underbrace{4}_{S} + \underbrace{2+2}_{MS} \,. \tag{2.137}$$

As already discussed, irreducible representations of permutation symmetry carry definite spin. For instance, $s = 3/2$ states have four components labeled by s_z. They are symmetrized states of:

$$\uparrow\uparrow\uparrow, \ \uparrow\uparrow\downarrow, \ \uparrow\downarrow\downarrow, \ \downarrow\downarrow\downarrow, \tag{2.138}$$

corresponding to 4 in Eq. (2.137).

Applying λ and ρ symmetrization to $\uparrow\uparrow\downarrow$ and $\uparrow\downarrow\downarrow$ (application to $\uparrow\uparrow\uparrow$ and $\downarrow\downarrow\downarrow$ states vanishes), we obtain four mixed-symmetric states. Two of them are of λ-type and the other two are of ρ-type. The equivalence between the irreducible representations of permutation and $SU(2)$ symmetry is verified by the fact that λ- and ρ-type functions are generated by the coupling

$$\lambda \ : \ [[1/2, \ 1/2]^1, \ 1/2]^{1/2}, \tag{2.139}$$

$$\rho \ : \ [[1/2, \ 1/2]^0, \ 1/2]^{1/2}, \tag{2.140}$$

where $[s_1, \ s_2]^{s_3}$ represents a coupling scheme of spin angular momenta.

Example 2: SU(3)

The irreducible decomposition for $SU(3)$ is expressed as (see Eq. (2.104))

$$3 \times 3 \times 3 = \underbrace{10}_{S} + \underbrace{8+8}_{MS} + \underbrace{1}_{A} . \tag{2.141}$$

We can generate ten totally symmetric (S) states (decuplet) by symmetrizing elements

$$\begin{aligned} S \quad \sim \quad & uuu\,,\ ddd\,,\ sss\,,\ uud\,,\ uus\,, \\ & ddu\,,\ dds\,,\ ssu\,,\ ssd\,,\ usd\,. \end{aligned} \tag{2.142}$$

In the notation given in section 2.4, they form the representation

$$(3,0) = 10\,. \tag{2.143}$$

There is one totally antisymmetric combination

$$A \sim uds\,. \tag{2.144}$$

The remaining 16 elements are split into two mixed symmetric states of the λ and ρ types. Formally they can be generated by applying the λ and ρ projection operators onto the states

$$MS \sim uud,\ uus,\ ddu,\ dds,\ ssu,\ ssd,\ uds,\ dsu\,.$$

The last two elements uds and dsu should be distinguished. But the MS states formed by (in terms of the two projection operators) sud are linear combinations of those of uds and dsu.

Two-dimensional representation of MS symmetry

Let us investigate more the mixed-symmetric states. We show explicitly that they form two-dimensional representations of permutation symmetry. Originally we had six functions of the type $\alpha\beta\gamma$. Out of them, two linear combinations are irreducible symmetric and antisymmetric representations. Therefore, four functions remain. They are reducible and split into two two-dimensional representations. The method of this decomposition is discussed in Problem 2.5.

Here we discuss the case $\beta = \gamma$, which is less general but practically of great importance. In this case we have three functions, $\alpha\beta\beta$, $\beta\alpha\beta$ and

$\beta\beta\alpha$. Since the totally antisymmetric combination vanishes, the only one-dimensional representation is the totally symmetric combination

$$\psi_S = \frac{1}{\sqrt{3}}(\alpha\beta\beta + \beta\alpha\beta + \beta\beta\alpha)\,. \tag{2.145}$$

The remaining two functions form uniquely the two-dimensional representation:

$$\psi^{\lambda}_{MS} = \frac{1}{\sqrt{6}}(\alpha\beta\beta + \beta\alpha\beta - 2\beta\beta\alpha)\,, \tag{2.146}$$

$$\psi^{\rho}_{MS} = \frac{1}{\sqrt{2}}(\alpha\beta\beta - \beta\alpha\beta)\,. \tag{2.147}$$

A familiar example is obtained by setting $\alpha = \boldsymbol{r}, \beta = \gamma = 1$, where we have

$$\begin{aligned} \psi_S(1,2,3) &\sim \boldsymbol{R} = \frac{1}{\sqrt{3}}(\boldsymbol{r}_1 + \boldsymbol{r}_2 + \boldsymbol{r}_3)\,, \\ \psi_\lambda(1,2,3) &\sim \boldsymbol{\lambda} = \frac{1}{\sqrt{6}}(\boldsymbol{r}_1 + \boldsymbol{r}_2 - 2\boldsymbol{r}_3)\,, \\ \psi_\rho(1,2,3) &\sim \boldsymbol{\rho} = \frac{1}{\sqrt{2}}(\boldsymbol{r}_1 - \boldsymbol{r}_2)\,, \end{aligned} \tag{2.148}$$

which are the Jacobi coordinates for the three-particle system.

In this basis, the six permutation operators are represented by the following 2×2 matrices:

$$\begin{aligned} \mathcal{E}_1 &= \begin{pmatrix} 1 & 0 \\ 0 & 1 \end{pmatrix} & \mathcal{O}_1 &= \begin{pmatrix} 1 & 0 \\ 0 & -1 \end{pmatrix} \\ \mathcal{E}_2 &= \begin{pmatrix} -1/2 & \sqrt{3}/2 \\ -\sqrt{3}/2 & -1/2 \end{pmatrix} & \mathcal{O}_2 &= \begin{pmatrix} -1/2 & \sqrt{3}/2 \\ \sqrt{3}/2 & 1/2 \end{pmatrix} \\ \mathcal{E}_3 &= \begin{pmatrix} -1/2 & -\sqrt{3}/2 \\ \sqrt{3}/2 & -1/2 \end{pmatrix} & \mathcal{O}_3 &= \begin{pmatrix} -1/2 & -\sqrt{3}/2 \\ -\sqrt{3}/2 & 1/2 \end{pmatrix} \end{aligned} \tag{2.149}$$

Interpretation of these matrices is discussed in Problem 2.4.

2.6 Product-States

In the quark model, we face a problem of constructing product wave functions of spin and isospin wave functions with definite permutation symmetry for N (especially three)-particle system. We investigate symmetry properties of such product-states. In a group theoretical language, it is called inner products, and the decomposition of

$$\chi_{\sigma_1}(1,2,3)\,\phi_{\sigma_2}(1,2,3) = \sum_{\sigma} C^{\sigma}_{\sigma_1\sigma_2}\Psi_\sigma(1,2,3) \tag{2.150}$$

follows. Here Ψ_σ is referred to as a product-state of χ and ϕ which carry the symmetry σ. The coefficients $C^{\sigma}_{\sigma_1\sigma_2}$ are the Clebsch-Gordan coefficients of the permutation symmetry.

Products of two-particle states

Let us first consider, as the simplest case, products of χ and ϕ wave functions for two-particle states. For this purpose, we consider products of projection operators. For a two-particle system, there are only two possibilities, either symmetric (S) or antisymmetric (A), as in Eq. (2.107). For instance, the product of symmetric spin and antisymmetric isospin states is represented by the projection operators as

$$\Pi^{\chi}_{S}\Pi^{\phi}_{A}\,,$$

where

$$\Pi^{\chi}_{S} = \frac{1}{2}(1+\mathcal{O}^{\chi})\,,\quad \Pi^{\phi}_{A} = \frac{1}{2}(1-\mathcal{O}^{\phi})\,. \tag{2.151}$$

They operate either spin or isospin wave function as denoted in the superscripts, e.g.,

$$\begin{aligned} \Pi^{\phi}_{S}\chi\phi &= \chi(\Pi^{\phi}_{S}\phi)\,, \\ \Pi^{\chi}_{S}\Pi^{\phi}_{A}\chi\phi &= (\Pi^{\chi}_{S}\chi)(\Pi^{\phi}_{S}\phi)\,. \end{aligned}$$

Projection operators for product states are defined by

$$\Pi^{\chi\phi}_{S} = \frac{1}{2}(1+\mathcal{O}^{\chi}\mathcal{O}^{\phi})\,, \tag{2.152}$$

$$\Pi^{\chi\phi}_{A} = \frac{1}{2}(1-\mathcal{O}^{\chi}\mathcal{O}^{\phi})\,. \tag{2.153}$$

It is then straightforward to verify, for instance, that

$$\begin{aligned} \Pi^{\chi\phi}_{S}(\Pi^{\chi}_{S}\Pi^{\phi}_{A}) &= 0\,, \\ \Pi^{\chi\phi}_{A}(\Pi^{\chi}_{S}\Pi^{\phi}_{A}) &= \Pi^{\chi}_{S}\Pi^{\phi}_{A}\,, \end{aligned}$$

Table 2.6 Products of two wave functions, $\chi_{\sigma_1}\phi_{\sigma_2}$.

	ϕ_S	ϕ_A
χ_S	Ψ_S	Ψ_A
χ_A	Ψ_A	Ψ_S

which implies that the product of symmetric and antisymmetric wave functions is antisymmetric. Similarly, one can easily verify other products. Results of the products of two wave functions are summarized in Table 2.6.

Products of three-particle states

We can extend the discussion in the previous subsection to three-particle systems. Trivial case is for products where one of the two is either symmetric or antisymmetric. We present the results in Table 2.7 directly.

A somewhat nontrivial case is the product of mixed-symmetric states. We consider products of functions of the type in Eqs. (2.146) and (2.147). We investigate decomposition of the product

$$\chi_\lambda \phi_\lambda \tag{2.154}$$

into various symmetry states S, A, λ and ρ. We apply the projection operator for the symmetric state

$$\Pi_S \equiv \frac{1}{6}\left(\mathcal{E}_1^\chi \mathcal{E}_1^\phi + \mathcal{E}_2^\chi \mathcal{E}_2^\phi + \mathcal{E}_3^\chi \mathcal{E}_3^\phi + \mathcal{O}_4^\chi \mathcal{O}_4^\phi + \mathcal{O}_5^\chi \mathcal{O}_5^\phi + \mathcal{O}_6^\chi \mathcal{O}_6^\phi\right) . \tag{2.155}$$

From the representation matrices of Eq. (2.149), we obtain

$$\begin{aligned} \Pi_S\, \chi_\lambda \phi_\lambda &= \frac{1}{6}\left((\mathcal{E}_1^\chi \chi_\lambda)(\mathcal{E}_1^\phi \phi_\lambda) + (\mathcal{E}_2^\chi \chi_\lambda)(\mathcal{E}_2^\phi \phi_\lambda) + \cdots\right) \\ &= \frac{1}{2}\left(\chi_\lambda \phi_\lambda + \chi_\rho \phi_\rho\right) . \end{aligned} \tag{2.156}$$

Similarly, one can verify that

$$\begin{aligned} \Pi_A \chi_\lambda \phi_\lambda &= 0 \\ \Pi_\lambda \chi_\lambda \phi_\lambda &= \frac{1}{2}\left(\chi_\lambda \phi_\lambda - \chi_\rho \phi_\rho\right) . \\ \Pi_\rho \chi_\lambda \phi_\lambda &= 0 . \end{aligned}$$

Table 2.7 Decomposition of products into product-states Ψ with definite symmetry.

	ϕ_S	ϕ_λ	ϕ_ρ	ϕ_A
χ_S	Ψ_S	Ψ_λ	Ψ_ρ	Ψ_A
χ_λ	Ψ_λ	$\frac{1}{\sqrt{2}}(\Psi_S+\Psi_\lambda)$	$\frac{1}{\sqrt{2}}(\Psi_\rho+\Psi_A)$	Ψ_ρ
χ_ρ	Ψ_ρ	$\frac{1}{\sqrt{2}}(\Psi_\rho-\Psi_A)$	$\frac{1}{\sqrt{2}}(\Psi_S-\Psi_\lambda)$	Ψ_λ
χ_A	Ψ_A	Ψ_ρ	Ψ_λ	Ψ_S

These equations imply the decomposition

$$\chi_\lambda\phi_\lambda = \frac{1}{\sqrt{2}}(\Psi_S + \Psi_\lambda)\,, \tag{2.157}$$

where the symmetric and λ states of product-states are defined by

$$\Psi_S = \frac{1}{\sqrt{2}}(\chi_\lambda\phi_\lambda + \chi_\rho\phi_\rho)\,, \tag{2.158}$$

$$\Psi_\lambda = \frac{1}{\sqrt{2}}(\chi_\lambda\phi_\lambda - \chi_\rho\phi_\rho)\,. \tag{2.159}$$

One can check other cases and find that

$$\chi_\rho\phi_\rho = \frac{1}{\sqrt{2}}(\Psi_S - \Psi_\lambda)\,, \tag{2.160}$$

$$\chi_\lambda\phi_\rho = \frac{1}{\sqrt{2}}(\Psi_\rho + \Psi_A)\,, \tag{2.161}$$

$$\chi_\rho\phi_\lambda = \frac{1}{\sqrt{2}}(\Psi_\rho - \Psi_A)\,, \tag{2.162}$$

where the ρ and A symmetric functions are defined by

$$\Psi_\rho = \frac{1}{\sqrt{2}}(\chi_\lambda\phi_\rho + \chi_\rho\phi_\lambda)\,, \tag{2.163}$$

$$\Psi_A = \frac{1}{\sqrt{2}}(\chi_\lambda\phi_\rho - \chi_\rho\phi_\lambda)\,. \tag{2.164}$$

Results of Eqs. (2.157), (2.160) - (2.162) are summarized in Table 2.7. For practical uses, inverse relations for the decomposition of product-states Ψ into products of two functions $\chi\phi$ is also summarized in Table 2.8.

Table 2.8 Decomposition of product-states Ψ into products $\chi\phi$ with definite symmetry.

Ψ_S	$\chi_S\phi_S,\ \chi_A\phi_A,\ \frac{1}{\sqrt{2}}(\chi_\lambda\phi_\lambda+\chi_\rho\phi_\rho)$
Ψ_λ	$\chi_S\phi_\lambda,\ \chi_\lambda\phi_S,\ \chi_A\phi_\rho,\ \chi_\rho\phi_A,\ \frac{1}{\sqrt{2}}(\chi_\lambda\phi_\lambda-\chi_\rho\phi_\rho)$
Ψ_ρ	$\chi_S\phi_\rho,\ \chi_\rho\phi_S,\ \chi_A\phi_\lambda,\ \chi_\lambda\phi_A,\ \frac{1}{\sqrt{2}}(\chi_\lambda\phi_\rho+\chi_\rho\phi_\lambda)$
Ψ_A	$\chi_A\phi_S,\ \chi_S\phi_A,\ \frac{1}{\sqrt{2}}(\chi_\lambda\phi_\rho-\chi_\rho\phi_\lambda)$

Table 2.9 Notations for quark model wave functions.

	Single-particle	Three-particle
Spin	$\uparrow,\downarrow$	χ
Flavor	u, d, s	ϕ
Orbital	$f(\boldsymbol{x}_i)$	$\psi(\boldsymbol{\lambda},\boldsymbol{\rho})$

2.7 Quark Model Wave Functions

In the previous section we have established various symmetry properties under permutations of three particle systems. Classification of products of spin and flavor wave functions is expressed symbolically as:

$$\begin{aligned}
&(2_\lambda+2_\rho+4_S)\times(1_A+8_\lambda+8_\rho+10_S)\\
&= \ 2_\lambda 1_A+2_\lambda 8_\lambda+2_\lambda 8_\rho+2_\lambda 10_S\\
&+ \ 2_\rho 1_A+2_\rho 8_\lambda+2_\rho 8_\rho+2_\rho 10_S\\
&+ \ 4_S 1_A+4_S 8_\lambda+4_S 8_\rho+4_S 10_S\,. \qquad (2.165)
\end{aligned}$$

Then it follows from Table 2.7 that every term on the right hand side is decomposed into terms of certain symmetries.

In the quark model, baryon wave functions are totally symmetric product-states of spin, flavor and orbital parts. Therefore, if we look at the spin-flavor part, it must carry an appropriate symmetry depending on the symmetry of the orbital part such that their product can form a totally symmetric wave function. In many quark model studies, nomenclature from the spin-flavor multiplets is adopted to classify physical baryon states. We see some examples in this section. Our notation is given in Table 2.9.

Orbital part – Harmonic oscillator

In non-relativistic quark models, orbital motion is often described by the three-dimensional harmonic oscillator. It is also adopted here. Many consequences which are derived from symmetry arguments hold in a wider class of potential models.

The hamiltonian of the harmonic oscillator model is assumed to be

$$H_{HO} = \sum_{i=1,2,3} \left(\frac{p_i^2}{2m_i} + \frac{m_i \omega^2}{2} \boldsymbol{x}_i^2 \right) , \tag{2.166}$$

where ω is an oscillator parameter (frequency) of a spherical oscillator potential for quark confinement. In many potential models, there is a problem of spurious center of mass motion. This will be discussed as it becomes necessary.

Harmonic oscillator states may be labeled by oscillator quantum numbers of three quarks in the spherical base [‡]:

$$\begin{aligned} n &: \text{principal quantum number} \\ l &: \text{orbital angular momentum} \end{aligned} \tag{2.167}$$

The single-particle wave function of the i-th quark is then denoted as

$$f_{n_i l_i}(\boldsymbol{x}_i) . \tag{2.168}$$

Explicit forms of the wave functions $f_{n_i l_i}$ are presented in Appendix A. Then three-quark states are linear combinations of products:

$$\psi^N = f_{n_1 l_1}(\boldsymbol{x}_1) f_{n_2 l_2}(\boldsymbol{x}_2) f_{n_3 l_3}(\boldsymbol{x}_3) \tag{2.169}$$

with definite permutation symmetry. If necessary, one can form a state of definite total angular momentum by the coupling:

$$[[l_1, \ l_2]^{l_{12}}, \ l_3]^{l_{tot}} \sim [l_1, \ [l_2, \ l_3]^{l_{23}}]^{l_{tot}} . \tag{2.170}$$

The total energy for a three-particle state is given by

$$\begin{aligned} E_N &= (2n + l + 3)\,\omega , \\ n &= \sum_i n_i , \quad l = \sum_i l_i , \end{aligned} \tag{2.171}$$

[‡]Here the third component of the angular momentum m is not considered explicitly.

where $3 = \frac{1}{2} \times 6$ (for the zero point energy) is half of the number of internal degrees of freedom with the center of mass degree of freedom excluded.

Spin-flavor part

Spin and flavor wave functions are product-states of a three-particle spin state and a three-particle flavor state. The method to construct such product-states was already discussed.

To start with, we notice that each quark may be labeled by spin and flavor quantum numbers,

$$u\uparrow, \quad u\downarrow, \quad d\uparrow, \quad d\downarrow, \quad s\uparrow, \quad s\downarrow . \tag{2.172}$$

They form the basis of the fundamental (six-dimensional) representation of the $SU(6)$ group which has $SU(2)$ and $SU(3)$ in its subgroups. For three particle systems, there are

$$6 \times 6 \times 6 = 216 \tag{2.173}$$

states forming the 216-dimensional irreducible representation. Notice that this equals to the total number of Eq. (2.165). If spin and/or flavor dependent interactions are turned off, all these states appear as members of a degenerate multiplet. The question now is how to classify these 216 states in terms of spin and flavor quantum numbers.

A spin-flavor state is often denoted as

$$^{2s+1}D \tag{2.174}$$

where D is the dimension of a flavor $SU(3)$ representation, and s the total spin. This notation mimics the familiar spectroscopic notation $^{2s+1}L_J$; however the spin and flavor can not be coupled, and so there is no subscript in Eq. (2.174). Examples are

$$^{2}8\,, \quad ^{4}8\,, \quad ^{4}10\,, \quad \text{etc.} \tag{2.175}$$

Combining those spin-flavor states with an orbital wave function, let us now construct totally symmetric states.

Table 2.10 Physical baryons as members of $SU(2) \times SU(3)$ (spin-flavor) multiplets.

Multiplet	Baryons	s	I_3	S
	N	1/2	1/2	0
$^2 8$	Λ (Octet)	1/2	0	−1
	Σ (Octet)	1/2	1	−1
	Ξ	1/2	1/2	−2
	Δ	3/2	3/2	0
	Σ^* (Decuplet)	3/2	1	−1
$^4 10$	Ξ^* (Decuplet)	3/2	1/2	−2
	Ω (Decuplet)	3/2	0	−3

The ground state, $N = 0$

All three quarks are in the ground states. Therefore, the only allowed symmetry for the three-particle ground state is total symmetry S:

$$\psi_S^{N=0} = f_{00}(\boldsymbol{x}_1) f_{00}(\boldsymbol{x}_2) f_{00}(\boldsymbol{x}_2) \,. \tag{2.176}$$

Consequently, spin-flavor wave functions are also totally symmetric. From Table 2.8, following combinations are available for totally symmetric spin-flavor wave functions:

$$\Psi_S^{N=0} = \begin{cases} \chi_{MS} \times \phi_{MS} & \sim \quad {}^2 8 = 16 \text{ dim} \\ \chi_S \times \phi_S & \sim \quad {}^4 10 = 40 \text{ dim} \end{cases} \tag{2.177}$$

Therefore, the ground states form a 56-dimensional representation. Identification of these states with physical baryons is shown in Table 2.10.

First excited state, $N = 1$

The $N = 1$ state is realized when one of the three quarks is excited to the p-state, $(n, l) = (0, 1)$. Then three single-particle states are

$$f_{00} \,, \quad f_{00} \,, \quad f_{01} \,. \tag{2.178}$$

Therefore, there are two classes of symmetries, total symmetry S and mixed symmetry MS. The totally symmetric state, however, corresponds to the excitation of the center of mass motion (see Eq. (2.148) also) and is irrelevant when discussing internal excitations. The relevant wave function for the first excited state in the quark model is then of mixed symmetry.

An explicit form of the p-wave state is given by the rank one spherical harmonics (see also Appendix H, Eq. (H.3))

$$Y_{1\,\mu} = \frac{1}{r}\sqrt{\frac{3}{4\pi}}\,x_\mu\,, \tag{2.179}$$

where the spherical component of a position vector $\boldsymbol{x} = (x, y, z)$ are given by

$$x_\mu = \begin{pmatrix} -\frac{1}{\sqrt{2}}(x+iy) \\ z \\ \frac{1}{\sqrt{2}}(x-iy) \end{pmatrix}. \tag{2.180}$$

Apart from the radial part, properties of the p state wave functions under rotation is determined by the unit radial vectors, and so we set

$$f_{00}(\boldsymbol{x}_i) \sim 1 \quad f_{01}(\boldsymbol{x}_i) \sim \boldsymbol{x}_i\,. \tag{2.181}$$

Therefore, taking mixed-symmetry combinations of the three functions 1, 1 and $\boldsymbol{x}$, we see that

$$\begin{aligned} \psi_\lambda &\sim \boldsymbol{x}_1 + \boldsymbol{x}_2 - 2\boldsymbol{x}_3 \sim \boldsymbol{\lambda}\,, \\ \psi_\rho &\sim \boldsymbol{x}_1 - \boldsymbol{x}_2 \sim \boldsymbol{\rho}\,. \end{aligned}$$

Associated with these mixed-symmetric orbital part, the spin-flavor part must be mixed symmetric. From Table 2.8, possible combinations are:

$$\Psi_{MS}^{N=1} = \begin{cases} \chi_{MS} \times \phi_{MS} & \sim \quad {}^2 8 = 16 \text{ dim} \\ \chi_{MS} \times \phi_{S} & \sim \quad {}^2 10 = 20 \text{ dim} \\ \chi_{S} \times \phi_{MS} & \sim \quad {}^4 8 = 32 \text{ dim} \\ \chi_{MS} \times \phi_{A} & \sim \quad {}^2 1 = 2 \text{ dim} \end{cases} \tag{2.182}$$

Hence total dimension of this multiplet is 70. The correspondence of these quark model states with physical baryons (excited states) are partially recognized. This will be discussed more in detail in Chapter 8

Second excited state, $N = 2$

States of $N = 2$ are provided by either two $l = 1$ states, one $l = 2$ state or one $n = 1$ state:

$$\begin{aligned} &(f_{00}\,,\ f_{01}\,,\ f_{01})\,, \\ &(f_{00}\,,\ f_{00}\,,\ f_{02})\,, \end{aligned}$$

$$(f_{00}\,,\ f_{00}\,,\ f_{10})\,. \tag{2.183}$$

If only one quark is excited into an $N = 2$ (either $l = 2$ or $n = 1$) orbit, the orbital wave function is either S or MS. If, in contrast, two quarks are excited into $N = 1$ ($l = 1$) orbits, the orbital wave function can take any symmetry (Note that $l = 1$ states are distinguished by the third component l_z.).

As in Eq. (2.179), these excitations are described, in this case, by five components of the second rank spherical harmonics Y_{2m}, and one component of square radius r^2. They are linear combinations of second order polynomials of x, y and z: There are six of them:

$$x^2\,,\ y^2\,,\ z^2\,,\ xy\,,\ yz\,,\ zx\,. \tag{2.184}$$

Three quark states of $N = 2$ and their irreducible decomposition is rather tedious and is discussed in Appendix A.

Depending on the symmetries of the orbital wave functions, spin-flavor part can take any type of symmetry. Hence there are totally symmetric 56-dimensional and mixed symmetric 70-dimensional representations. The totally anti-symmetric orbital part is also possible and accompanied by the totally anti-symmetric spin-flavor parts. They are

$$\Psi_A^{N=2} = \begin{cases} \chi_S \times \phi_A & \sim \quad {}^4 1 = 4\ \text{dim} \\ \chi_{MS} \times \phi_{MS} & \sim \quad {}^2 8 = 16\ \text{dim} \end{cases} \tag{2.185}$$

Therefore, they form a 20 dimensional representation.

Summary tables

Based on the discussions on the permutation symmetry for multi-particle system, one can, in principle, generate all wave functions for baryons in the quark model. Somewhat non-trivial is the derivation of orbital wave functions when written by the Jacobi coordinates $\boldsymbol{\rho}$ and $\boldsymbol{\lambda}$, after center of mass motion is removed. We discuss this problem in Appendix A. Here we summarize explicit forms in the following Tables. They include:

(1) Table 2.11: Orbital wave functions. In tabulating this, we introduce single-particle wave function f^N inspired by the harmonic oscillator wave function, Eq. (2.169).

(2) Table 2.12: Spin wave functions.

(3) Table 2.13: Flavor wave functions.
(4) Table 2.14: Totally symmetric baryon wave functions.

Table 2.11 Orbital wave functions written in terms of the Jacobi coordinates.

$N=0$	$\psi_S^0 = f^0(\boldsymbol{\lambda})f^0(\boldsymbol{\rho})$
$N=1$	$\psi_\lambda^1 = f^1(\boldsymbol{\lambda})f^0(\boldsymbol{\rho})$
	$\psi_\rho^1 = f^0(\boldsymbol{\lambda})f^1(\boldsymbol{\rho})$
$N=2$	$\psi_S^2 = \frac{1}{\sqrt{2}}(f^2(\boldsymbol{\lambda})f^0(\boldsymbol{\rho}) + f^0(\boldsymbol{\lambda})f^2(\boldsymbol{\rho}))$
	$\psi_\lambda^2 = \frac{1}{\sqrt{2}}(f^2(\boldsymbol{\lambda})f^0(\boldsymbol{\rho}) - f^0(\boldsymbol{\lambda})f^2(\boldsymbol{\rho}))$
	$\psi_\rho^2 = f^1(\boldsymbol{\lambda})f^1(\boldsymbol{\rho})$

Table 2.12 Spin wave functions χ.

Spin3/2	Spin1/2
$\chi_{3/2}^S = \mid\uparrow\uparrow\uparrow\rangle$	$\chi_{1/2}^\rho = \frac{1}{\sqrt{2}}(\mid\uparrow\downarrow\uparrow\rangle - \mid\downarrow\uparrow\uparrow\rangle)$
$\chi_{1/2}^S = \frac{1}{\sqrt{3}}(\mid\uparrow\uparrow\downarrow\rangle + \mid\uparrow\downarrow\uparrow\rangle + \mid\uparrow\uparrow\downarrow\rangle)$	$\chi_{-1/2}^\rho = \frac{1}{\sqrt{2}}(\mid\uparrow\downarrow\downarrow\rangle - \mid\downarrow\uparrow\downarrow\rangle)$
$\chi_{-1/2}^S = \frac{1}{\sqrt{3}}(\mid\uparrow\downarrow\downarrow\rangle + \mid\downarrow\downarrow\uparrow\rangle + \mid\downarrow\uparrow\downarrow\rangle)$	$\chi_{1/2}^\lambda = \frac{-1}{\sqrt{6}}(\mid\uparrow\downarrow\uparrow\rangle + \mid\downarrow\uparrow\uparrow\rangle - 2\mid\uparrow\uparrow\downarrow\rangle)$
$\chi_{-3/2}^S = \mid\downarrow\downarrow\downarrow\rangle$	$\chi_{-1/2}^\lambda = \frac{1}{\sqrt{6}}(\mid\downarrow\uparrow\downarrow\rangle + \mid\uparrow\downarrow\downarrow\rangle - 2\mid\downarrow\downarrow\uparrow\rangle)$

Table 2.13 Flavor wave functions $\phi^{\sigma}_{S\,I\,I_z}$.

Wave function	Baryons
$\phi^{\rho}_{0\,1/2\,1/2} = \frac{1}{\sqrt{2}}(udu - duu)$	N
$\phi^{\rho}_{0\,1/2\,-1/2} = \frac{1}{\sqrt{2}}(udd - dud)$	
$\phi^{\rho}_{-1\,1\,1} = \frac{1}{\sqrt{2}}(usu - suu)$	Σ
$\phi^{\rho}_{-1\,1\,0} = \frac{1}{2}(sud - usd + sdu - dsu)$	
$\phi^{\rho}_{-1\,1\,-1} = \frac{1}{\sqrt{2}}(dsd - sdd)$	
$\phi^{\rho}_{-1\,0\,0} = \frac{1}{2\sqrt{3}}(2uds - 2dus + usd - dsu - sud + sdu)$	Λ
$\phi^{\rho}_{-2\,1/2\,1/2} = \frac{1}{\sqrt{2}}(sus - uss)$	Ξ
$\phi^{\rho}_{-2\,1/2\,1/2} = \frac{1}{\sqrt{2}}(sds - dss)$	
$\phi^{\lambda}_{0\,1/2\,1/2} = -\frac{1}{\sqrt{6}}(udu + duu - 2uud)$	N
$\phi^{\lambda}_{0\,1/2\,-1/2} = \frac{1}{\sqrt{6}}(dud + udd - 2ddu)$	
$\phi^{\lambda}_{-1\,1\,1} = \frac{1}{\sqrt{6}}(suu + usu - 2uus)$	Σ
$\phi^{\lambda}_{-1\,1\,0} = \frac{1}{2\sqrt{3}}(sud + usd + sdu + dsu - 2uds - 2dus)$	
$\phi^{\lambda}_{-1\,1\,-1} = \frac{1}{\sqrt{6}}(sdd + dsd - 2dds)$	
$\phi^{\lambda}_{-1\,0\,0} = \frac{1}{2}(usd - dsu + sud - sdu)$	Λ
$\phi^{\lambda}_{-2\,1/2\,1/2} = -\frac{1}{\sqrt{6}}(sus + uss - 2ssu)$	Ξ
$\phi^{\lambda}_{-2\,1/2\,-1/2} = \frac{1}{\sqrt{6}}(sds + dss - 2ssd)$	
$\phi^{S}_{0\,3/2\,3/2} = uuu$	Δ
$\phi^{S}_{0\,3/2\,1/2} = \frac{1}{\sqrt{3}}(uud + udu + duu)$	
$\phi^{S}_{0\,3/2\,-1/2} = \frac{1}{\sqrt{3}}(udd + ddu + dud)$	
$\phi^{S}_{0\,3/2\,-3/2} = ddd$	
$\phi^{S}_{-1\,1\,1} = \frac{1}{\sqrt{3}}(uus + usu + suu)$	Σ
$\phi^{S}_{-1\,1\,0} = \frac{1}{\sqrt{6}}(uds + dus + usd + dsu + sud + sdu)$	
$\phi^{S}_{-1\,1\,-1} = \frac{1}{\sqrt{3}}(dds + dsd + sdd)$	
$\phi^{S}_{-2\,1/2\,1/2} = \frac{1}{\sqrt{3}}(uss + ssu + sus)$	Ξ
$\phi^{S}_{-2\,1/2\,-1/2} = \frac{1}{\sqrt{3}}(dss + ssd + sds)$	
$\phi^{S}_{-3\,0\,0} = sss$	Ω

Table 2.14 Construction of total wave functions.

56 dim	$^{2}8$	$\psi_S^N \frac{1}{\sqrt{2}}(\chi_\lambda\phi_\lambda + \chi_\rho\phi_\rho)$
	$^{4}10$	$\psi_S^N \chi_S\phi_S$
70 dim	$^{2}8$	$\frac{1}{2}(\psi_\rho^N\chi_\rho\phi_\lambda + \psi_\rho^N\chi_\lambda\phi_\rho + \psi_\lambda^N\chi_\rho\phi_\rho - \psi_\lambda^N\chi_\lambda\phi_\lambda)$
	$^{4}8$	$\frac{1}{\sqrt{2}}(\psi_\lambda^N\chi_\lambda + \psi_\rho^N\chi_\rho)\phi_S$
	$^{2}10$	$\frac{1}{\sqrt{2}}(\psi_\lambda^N\chi_S\phi_\lambda + \psi_\rho^N\chi_S\phi_\rho)$
	$^{2}1$	$\frac{1}{\sqrt{2}}(\psi_\lambda^N\chi_\rho - \psi_\rho^N\chi_\lambda)\phi_A$
20 dim	$^{2}8$	$\psi_A^N \frac{1}{\sqrt{2}}(\chi_\lambda^N\phi_\rho - \chi_\rho^N\phi_\lambda)$
	$^{4}1$	$\psi_A^N\chi_S\phi_A$

Problems

2.1

(1) Verify the invariance of the determinant $\phi^1\psi^2 - \phi^2\psi^1$ under $SU(2)$ transformations (see Eq. (2.42)).
(2) In the Euler angle representation, an $SU(2)$ matrix is parameterized as

$$U(\alpha, \beta, \gamma) = \exp(it_z\alpha)\exp(it_y\beta)\exp(it_z\gamma)\,. \tag{2.186}$$

Compute the trace

$$D_{ij} = \frac{1}{2}\,\mathrm{tr}\,(\tau_i U \tau_j U^\dagger) \tag{2.187}$$

and find the rotation matrix for an ordinary vectors.

2.2

Prove that the dimension of the (p_1, p_2) representation of $SU(3)$ is given by Eq. (2.86) by explicitly counting the number of sites of the hexagon for the representation including degeneracy.

2.3

Consider a tensor of the type $(m, m'; n, n')$ (see Eq. (2.92) and explanation below)

$$T^{i_1,\cdots,i_m,j_1,\cdots,j_{m'}}_{k_1,\cdots,k_n,l_1,\cdots,l_{n'}}\,. \tag{2.188}$$

The anti-symmetric combination of this tensor is written using the ϵ tensor as in Eq. (2.95):

$$A^{i_2,\cdots,j_2,\cdots}_{p,k_1,\cdots,l_1,\cdots} = \epsilon_{pi_1j_1}A^{i_1,i_2\cdots,j_1,j_2\cdots}_{k_1,\cdots,l_1,\cdots}\,. \tag{2.189}$$

In order to show that it is symmetric among the lower indices $k_1, \cdots, l_1, \cdots$ (the first index p is irrelevant!), form an antisymmetric combination with respect to the lower two indices, say, k_1 and l_1:

$$\epsilon^{qk_1l_1}A^{i_2,\cdots,j_2,\cdots}_{p,k_1,k_2,\cdots,l_1,l_2,\cdots} = \epsilon^{qk_1l_1}\epsilon_{pi_1j_1}A^{i_1,i_2\cdots,j_1,j_2\cdots}_{p,k_1,k_2,\cdots,l_1,l_2,\cdots}\,. \tag{2.190}$$

Using

$$\epsilon^{qk_1l_1}\epsilon_{pi_1j_1} = \delta^q_p\delta^{k_1}_{i_1}\delta^{l_1}_{j_1} - \delta^q_p\delta^{l_1}_{i_1}\delta^{k_1}_{j_1} + (\text{permutation}) \tag{2.191}$$

and the traceless property, verify that this quantity vanishes.

2.4

(1) Verify that the six elements of the permutation symmetry group S^3 are represented by the six 2×2 matrices of Eq. (2.149).

(2) Show that these matrices are expressed by matrices of rotation and parity inversion:

$$R(\theta) = \begin{pmatrix} \cos\theta & -\sin\theta \\ \sin\theta & \cos\theta \end{pmatrix}, \quad P = \begin{pmatrix} 1 & 0 \\ 0 & -1 \end{pmatrix}, \tag{2.192}$$

where rotation angle θ is a multiple of 120 degrees.

2.5

(1) Introducing six-dimensional vectors for three-particle states $\alpha\beta\gamma \ldots$ as

$$\begin{aligned}
\alpha\beta\gamma &= (1,\ 0,\ 0,\ 0,\ 0,\ 0)\,, & \beta\gamma\alpha &= (0,\ 1,\ 0,\ 0,\ 0,\ 0)\,,\\
\gamma\alpha\beta &= (0,\ 0,\ 1,\ 0,\ 0,\ 0)\,, & \beta\alpha\gamma &= (0,\ 0,\ 0,\ 1,\ 0,\ 0)\,,\\
\alpha\gamma\beta &= (0,\ 0,\ 0,\ 0,\ 1,\ 0)\,, & \gamma\beta\alpha &= (0,\ 0,\ 0,\ 0,\ 0,\ 1)\,.
\end{aligned}$$

we can write the λ and ρ functions as

$$\begin{aligned}
\phi_\lambda &= \frac{1}{6}(2,\ -1,\ -1,\ 2,\ -1,\ -1)\,,\\
\phi_\rho &= \frac{1}{6}(2,\ -1,\ -1,\ -2,\ 1,\ 1)\,.
\end{aligned}$$

Verify that six elements of permutation symmetry group S^3, $\mathcal{E}_i$ and $\mathcal{O}_i$ are represented in two different ways by two-dimensional representations in the bases of, for instance, $(\phi_\lambda,\ \mathcal{E}_2\phi_\lambda)$ and $(\phi_\rho,\ \mathcal{E}_2\phi_\rho)$.

(2) When $\beta = \gamma$, verify that the above two representations reduces to a single representation as shown in section 2.5 (see Eqs. (2.146), (2.147) and (2.149)).

Chapter 3

CHIRAL SYMMETRY

Chiral symmetry is one of the main topics of this book. It is very important in low energy hadron physics, since it generates the pion, the lightest hadron, as the Nambu-Goldstone mode associated with its spontaneous breaking. Then chiral symmetry dictates the interactions of the pions and baryons.

Since chiral symmetry is related to the right and left handed decomposition of relativistic spinors, we start with the discussion of the Lorentz group and its representations *. Chiral symmetry is then introduced in section 3.2 as internal symmetry of the right and left handed spinors. We have made a rather detailed description for these two topics, not only because they are important but also because students are often confused by the concept of chirality.

3.1 Lorentz Group and Chiral Fermions

Commutation relations

Lorentz transformations are defined as linear transformations which preserve the length of four vectors in the Minkowski space,

$$x_\mu \to x'_\mu \quad = \quad a_\mu^{\ \nu} x_\nu \,, \tag{3.1}$$

*Part of the discussion in this and next chapters owe the excellent textbook by T. Kugo which, however, is written only in Japanese [Kugo 1989].

where the infinitesimal form of the transformation matrices $a_\mu^{\ \nu}$ are given by

$$a_\mu^{\ \nu} = \delta_\mu^{\ \nu} + \epsilon_\mu^{\ \nu} \,. \tag{3.2}$$

Here $\epsilon_\mu^{\ \nu}$ are antisymmetric real parameters. The transformation of Eq. (3.1) can be written in the form of group representation as

$$x'_\mu = (1 + \frac{i}{2}\epsilon_{\alpha\beta} M^{\alpha\beta})_\mu^{\ \nu}\, x_\nu \,, \tag{3.3}$$

Here $M^{\alpha\beta}$ are the six ($= 4 \times 3/2$) generators of the Lorentz group, which are given in 4×4 matrix form by

$$(M^{\alpha\beta})_\mu^{\ \nu} = i(\delta_\mu^\alpha g^{\beta\nu} - \delta_\mu^\beta g^{\alpha\nu}) \,. \tag{3.4}$$

Note that $M^{\alpha\beta}$ are four dimensional extensions of the generators of three dimensional rotation,

$$\begin{aligned} (M^{ij})_{lm} &\to i\epsilon^{ijk}(t_k)_{lm} \\ &= i\epsilon^{ijk}\epsilon_{klm} = i(\delta_l^i \delta_m^j - \delta_m^i \delta_l^j) \,, \end{aligned}$$

where $(t_k)_{lm} = \epsilon_{klm}$ are the adjoint (spin one) representations of $SU(2)$ generators. The matrix form of Eq. (3.4) determines commutation relations among the generators $M^{\alpha\beta}$ which are of more fundamental nature;

$$\begin{aligned} &[M^{\alpha\beta}, M^{\gamma\delta}] \\ &= -i(g^{\alpha\gamma}M^{\beta\delta} - g^{\alpha\delta}M^{\beta\gamma} - g^{\beta\gamma}M^{\alpha\delta} + g^{\beta\delta}M^{\alpha\gamma}) \,. \end{aligned} \tag{3.5}$$

Let the angular momentum and boost operators be

$$J_i = \frac{1}{2}\epsilon_{ijk}M^{jk} \,, \tag{3.6}$$

$$K_i = M^{0i} \,. \tag{3.7}$$

Then commutation relations among these generators are given by

$$\begin{aligned} [J_i, J_j] &= i\epsilon_{ijk}J_k \,, \\ [J_i, K_j] &= i\epsilon_{ijk}K_k \,, \\ [K_i, K_j] &= -i\epsilon_{ijk}J_k \,. \end{aligned} \tag{3.8}$$

Defining the right and left generators by

$$\begin{aligned} R_i &= \frac{1}{2}(J_i + iK_i)\,, \\ L_i &= \frac{1}{2}(J_i - iK_i)\,, \end{aligned} \tag{3.9}$$

we find two independent commutation relations,

$$\begin{aligned} [R_i, R_j] &= i\epsilon_{ijk}R_k\,, \\ [L_i, L_j] &= i\epsilon_{ijk}L_k\,, \\ [L_i, R_j] &= 0\,. \end{aligned} \tag{3.10}$$

These form the algebra of the Lorentz group.

Several remarks are in order.

- The right and left generators satisfy separately the angular momentum algebra. Hence the irreducible representations are easily constructed. In order to write Eq. (3.10), we have formed combinations $J_i \pm iK_i$. The i factor in front of the boost generators K_i serves to transform them to rotation generators; the former are equivalent to the latter in the coordinate system of imaginary time.

- We can introduce parity operation. Under that operation, the boost generators change their signs, while the rotation generators do not;

$$K_i \overset{\text{parity}}{\longrightarrow} -K_i\,, \quad J_i \overset{\text{parity}}{\longrightarrow} J_i\,. \tag{3.11}$$

  Then it follows that the right and left generators interchange their roles under the parity transformation:

$$R_i \longleftrightarrow L_i\,. \tag{3.12}$$

- It is known that the algebra (3.10) of the Lorentz group is equivalent to that of $SL(2, C)$, a group of special linear transformations of two complex numbers, whose elements are 2×2 complex valued matrices,

$$A = \begin{pmatrix} a & b \\ c & d \end{pmatrix}\,, \quad \text{with} \quad \det A = 1\,. \tag{3.13}$$

Due to the constraint $\det A = 1$, A contains six real parameters corresponding to those of the Lorentz group. To be precise, the original Lorentz transformations for four-vectors as given in Eq. (3.1) are not completely the same as the transformations of $SL(2, C)$ (see footnote on page on page 70).

Representations

We consider the representations of the Lorentz group dictated by the commutation relations, Eq. (3.10). Since the R and L generators form two independent $SU(2)$ algebras, the group may be denoted as $SU(2)_R \times SU(2)_L$, where $SU(2)_R$ and $SU(2)_L$ are spin (rotational) groups for the right and left handed fermions. The representations may be labeled by two spin values, one for the right $SU(2)_R$ and another for left $SU(2)_L$ representations: (s_R, s_L). The lowest (fundamental) representations are the spinor representations $(\frac{1}{2}, 0)$ and $(0, \frac{1}{2})$. Higher dimensional representations are constructed from their tensor products.

Fundamental representations

The spinor representation $(\frac{1}{2}, 0)$ is called the right-handed fundamental representation of the Lorentz group $SL(2, C)$. It is represented by a two component complex spinor χ_R

$$(\frac{1}{2}, 0) \sim \chi_R \,. \tag{3.14}$$

In this basis, the right generators R_i are represented by the Pauli matrices

$$R_i \sim \sigma_i / 2 \,. \tag{3.15}$$

Accordingly the rotation and boost generators are given by

$$\begin{aligned} J_i &= R_i + L_i = \sigma_i/2 \equiv s_i \,, \\ K_i &= i(-R_i + L_i) = -i\sigma_i/2 \equiv -is_i \,. \end{aligned} \tag{3.16}$$

Elements of the Lorentz group are then represented as

$$U_R(\boldsymbol{a}, \boldsymbol{b}) = \exp\left(i\boldsymbol{s} \cdot \boldsymbol{a} + \boldsymbol{s} \cdot \boldsymbol{b}\right) \,, \tag{3.17}$$

where $\boldsymbol{a} = (a_1, a_2, a_3)$ and $\boldsymbol{b} = (b_1, b_2, b_3)$ are real parameters for rotation and boost. These matrices act on right-handed spin 1/2 spinors (fermions)

$\chi_R \in (\frac{1}{2}, 0)$ as †,

$$\chi'_R = U_R(\boldsymbol{a}, \boldsymbol{b})\chi_R\,. \tag{3.18}$$

Similarly, one can construct the left-handed spinor $\chi_L \sim (0, \frac{1}{2})$, where transformation matrices are given by

$$U_L(\boldsymbol{a}, \boldsymbol{b}) = \exp\left(i\boldsymbol{s}\cdot\boldsymbol{a} - \boldsymbol{s}\cdot\boldsymbol{b}\right)\,, \tag{3.19}$$

acting on the left-handed spinor χ_L,

$$\chi'_L = U_L(\boldsymbol{a}, \boldsymbol{b})\chi_L\,. \tag{3.20}$$

The right and left handed representations are related to each other by complex conjugation. This is expected from Eq. (3.9):

$$\begin{aligned} R_i^* &= \frac{1}{2}(J_i + iK_i)^* \\ &= \frac{1}{2}(J_i - iK_i) = L_i\,. \end{aligned} \tag{3.21}$$

In order to see this more precisely, let us take a complex conjugate of Eq. (3.20):

$$\begin{aligned} (\chi'_L)^* &= U_L^*(\boldsymbol{a}, \boldsymbol{b})\chi_L^* \\ &= \exp\left(-i\boldsymbol{s}^*\cdot\boldsymbol{a} - \boldsymbol{s}^*\cdot\boldsymbol{b}\right)\chi_L^*\,. \end{aligned} \tag{3.22}$$

Multiplying the antisymmetric tensor

$$\epsilon = \begin{pmatrix} 0 & 1 \\ -1 & 0 \end{pmatrix}, \quad \epsilon^{-1} = \epsilon^T \ (T:\ \text{transpose}) \tag{3.23}$$

from the left, we obtain

$$\begin{aligned} (\epsilon\chi_L^*)' &= \exp\left(-i\epsilon\boldsymbol{s}^*\epsilon^{-1}\cdot\boldsymbol{a} - \epsilon\boldsymbol{s}^*\epsilon^{-1}\cdot\boldsymbol{b}\right)(\epsilon\chi_L^*) \\ &= \exp\left(i\boldsymbol{s}\cdot\boldsymbol{a} + \epsilon\boldsymbol{s}\cdot\boldsymbol{b}\right)(\epsilon\chi_L^*) \end{aligned} \tag{3.24}$$

From this we observe the correspondence

$$\epsilon\chi_L^* \sim \chi_R\,. \tag{3.25}$$

†In literatures, the right-handed spinors are often denoted by ξ and the left handed-spinors as η [Lifschitz and Pitaevskii 1971; Kugo 1989; Miransky 1995].

Likewise,

$$\epsilon\chi_R^* \sim \chi_L\,. \tag{3.26}$$

The right and left-handed fermions $\chi_R \sim (\frac{1}{2},0)$ and $\chi_L \sim (0,\frac{1}{2})$ are called chiral fermions and are the building blocks of the relativistic description of spin 1/2 particles.

Scalar

Lorentz scalars are formed as bi-spinors with the ϵ-tensor inserted,

$$\begin{aligned} S \;\sim\;& (\epsilon\chi_R)^T\phi_R = \chi_R^T\epsilon^T\phi_R = -\chi_R^T\epsilon\phi_R\,, \\ \text{and}\quad & (\epsilon\chi_L)^T\phi_L = \chi_L^T\epsilon^T\phi_L = -\chi_L^T\epsilon\phi_L\,. \end{aligned} \tag{3.27}$$

The invariance of these quantities can be verified if we notice the property of the 2×2 Lorentz transformation matrix

$$\det U_{R,L}(\boldsymbol{a},\boldsymbol{b}) = 1\,. \tag{3.28}$$

Also from Eq. (3.25), we obtain other types of Lorentz scalars

$$(\chi_L)^\dagger\phi_R\,,\quad \text{and}\quad (\chi_R)^\dagger\phi_L\,. \tag{3.29}$$

Four-vector

The rule to form a four-vector from spinors is [‡]

$$V^\mu \sim \chi_R^\dagger\sigma^\mu\phi_R\,, \tag{3.30}$$

where

$$\sigma^\mu = (1,\boldsymbol{\sigma})\,. \tag{3.31}$$

It is an elementary exercise to check that $SL(2,C)$ transformations on spinors as given in Eqs. (3.18) and (3.20) yield familiar Lorentz transformations of four-vectors. By using the property (3.25), one can write other types of four-vectors

$$\chi_L^T\epsilon^T\sigma^\mu\phi_R\,,\quad \chi_R^\dagger\sigma^\mu\epsilon\phi_L^*\,,\quad \chi_L^T\epsilon^T\sigma^\mu\epsilon\phi_L^*\,. \tag{3.32}$$

[‡]Due to the bilinear nature, the four-vector transforms in the same way under the two distinctive transformations: $\chi\to A\chi$ and $\chi\to -A\chi$.

Table 3.1 Lorentz scalar and four vectors

scalar	$\chi_R^T\epsilon^T\chi_R$	$\chi_L^T\epsilon^T\chi_L$	$\chi_L^\dagger\chi_R$	$\chi_R^\dagger\chi_L$
vector	$\chi_R^\dagger\sigma^\mu\chi_R$	$\chi_L^\dagger\bar{\sigma}^\mu\chi_L$	$\chi_L^T\epsilon^T\sigma^\mu\chi_R$	$\chi_L^\dagger\bar{\sigma}^\mu\epsilon^T\chi_R^*$

It is not difficult to verify that the last combination can be written as

$$\phi_L^\dagger\bar{\sigma}^\mu\chi_L\,, \tag{3.33}$$

where

$$\bar{\sigma}^\mu = (\epsilon^T\sigma^\mu\epsilon)^T = (1, -\boldsymbol{\sigma})\,. \tag{3.34}$$

In Table 3.1, we summarize bi-spinor forms of Lorentz scalar and vectors for $\phi = \chi$.

Lagrangian

The Lagrangian $\mathcal{L}$ may be constructed by imposing the conditions of (1) Hermitian, $\mathcal{L}^\dagger = \mathcal{L}$, (2) Lorentz invariance, and (3) fermion number conservation. Let us first consider kinetic terms. They contain the derivative ∂_μ of four-vector. This can be contracted with another four-vector, either $\chi_R^\dagger\sigma^\mu\chi_R$ or $\chi_L^\dagger\bar{\sigma}^\mu\chi_L$. Therefore, we have the following candidates for the kinetic terms:§

$$\mathcal{L}_K = i\chi_R^\dagger\sigma^\mu\partial_\mu\chi_R + i\chi_L^\dagger\bar{\sigma}^\mu\partial_\mu\chi_L\,, \tag{3.35}$$

Next we consider the mass term. It should be a scalar, and therefore, a candidate would be,

$$\mathcal{L}_M = m(\chi_L^\dagger\chi_R + \chi_R^\dagger\chi_L)\,. \tag{3.36}$$

This is the Dirac mass term which conserves the fermion number (which is guaranteed by the combination $\chi^\dagger\chi$). If we allow non-conservation of the fermion number, the Majorana mass term is also possible using the first two scalars in Table 3.1. The physical meaning of the Dirac and Majorana mass terms is nicely explained in the book [Miransky 1995]. Here and in what follows we consider only the Dirac mass, since in strong interaction

§A symmetric form is also possible such as $\mathcal{L}_K^l = (i/2)(\chi_R^\dagger\sigma^\mu(\partial_\mu\chi_R) - (\partial_\mu\chi_R^\dagger)\sigma^\mu\chi_R)$.

processes, the fermion number is conserved. Then the lagrangian may be written as

$$\mathcal{L} = i\chi_R^\dagger \sigma^\mu \partial_\mu \chi_R + i\chi_L^\dagger \bar{\sigma}^\mu \partial_\mu \chi_L + m(\chi_L^\dagger \chi_R + \chi_R^\dagger \chi_L) \,. \tag{3.37}$$

Weyl representation

We have just seen that we need both χ_R and χ_L to describe a massive spinor in a relativistic theory. We combine them into one four component spinor ψ. In the so called Weyl representation, the right and left handed spinors are located in the upper and lower parts,

$$\psi = \psi_R + \psi_L = \begin{pmatrix} \chi_R \\ \chi_L \end{pmatrix} , \tag{3.38}$$

where we have introduced formally four-component chiral fermions

$$\psi_R = \begin{pmatrix} \chi_R \\ 0 \end{pmatrix} , \quad \psi_L = \begin{pmatrix} 0 \\ \chi_L \end{pmatrix} . \tag{3.39}$$

Then we can write the Lagrangian (3.37) in the standard form as

$$\mathcal{L} = i\bar{\psi}\gamma^\mu \partial_\mu \psi + m\bar{\psi}\psi \,. \tag{3.40}$$

Here we have introduced the gamma matrices in the Weyl representation

$$\gamma^0 = \begin{pmatrix} 0 & 1 \\ 1 & 0 \end{pmatrix} , \quad \gamma^i = \begin{pmatrix} 0 & -\sigma_i \\ \sigma_i & 0 \end{pmatrix} \tag{3.41}$$

and the Dirac conjugate

$$\bar{\psi} = \psi^\dagger \gamma_0 \,. \tag{3.42}$$

In particular, we have

$$\bar{\psi}_R = \begin{pmatrix} 0 \\ \chi_R^\dagger \end{pmatrix} , \quad \bar{\psi}_L = \begin{pmatrix} \chi_L^\dagger \\ 0 \end{pmatrix} . \tag{3.43}$$

Using the four gamma matrices we can form the gamma-five matrix which is defined by

$$\gamma_5 \equiv i\gamma^0\gamma^1\gamma^2\gamma^3 = \begin{pmatrix} 1 & 0 \\ 0 & -1 \end{pmatrix} . \tag{3.44}$$

Then it follows that the right and left handed fermions are eigenstates of γ_5,

$$\gamma_5\psi_R = +\psi_R\,, \quad \gamma_5\psi_L = -\psi_L\,. \tag{3.45}$$

The right and left handed fermions are projected out by the projection operators

$$\begin{aligned} P_R &= \frac{1+\gamma_5}{2} = \begin{pmatrix} 1 & 0 \\ 0 & 0 \end{pmatrix}, \\ P_L &= \frac{1-\gamma_5}{2} = \begin{pmatrix} 0 & 0 \\ 0 & 1 \end{pmatrix}. \end{aligned} \tag{3.46}$$

The operators P_R and P_L satisfy the relations

$$\begin{aligned} &P_R + P_L = 1\,, \\ &P_R^2 = P_R\,, \quad P_L^2 = P_L\,, \quad P_R P_L = 0\,. \end{aligned} \tag{3.47}$$

Parity

Under the parity transformation, the role of the right and left handedness is interchanged as implied by Eq. (3.12), and is shown in Fig. 3.1. It implies an interchange of states

$$\chi_R(x) \leftrightarrow \eta\chi_L(\bar{x})\,, \quad x = (t, \boldsymbol{x})\,, \tag{3.48}$$

$$\bar{x} = (t, -\boldsymbol{x})\,, \tag{3.49}$$

where $\eta = \pm 1$ is a phase which determines the eigenvalue of the parity. In the four component Weyl representation, this interchange alone is not sufficient, since for instance

$$\psi_R(x) = \begin{pmatrix} \chi_R(x) \\ 0 \end{pmatrix} \to \begin{pmatrix} \eta\chi_L(\bar{x}) \\ 0 \end{pmatrix}. \tag{3.50}$$

But the left handed component should be placed in the lower part of the four component spinor. This can be done by multiplying γ_0 to the above equation. Thus the correct parity transformation is defined by (for the right handed fermion)

$$\psi_R(x) = \begin{pmatrix} \chi_R(x) \\ 0 \end{pmatrix}$$

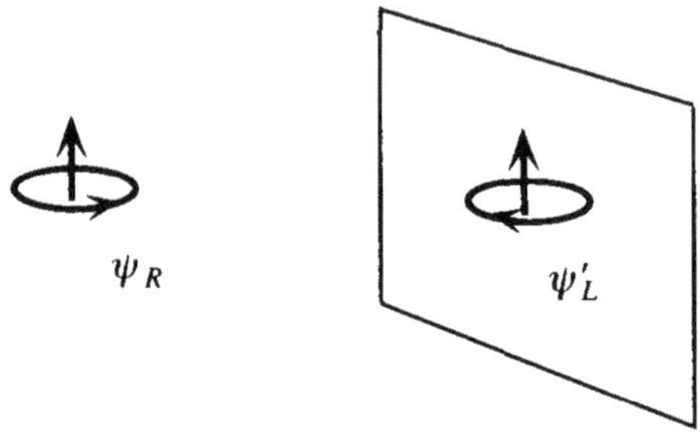

Fig. 3.1 Parity transformation of ψ_L and ψ_R.

$$\rightarrow \begin{pmatrix} 0 & 1 \\ 1 & 0 \end{pmatrix} \begin{pmatrix} \eta\chi_L(\bar{x}) \\ 0 \end{pmatrix} = \begin{pmatrix} 0 \\ \eta\chi_L(\bar{x}) \end{pmatrix} . \tag{3.51}$$

It is apparent that we can not form a parity eigenstate from ψ_R or ψ_L alone. In the Weyl representation it is given by

$$\begin{aligned} \psi(x) &= \begin{pmatrix} \chi_R(x) \\ \chi_L(x) \end{pmatrix} \\ &\rightarrow \gamma_0 \begin{pmatrix} \eta\chi_L(\bar{x}) \\ \eta\chi_R(\bar{x}) \end{pmatrix} = \eta\psi(\bar{x}) , \end{aligned} \tag{3.52}$$

where η is the eigenvalue of the parity.

Dirac representation

When discussing the non-relativistic limit of fermions, it is convenient to introduce the Dirac representation, where ψ is defined by

$$\psi = \begin{pmatrix} \frac{1}{\sqrt{2}}(\chi_R + \chi_L) \\ \frac{1}{\sqrt{2}}(\chi_R - \chi_L) \end{pmatrix} . \tag{3.53}$$

In this basis, the gamma matrices are given by

$$\begin{aligned} \gamma_0 &= \begin{pmatrix} 1 & 0 \\ 0 & -1 \end{pmatrix} , \quad \gamma^i = \begin{pmatrix} 0 & \sigma_i \\ -\sigma_i & 0 \end{pmatrix} , \\ \gamma_5 &= \begin{pmatrix} 0 & 1 \\ 1 & 0 \end{pmatrix} . \end{aligned} \tag{3.54}$$

Note that Eq. (3.53) is also an eigenstate of the parity. The transformation matrix between the Weyl and Dirac representations,

$$T = \frac{1}{\sqrt{2}} \begin{pmatrix} 1 & 1 \\ 1 & -1 \end{pmatrix} \tag{3.55}$$

relates the parity eigenstate in the Weyl representation, Eq. (3.52), to that of the Dirac representation, Eq. (3.53).

Helicity states

At the end of this section, we comment on the relation between the helicity states with the chiral fermions ψ_R or ψ_L. The helicity is defined by

$$h = \boldsymbol{\sigma} \cdot \hat{\boldsymbol{p}} . \tag{3.56}$$

Eigenstates of the energy (either positive or negative), momentum ($\boldsymbol{p}$) and the helicity ($h = \pm$) satisfy (u for positive energy and v for negative energy states)

$$\begin{aligned} (\not{p} - m)u(\boldsymbol{p}, h) &= 0 , \\ (\not{p} + m)v(\boldsymbol{p}, h) &= 0 . \end{aligned} \tag{3.57}$$

The solutions are given by

$$\begin{aligned} u(\boldsymbol{p}, h) &= \sqrt{\frac{E+m}{2E}} \begin{pmatrix} 1 \\ \frac{\boldsymbol{\sigma} \cdot \boldsymbol{p}}{E+m} \end{pmatrix} \chi_h , \\ v(\boldsymbol{p}, h) &= \sqrt{\frac{E+m}{2E}} \begin{pmatrix} \frac{\boldsymbol{\sigma} \cdot \boldsymbol{p}}{E+m} \\ 1 \end{pmatrix} \chi_h , \end{aligned} \tag{3.58}$$

where χ_h is the two component spinor quantized along the $\boldsymbol{p}$ direction. Choose the momentum $\boldsymbol{p}$ along the z-direction. Then, for instance,

$$u(\boldsymbol{p}, +) = \sqrt{\frac{E+m}{2E}} \begin{pmatrix} 1 \\ \frac{\sigma_z}{E+m} \end{pmatrix} \chi_+ , \tag{3.59}$$

where $\chi_+ = (1, 0)$. In the massless limit $m = 0$,

$$u(\boldsymbol{p}, +) = \begin{pmatrix} 1 \\ 0 \\ 1 \\ 0 \end{pmatrix} \tag{3.60}$$

is an eigenstate of γ_5 with the eigenvalue +1. Similarly, one can verify that $u(\boldsymbol{p}, -)$ is an eigenstate of $\gamma_5 = -1$. Therefore, in the massless limit, the helicity states coincide with the chiral fermions. It is clear that in the massive case, this does not hold.

3.2 Chiral Group

$U(1)_R \times U(1)_L$

In the massless limit the lagrangian (3.37) has symmetries under *internal* phase transformations of ψ_R and ψ_L separately:

$$\psi_R \to \exp(i\theta_R)\psi_R\,, \quad \psi_L \to \exp(i\theta_L)\psi_L\,. \tag{3.61}$$

The transformations of the right (left) handed fermion is called the right (left) transformation. Associated with these symmetries, right and left currents are conserved independently:

$$\partial^\mu R_\mu = 0\,, \quad \partial^\mu L_\mu = 0\,, \tag{3.62}$$

where

$$\begin{aligned} R_\mu &= \bar{\psi}_R \gamma_\mu \psi_R = \chi_R^\dagger \sigma_\mu \chi_R\,, \\ L_\mu &= \bar{\psi}_L \gamma_\mu \psi_L = \chi_L^\dagger \bar{\sigma}_\mu \chi_L\,. \end{aligned} \tag{3.63}$$

The conservation of the currents in Eq. (3.62) implies that the charges

$$Q_R = \int d^3x\, R_0\,, \quad Q_L = \int d^3x\, L_0 \tag{3.64}$$

are the constants of motion. The symmetry associated with the two independent phase transformations is called chiral symmetry, and is denoted as $U(1)_R \times U(1)_L$.

For massive fermions the right and left symmetries are not exact symmetries. There is, however, a residual symmetry in the diagonal part whose elements are parameterized by $\theta_R = \theta_L \equiv \theta_V$. Namely, the right and left handed fermions are transformed in the same way:

$$\psi \to \exp(i\theta_V)\psi = \begin{pmatrix} \exp(i\theta_V)\psi_R \\ \exp(i\theta_V)\psi_L \end{pmatrix}. \tag{3.65}$$

This symmetry is also referred to as the vector symmetry. It is straightforward to verify that the mass term of Eq. (3.36) is invariant under this transformation.

Another transformation which is independent of the vector transformation is the axial transformation where $\theta_L = -\theta_R = \theta_A$:

$$\psi \to \exp(i\theta_A\gamma_5)\psi = \begin{pmatrix} \exp(i\theta_A)\psi_R \\ \exp(-i\theta_A)\psi_L \end{pmatrix} . \tag{3.66}$$

Apparently, the mass term is no longer invariant under the axial transformation.

Associated with these transformations, the vector and axial-vector currents are given by

$$\begin{aligned} V_\mu &= R_\mu + L_\mu = \bar{\psi}\gamma_\mu\psi , \\ A_\mu &= R_\mu - L_\mu = \bar{\psi}\gamma_\mu\gamma_5\psi . \end{aligned} \tag{3.67}$$

For the massive case, only the vector current is conserved, but the axial-vector current is not. Thus the vector charge, the fermion number,

$$Q_V = Q_R + Q_L = \int d^3x(\chi_R^\dagger\chi_R + \chi_L^\dagger\chi_L) \tag{3.68}$$

is a constant of motion independent of the mass term. In contrast, the axial charge

$$Q_A = Q_R - Q_L = \int d^3x(\chi_R^\dagger\chi_R - \chi_L^\dagger\chi_L) \tag{3.69}$$

can be a constant of motion only when the fermion is massless.

$SU(2)_R \times SU(2)_L$

The $U(1)_R \times U(1)_L$ chiral symmetry is extended to $U(N)_R \times U(N)_L$, where $N \geq 2$. The parameter N is identified with the number of light flavors. In this chapter we discuss the chiral symmetry for the two flavor (isospin) case, $SU(2)_L \times SU(2)_R$. An extension to three-flavor case will be given in the next chapter.

For two flavors, a fermion forms a two-component iso-spinor

$$\psi = \begin{pmatrix} u \\ d \end{pmatrix} \tag{3.70}$$

which is the fundamental representation of isospin $SU(2)$. Both components, u and d, are four-component spinors. In the Weyl representation, they are

$$u = u_R + u_L \,, \tag{3.71}$$

and

$$u_R = \begin{pmatrix} \chi_R^u \\ 0 \end{pmatrix} , \quad u_L = \begin{pmatrix} 0 \\ \chi_L^u \end{pmatrix} ,$$

where χ_R^u and χ_L^u are two component spinors. We have a similar relation for the d component. Thus one can write

$$\psi = \psi_R + \psi_L = \begin{pmatrix} u_R + u_L \\ d_R + d_L \end{pmatrix} = \begin{pmatrix} \chi_R^u \\ \chi_L^u \\ \chi_R^d \\ \chi_L^d \end{pmatrix} . \tag{3.72}$$

The free lagrangian is given by

$$\mathcal{L} = \bar{\psi}(i\partial\!\!\!/ - m)\psi \tag{3.73}$$

where m is a diagonal 2×2 mass matrix in the isospin space,

$$m = \begin{pmatrix} m_u & 0 \\ 0 & m_d \end{pmatrix} . \tag{3.74}$$

$SU(2)_R \times SU(2)_L$ chiral transformations are isospin rotations for

$$\psi_R = \begin{pmatrix} u_R \\ d_R \end{pmatrix} , \quad \psi_L = \begin{pmatrix} u_L \\ d_L \end{pmatrix} . \tag{3.75}$$

Therefore, the right or left transformations are now written as

$$\psi_R \to \exp(i\, r)\psi_R \,, \quad \psi_L \to \exp(i\, l)\psi_L \,, \tag{3.76}$$

where

$$r = \boldsymbol{t} \cdot \boldsymbol{r} \,, \quad l = \boldsymbol{t} \cdot \boldsymbol{l} \,, \tag{3.77}$$

and $\boldsymbol{t} = \boldsymbol{\tau}/2$ are the generators of $SU(2)$.

Associated with these transformations, the isospin right and left currents are defined by

$$\begin{aligned} R^a_\mu &= \bar{\psi}_R \gamma_\mu t^a \psi_R = \chi^\dagger_R \sigma_\mu t^a \chi_R\,, \\ L^a_\mu &= \bar{\psi}_L \gamma_\mu t^a \psi_L = \chi^\dagger_L \bar{\sigma}_\mu t^a \chi_L\,. \end{aligned} \tag{3.78}$$

Here we have introduced isodoublet two-component spinors

$$\chi_R = \begin{pmatrix} \chi^u_R \\ \chi^d_R \end{pmatrix}, \quad \chi_L = \begin{pmatrix} \chi^u_L \\ \chi^u_L \end{pmatrix}. \tag{3.79}$$

As in Eq. (3.67), one can define the isospin vector and axial-vector currents:

$$\begin{aligned} V^a_\mu &= R^a_\mu + L^a_\mu = \bar{\psi}\gamma_\mu t^a \psi\,, \\ A^a_\mu &= R^a_\mu - L^a_\mu = \bar{\psi}\gamma_\mu \gamma_5 t^a \psi\,. \end{aligned} \tag{3.80}$$

In the massless limit they are both conserved under vector and axial-vector transformations:

$$\begin{aligned} \psi &\to \exp(iv)\psi = \begin{pmatrix} \exp(iv)u_R + \exp(iv)u_L \\ \exp(iv)d_R + \exp(iv)d_L \end{pmatrix}, \\ \psi &\to \exp(ia\gamma_5)\psi = \begin{pmatrix} \exp(ia)u_R - \exp(ia)u_L \\ \exp(ia)d_R - \exp(ia)d_L \end{pmatrix}. \end{aligned} \tag{3.81}$$

Here we have introduced

$$v = \boldsymbol{t}\cdot\boldsymbol{v}\,, \quad a = \boldsymbol{t}\cdot\boldsymbol{a} \tag{3.82}$$

with vector and axial parameters $\boldsymbol{v}$ and $\boldsymbol{a}$. As in the case of $U(1)_R \times U(1)_L$, for finite masses, both currents are not conserved. When $m_u = m_d$, the vector current is conserved, but the axial-vector current is not.

Let us now look at the transformation property of the currents. Under an infinitesimal transformations, the right current is subject to the variation

$$\begin{aligned} R^a_\mu &\to \bar{\psi}_R(1 + i\boldsymbol{t}\cdot\boldsymbol{r})\gamma_\mu t^a (1 - i\boldsymbol{t}\cdot\boldsymbol{r})\psi_R \\ &\sim \bar{\psi}_R \gamma_\mu t^a \psi_R - i\bar{\psi}_R \gamma_\mu [t^a, \boldsymbol{t}\cdot\boldsymbol{r}]\psi_R \\ &= R^a_\mu + \epsilon_{abc} r^b R^c_\mu\,, \end{aligned} \tag{3.83}$$

where we have used the commutation relations (2.36). Similarly

$$L^a_\mu \to L^a_\mu + \epsilon_{abc} l^b L^c_\mu\,. \tag{3.84}$$

These right and left transformations can be converted into the vector and axial transformations. The vector transformation is obtained when $r = l = v$. Therefore, by adding or subtracting the two equations (3.83) and (3.84), we find

$$\begin{aligned} V_\mu^a &\to V_\mu^a + \epsilon_{abc} v^b V_\mu^c \,, \\ A_\mu^a &\to A_\mu^a + \epsilon_{abc} v^b A_\mu^c \,. \end{aligned} \tag{3.85}$$

The axial transformation is for $r = -l = a$, and so

$$\begin{aligned} V_\mu^a &\to V_\mu^a + \epsilon_{abc} a^b A_\mu^c \,, \\ A_\mu^a &\to A_\mu^a + \epsilon_{abc} a^b V_\mu^c \,. \end{aligned} \tag{3.86}$$

Finally we would like to derive charge commutation relations corresponding to the current transformations shown in Eqs. (3.85) and (3.86) in the quantum theory. Applying the commutation relation (2.21) to the time components of the currents in Eq. (3.78), we obtain the two independent commutation relations

$$\begin{aligned} \left[Q_R^a, Q_R^b\right] &= i\epsilon_{abc} Q_R^c \,, \\ \left[Q_L^a, Q_L^b\right] &= i\epsilon_{abc} Q_L^c \,, \\ \left[Q_L^a, Q_R^b\right] &= 0 \,. \end{aligned} \tag{3.87}$$

As expected, the commutation relations for the right and left charges are closed within each sector. The algebra of Eq. (3.87) and the associated symmetry is called the chiral symmetry $SU(2)_R \times SU(2)_L$. Furthermore, the commutation relations among the vector and axial vector charges are given by

$$\begin{aligned} \left[Q_V^a, Q_V^b\right] &= i\epsilon_{abc} Q_V^c \,, \\ \left[Q_V^a, Q_A^b\right] &= i\epsilon_{abc} Q_A^c \,, \\ \left[Q_A^a, Q_A^b\right] &= i\epsilon_{abc} Q_V^c \,. \end{aligned} \tag{3.88}$$

These show that the vector and axial-vector transformations mix with each other. Although these commutation relations look very similar to the algebra of the Lorentz group, Eqs. (3.8) and (3.10), they should be clearly distinguished.

Chiral vs vector

So far we have considered only one fermion. If there are two (or more) fermions, an interesting feature emerges. This is essentially because chiral symmetry is an independent internal symmetry for the right and left handed fermions. To look at the relevant point, let us recall the naming of chiral symmetry. We have called the symmetry of the right (left) handed fermion as right (left) and denoted the corresponding group by $SU(N_f)_{R(L)}$. However, this naming is just a convention; one could have called them as $A(B)$ and denoted it by $SU(N_f)_{A(B)}$.

Consider two fermions which have right and left handed components. Thus we need two right handed and two left handed components:

$$a_R\,,\ b_R\,,\ c_L\,,\ d_L\,. \tag{3.89}$$

Naively, they are assumed to transform as

$$\begin{aligned} a_R\,,\ b_R &\to Aa_R\ Ab_R \\ c_L\,,\ d_L &\to Bc_L\ Bd_L\,, \end{aligned} \tag{3.90}$$

where $A \in SU(2)_A$ and $B \in SU(2)_B$. However, another transformation is also possible:

$$\begin{aligned} a_R\,,\ b_R &\to Aa_R\ Bb_R \\ c_L\,,\ d_L &\to Bc_L\ Ad_L\,. \end{aligned} \tag{3.91}$$

We call the first assignment of Eqs. (3.90) chiral, and the second one of Eqs. (3.91) vector.

In order to form parity eigenstates, right handed and left handed components must be paired (see Eq. (3.52)). One can form such pairs in two different ways:

$$\psi^1 = \begin{pmatrix} a_R \\ c_L \end{pmatrix} \equiv \begin{pmatrix} \psi_R^1 \\ \psi_L^1 \end{pmatrix}, \quad \phi^1 = \begin{pmatrix} b_R \\ d_L \end{pmatrix} \equiv \begin{pmatrix} \phi_R^1 \\ \phi_L^1 \end{pmatrix}, \tag{3.92}$$

or

$$\psi^2 = \begin{pmatrix} a_R \\ d_L \end{pmatrix} \equiv \begin{pmatrix} \psi_R^2 \\ \psi_L^2 \end{pmatrix}, \quad \phi^2 = \begin{pmatrix} b_R \\ c_L \end{pmatrix} \equiv \begin{pmatrix} \phi_R^2 \\ \phi_L^2 \end{pmatrix}. \tag{3.93}$$

In the chiral assignment of Eqs. (3.90), both $\{\psi^1, \phi^1\}$ and $\{\psi^2, \phi^2\}$ transform in the same way:

$$\begin{pmatrix} \psi_R^{1,2} \\ \psi_L^{1,2} \end{pmatrix} \to \begin{pmatrix} A\psi_R^{1,2} \\ B\psi_L^{1,2} \end{pmatrix}, \quad \begin{pmatrix} \phi_R^{1,2} \\ \phi_L^{1,2} \end{pmatrix} \to \begin{pmatrix} A\phi_R^{1,2} \\ B\phi_L^{1,2} \end{pmatrix}. \tag{3.94}$$

But in the vector assignment of Eqs. (3.91), $\{\psi^1, \phi^1\}$ and $\{\psi^2, \phi^2\}$ transform differently:

$$\begin{pmatrix} \psi_R^{1} \\ \psi_L^{1} \end{pmatrix} \to \begin{pmatrix} A\psi_R^{1} \\ A\psi_L^{1} \end{pmatrix}, \quad \begin{pmatrix} \phi_R^{1} \\ \phi_L^{1} \end{pmatrix} \to \begin{pmatrix} B\phi_R^{1} \\ B\phi_L^{1} \end{pmatrix} \tag{3.95}$$

while

$$\begin{pmatrix} \psi_R^{2} \\ \psi_L^{2} \end{pmatrix} \to \begin{pmatrix} A\psi_R^{2} \\ B\psi_L^{2} \end{pmatrix}, \quad \begin{pmatrix} \phi_R^{2} \\ \phi_L^{2} \end{pmatrix} \to \begin{pmatrix} B\phi_R^{2} \\ A\phi_L^{2} \end{pmatrix}. \tag{3.96}$$

The chiral assignment was called the naive assignment in Ref. [Jido et al. 2000b], while the second choice of the vector assignment, Eq. (3.96), was called the mirror assignment, since the transformation rule of ϕ looks a mirror image of that of ψ. In the vector assignment (including the mirror assignment), fermions can be massive without destroying chiral symmetry [Lee 1970; DeTar and Kunihiro 1989; Jido et al. 2000a].

3.3 Spontaneous Breaking of Chiral Symmetry

Degeneracy in multiplets

As we have seen, in the limit of massless QCD chiral symmetry becomes a good symmetry. Strictly, this holds only approximately in the real world due to small but nonzero masses of current quarks. In practice, chiral symmetry is considered to be good for the light flavor sector of u, d and s, in particular for u and d quarks.

Naively, the symmetry of the lagrangian (or hamiltonian) would lead to the symmetry of particles. In particular, since chiral symmetry (actually, γ_5 transformation) relates positive and negative parity states, they are expected to appear as degenerate particles. In reality, however, this is not the case as shown in Fig. 3.2, where masses of mesons and baryons with opposite parity are plotted. In the figure the still controversial state σ (It is denoted as $f_0(400-1200)$ by the Particle Data Group [PDG 2000]) is depicted with the large width of several hundred MeV. We see that in many

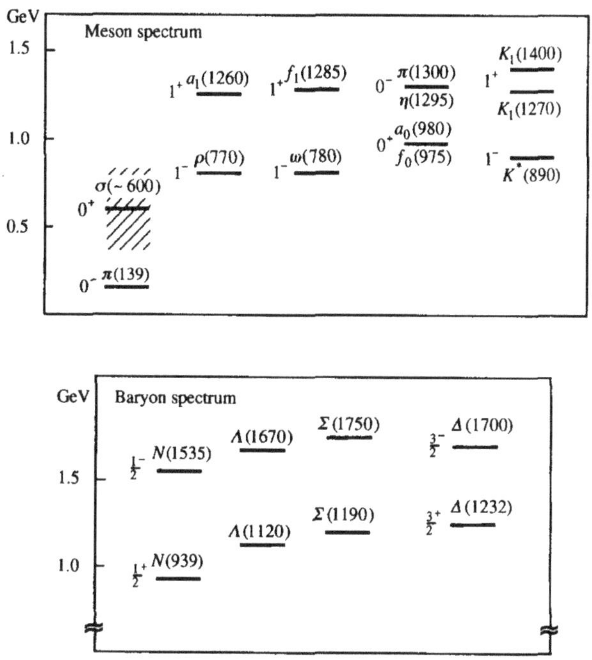

Fig. 3.2 Masses of mesons and baryons of opposite parity.

low lying channels, mass splittings between the positive (negative) parity ground states and first excited states of negative (positive) parity are about 500 MeV.

The symmetry of the system implies that the hamiltonian is invariant under the symmetry transformation U

$$UHU^\dagger = H\,. \tag{3.97}$$

Consider two states $|A\rangle$ and $|B\rangle$ which are related by the transformation U. Suppose that they are created by the creation operators $a^\dagger$ and $b^\dagger$ as $|A\rangle = a^\dagger|0\rangle, |B\rangle = b^\dagger|0\rangle$. Then U relates a and b through $UaU^\dagger = b$. Now the transformation of the state $|A\rangle$ is

$$\begin{aligned} |A\rangle \to U|A\rangle &= Ua^\dagger U^\dagger U|0\rangle \\ &= b^\dagger \; U|0\rangle \end{aligned} \tag{3.98}$$

Naively, if the vacuum was invariant

$$U|0\rangle = |0\rangle \tag{3.99}$$

then we would expect that

$$E_A = \langle A|H|A\rangle = \langle B|UHU^\dagger|B\rangle = \langle B|H|B\rangle = E_B\,. \tag{3.100}$$

In other words, the states $|A\rangle$ and $|B\rangle$ are degenerate.

The key assumption in the above derivation is the invariance of the vacuum as expressed in Eq. (3.99). If the symmetry of the hamiltonian is also the symmetry of the state, it is said that the symmetry is realized explicitly. Contrary, if the symmetry of the hamiltonian is not realized in states, it is said that the symmetry is broken spontaneously [Nambu and Jona-Lasinio 1961a; Nambu and Jona-Lasinio 1961b; Goldstone 1961]. The former is also called the Wigner-Weyl phase (mode), while the latter the Nambu-Goldstone phase (mode). In the present real world, there are many indications that chiral symmetry is spontaneously broken. More precisely, the full chiral symmetry of $SU(N_f)_L \times SU(N_f)_R$ is spontaneously broken down to the diagonal vector symmetry, $SU(N_f)_V$.

Of particular importance is the presence of light pseudoscalar mesons as the Nambu-Goldstone bosons; they are identified with the pions. Then, significant part of pion dynamics is described by chiral symmetry, which has been formulated by the current algebra and low energy theorems.

The Nambu-Goldstone theorem

This theorem is associated with the appearance of the zero mode when chiral symmetry is spontaneously broken [Nambu and Jona-Lasinio 1961a; Goldstone 1961]. Sometimes the spontaneous breaking of a symmetry is expressed as the nonvanishing of the vacuum when operated by the charge Q, as the invariance of $U|0\rangle$ implies,

$$Q|0\rangle = 0\,. \tag{3.101}$$

However, there is a problem of principle that in the broken phase, the charge Q does not exist. One can show this by proving the opposite proposition; if Q exists, the symmetry is realized in the Wigner phase.

If Q does not exist in the spontaneously broken (Nambu-Goldstone) mode, the proof of the Nambu-Goldstone theorem should be done carefully.

It turns out that although the charge Q is not well-defined, the commutation relation

$$i[Q,\phi(x)] = \int d^3y[J_0(y),\phi(x)] \equiv \delta\phi(x) \tag{3.102}$$

does make sense. One can prove that if there exists an operator which satisfies $\langle 0|\delta\phi(x)|0\rangle \neq 0$, then the symmetry is spontaneously broken.

In order to compute the commutation relation

$$0 \neq \langle 0|\delta\phi(0)|0\rangle = \langle 0|\int d^3x[J_0(x),\phi(0)]|0\rangle\,, \tag{3.103}$$

first we compute the time ordered product:

$$\langle 0|T[J_\mu(x)\phi(0)]|0\rangle = \partial_\mu \int d\sigma^2\rho(\sigma^2)\Delta_F(x;\sigma)\,. \tag{3.104}$$

Here $\Delta_F(x;\sigma)$ is the Feynman's scalar propagator:

$$\Delta_F(x;\sigma) = -\int \frac{d^4p}{(2\pi)^4}\frac{1}{p^2-\sigma^2+i\epsilon}e^{-ipx}\,, \tag{3.105}$$

and $\rho(\sigma)$ is a scalar spectral density. Then Eq. (3.104) is obtained by performing the standard spectral representation for the Green's functions. Consider the total derivative

$$\begin{aligned}\partial^\mu[TJ_\mu(x)\phi(0)] &= \delta(t)[J_0(x)\phi(0)] + T\partial^\mu J_\mu(x)\phi(0)\\ &= \delta(t)[J_0(x)\phi(0)]\,,\end{aligned} \tag{3.106}$$

where we have used the current conservation. Then the quantity which we are interested in is expressed as the total derivative:

$$\begin{aligned}\langle 0|\delta\phi(0)|0\rangle &= \langle 0|\int d^3x[J_0(x),\phi(0)]|0\rangle\\ &= \int d^4x\partial^\mu\langle 0|T[J_\mu(x),\phi(0)]|0\rangle\\ &= \int d^4x\int d\sigma^2\partial^2\rho(\sigma^2)\Delta_F(x)\,.\end{aligned} \tag{3.107}$$

Naively, this quantity would vanish. However in the momentum space, this equation can be written as

$$\lim_{q\to 0}\int d\sigma^2\rho(\sigma^2)\frac{-q^2}{q^2-\sigma^2+i\epsilon} \tag{3.108}$$

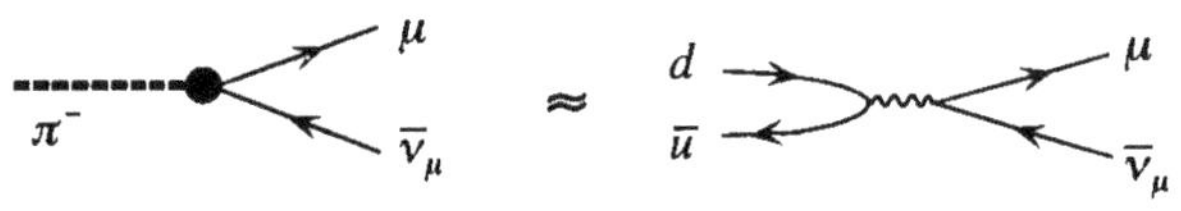

Fig. 3.3 Weak decay of the pion and its microscopic interpretation.

and survives if there exists a pole at the zero momentum $q \to 0$. This zero momentum state is the Nambu-Goldstone mode as corresponding to the pion. As it stands, this zero mode carries the quantum numbers which are the same as the charge Q. In reality, due to small masses of quarks, the pion acquires a small mass about 139 MeV.

Examples of low energy theorems

In this subsection we discuss three examples of low energy theorems. They are the weak decay of the pion, the beta decay of the neutron and the soft-pion theorem.

The weak decay of the pion and PCAC

The pion decays into the muon and neutrino through the weak interaction. This process is depicted in Fig. 3.3, where on the right hand side, the quark content of π^- and the weak boson (W^-) propagation are shown explicitly. At low energies as compared with the mass of W, the interaction reduces to a current-current form, $J_\mu J^\mu$. The currents contain both the leptonic and hadronic part: $J_\mu = J^h_\mu + J^l_\mu$. Thus the T-matrix of this process can be written as

$$T_{\pi\to\mu\nu} \sim \langle\mu\nu|J_\mu J^\mu|\pi(q)\rangle = \langle\mu\nu|J^l_\mu|0\rangle\langle 0|J^{h\,\mu}|\pi(q)\rangle \,. \tag{3.109}$$

The leptonic part is well understood and is given by $\bar{l}(1-\gamma_5)l$, while the hadronic part is not, due to the effect of strong interaction renormalization. Since the pion carries the negative parity, only the axial-vector current is relevant to the hadronic part:

$$\langle 0|J^{h\,\mu}|\pi(q)\rangle = \langle 0|A^\mu|\pi(q)\rangle \,. \tag{3.110}$$

The matrix element of the axial-vector current $A^a_\mu(\mathrm{x})$ can be written as

$$\langle 0|A^a_\mu(x)|\pi^b(q)\rangle = iq_\mu\delta_{ab}f_\pi(q^2)e^{-iqx} \,. \tag{3.111}$$

In writing this expression, we have used the translational invariance and the fact that the only relevant four vector is the pion momentum q_μ. A possible q^2 dependence is then included in the form factor $f_\pi(q^2)$. The on-mass-shell value of $f_\pi(q^2)$ is defined to be the pion decay constant. Taking the derivative of Eq. (3.111), we find

$$\langle 0|\partial^\mu A_\mu^a(x)|\pi^b(q)\rangle = q^2\delta_{ab}f_\pi(q^2)e^{-iqx} . \tag{3.112}$$

If the mass of the pion were zero ¶, $q^2 = m_\pi^2 = 0$, implying that the axial-vector current is conserved.

In reality, there is a small pion mass and $q^2 = m_\pi^2 \neq 0$ for an on-shell pion. If one introduces the pion field $\phi^a(x)$ such that

$$\langle 0|\phi^a(x)|\pi^b(q)\rangle = \delta_{ab}e^{-iqx} , \tag{3.113}$$

one can expect the following relation

$$\partial^\mu A_\mu^a(x) = m_\pi^2 f_\pi \phi^a(x) . \tag{3.114}$$

This is the PCAC (partially conserved axial-vector current) relation and is very powerful in deriving various low energy theorems [Nambu 1960; Gell-Mann and Levy 1960] (for critical comments, see Ref. [Coleman 1985], Chapter 2). One can also formulate many relations in the exact chiral limit of $m_\pi = 0$ and CAC (conserved axial-vector current). In this case, however, one has to be careful for the appearance of the massless pion pole (see the next subsection).

The Goldberger-Treiman relation

The neutron decays into the proton by emitting electron and electron-neutrino. The hadronic part contains both the vector and axial-vector current contributions. Here we discuss the contribution of the axial-vector current. Since the nucleon couples to the pion through the strong interaction, there are two contributions to the hadronic axial-vector current; one is the direct coupling and the other is the pion pole term as depicted in Fig. 3.4. The following discussion is done in the limit of CAC.

¶In fact, in this ideal case, the pion can not decay.

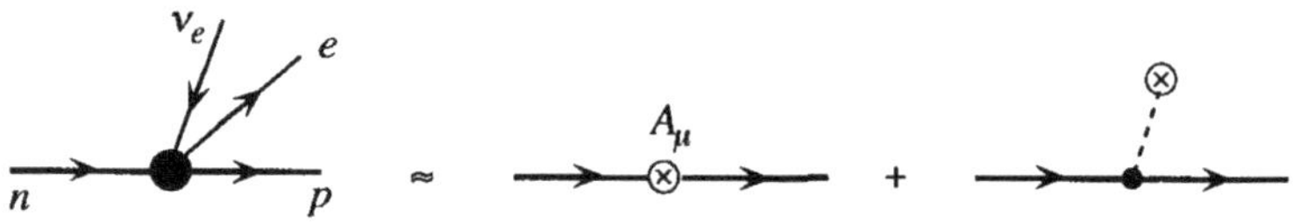

Fig. 3.4 Weak decay of the neutron.

Let us write the nucleon matrix element of the axial-vector current as

$$\begin{aligned}&\langle p(p_f)|A^a_\mu(0)|n(p_i)\rangle\\&=\bar{u}_p(p_f)\left(\gamma_\mu\gamma_5 g_A(q^2)+q_\mu\gamma_5 h_A(q^2)\right)t^a u_n(p_i)\,.\end{aligned}\tag{3.115}$$

Here $t^a = \tau^a/2$, $u_{p,n}$ are the Dirac spinors for the proton and neutron, g_A and h_A are the two form factors, and $q = p_f - p_i$. Taking derivative of Eq. (3.115) and using the conservation of the axial vector current, we obtain

$$\begin{aligned}0&=\langle p(p_f)|\partial^\mu A^a_\mu(0)|n(p_i)\rangle\\&=i\bar{u}_p(p_f)\left(\not{q}\gamma_5 g_A(q^2)+q^2\gamma_5 h_A(q^2)\right)t^a\ u_n(p_i)\,.\end{aligned}\tag{3.116}$$

Therefore,

$$h_A(q^2) = -2M_N\frac{g_A(q^2)}{q^2}\,,\tag{3.117}$$

where we have used the equation of motion for the nucleons,

$$(\not{p}_i - M_N)u_n(p_i) = 0\,,\quad \bar{u}_p(\not{p}_f + M_N) = 0\,.\tag{3.118}$$

This implies that the h_A is the pion pole term of Fig. 3.4.

Let us now compute the pion pole term explicitly. Eq. (3.111) implies that $A^a_\mu(x) = -f_\pi\partial_\mu\phi^a(x) + \cdots$ and therefore, we can write

$$\langle p(p_f)|A^a_\mu(0)|n(p_i)\rangle_{\pi-pole} = +i\ f_\pi q_\mu\langle p(p_f)|\phi^a(0)|n(p_i)\rangle\,.\tag{3.119}$$

We define the strong coupling constant $g_{\pi NN}$ by

$$\langle p(p_f)|\phi^a(0)|n(p_i)\rangle = \frac{i}{q^2}g_{\pi NN}(q^2)\bar{u}_p(p_f)\gamma_5\tau^a u_n(p_i)\,.\tag{3.120}$$

Then we have

$$\langle p(p_f)|A^a_\mu(0)|n(p_i)\rangle_{\pi-pole} = -f_\pi\frac{q_\mu}{q^2}g_{\pi NN}(q^2)\bar{u}_p(p_f)\gamma_5\tau^a u_n(p_i)\,.\tag{3.121}$$

By identifying this with the h_A term of Eq. (3.115) together with Eq. (3.117), we find the following Goldberger-Treiman relation [Goldberger and Treiman 1958]

$$M_N g_A = f_\pi g_{\pi NN} \,. \tag{3.122}$$

In reality, this relation holds only at 2 % level, by using $M_N = 939$ MeV, $g_A = 1.25$, $f_\pi = 92.5$ MeV and $g_{\pi NN} = 13.5$.

Soft pion theorem

This theorem relates a matrix element which involves n pions with the one that involves $n-1$ pions. Let us consider a process $\alpha \to \beta\pi$. We first apply the LSZ reduction formula [Bjorken and Drell 1965; Itzykson and Zuber 1980] to the matrix element

$$\begin{aligned} \langle \pi^a(q)\beta|\mathcal{O}|\alpha\rangle &= i\int d^4x\, e^{iqx}(\partial^2 + m_\pi^2)\langle\beta|T\phi^a(x)\mathcal{O}|\alpha\rangle \\ &= i\int d^4x\, e^{iqx}(-q^2 + m_\pi^2)\langle\beta|T\phi^a(x)\mathcal{O}|\alpha\rangle \,. \end{aligned} \tag{3.123}$$

We can replace the pion field $\phi^a(x)$ by the divergence of the axial-vector current through the PCAC (3.114);

$$= i\frac{m_\pi^2 - q^2}{f_\pi m_\pi^2}\int d^4x\, e^{iqx}\langle\beta|T\partial^\mu A_\mu^a \mathcal{O}|\alpha\rangle \,. \tag{3.124}$$

Integrating by parts, we obtain

$$\begin{aligned} = \; & i\frac{m_\pi^2 - q^2}{f_\pi m_\pi^2}\int d^4x e^{iqx}\left(-\langle\beta|[A_0^a, \mathcal{O}]|\alpha\rangle\delta(x_0)\right. \\ & \left. -iq^\mu\langle\beta|TA_\mu^a(x)\mathcal{O}(0)|\alpha\rangle\right) \,. \end{aligned} \tag{3.125}$$

If there is no pole contribution at $q \to 0$ when $m_\pi \neq 0$, the second term of this equation can be neglected and we find the following soft pion theorem[Nambu and Lurie 1962; Adler and Dashen 1968; J.F. Donoghue and Holstine 1992]:

$$\langle \pi^a(q)\beta|\mathcal{O}|\alpha\rangle = -\frac{i}{f_\pi}\langle\beta|[Q_A^a, \mathcal{O}]|\alpha\rangle \,. \tag{3.126}$$

As an application of the soft pion theorem, we discuss the pion photoproduction $\gamma N \to \pi N$ as depicted in Fig. 3.5. For simplicity we consider a

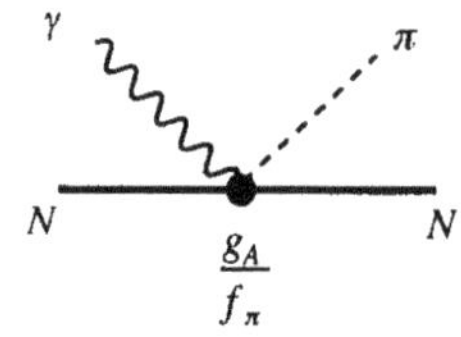

Fig. 3.5 Kroll-Ruderman term for the pion photoproduction.

charged pion state $\pi^2 = (1/\sqrt{2})(\pi^+ - i\pi^-)$. We apply the LSZ reduction formula for the photon state and then use the soft pion theorem:

$$\begin{aligned}\langle N\pi^2|\gamma N\rangle &\sim \int d^4x e^{-ikx}\epsilon^\mu \langle N\pi^2|J_\mu(x)|N\rangle \\ &= -\frac{1}{f_\pi}\int d^4x e^{-ikx}\langle N|[Q_A^2, J_\mu(x)]|N\rangle\epsilon^\mu . \quad (3.127)\end{aligned}$$

Here ϵ^μ is the polarization vector of the photon. The electromagnetic current contains the isoscalar and isovector part. But the commutator survives only for the isovector part

$$\begin{aligned}[Q_A^2, J_\mu^3(x)]\epsilon^\mu &= J_{5,\mu}^1(x)\epsilon^\mu \\ &= -\boldsymbol{J}_5^1(x)\cdot\boldsymbol{\epsilon} \sim g_A\tau^1\boldsymbol{\sigma}\cdot\boldsymbol{\epsilon} , \quad (3.128)\end{aligned}$$

where we have used the fact that the photon is real. Finally we have

$$\langle N\pi^2|\gamma N\rangle \sim \frac{g_A}{f_\pi}\tau^1\boldsymbol{\sigma}\cdot\boldsymbol{\epsilon} . \quad (3.129)$$

This is known as the Kroll-Ruderman theorem [Kroll and Ruderman 1954]. From this derivation the lowest order (and hence the dominant) term contributes only to the charged pion production.

Problems

3.1

Under the boost transformation, spinors $\chi_{R,L}$ are subject to transformation

$$\begin{aligned} \chi_R &\to \exp(\boldsymbol{v}\cdot\boldsymbol{\sigma})\chi_R\,, \\ \chi_L &\to \exp(-\boldsymbol{v}\cdot\boldsymbol{\sigma})\chi_L\,, \end{aligned} \tag{3.130}$$

where $\boldsymbol{v}$ corresponds to a velocity. Verify that the bispinors $V_\mu = \chi_R^\dagger \sigma_\mu \chi_R$ and $\chi_L^\dagger \bar{\sigma}_\mu \chi_L$ transform as a four-vector, where $\sigma_\mu = (\sigma_0, \boldsymbol{\sigma})$ and $\bar{\sigma}_\mu = (\sigma_0, -\boldsymbol{\sigma})$ (see Eqs. (3.31) and (3.34)).

3.2

Let us consider two fermions $\psi = (\psi_R, \psi_L)$ and $\phi = (\phi_R, \phi_L)$. As discussed in section 3.2, under the mirror assignment they transform as

$$\begin{aligned} (\psi_R, \psi_L) &\to (\exp(i\theta_R)\psi_R, \exp(i\theta_L)\psi_L)\,, \\ (\phi_R, \phi_L) &\to (\exp(i\theta_L)\phi_R, \exp(i\theta_R)\phi_L)\,. \end{aligned} \tag{3.131}$$

Find a chiral invariant mass term in the off-diagonal combination of ψ and ϕ. Show that it is written in the four-component representation as

$$\bar{\phi}\gamma_5\psi - \bar{\psi}\gamma_5\phi\,. \tag{3.132}$$

3.3

Derive the spectral representation of Eq. (3.104):

$$\langle 0|T[J_\mu(x)\phi(0)]|0\rangle = \partial_\mu \int d\sigma^2 \rho(\sigma^2)\Delta_F(x;\sigma)\,. \tag{3.133}$$

3.4

In Eq. (3.116), we need to take a (space-time) derivative $\partial_\mu A^\mu$. In the matrix element of the axial-vector current, however, there is no argument x. Interpreting $\partial_\mu A^\mu(0) = \lim_{x\to 0}\partial_\mu A^\mu(x)$, explain how to take the derivative of the matrix element and derive the second equation of Eq. (3.116).

3.5

In the text, we have derived the soft pion theorem of Eq. (3.126) for the physical pion with a small finite mass. Here we want to derive it for a massless pion. To do this we first write down the reduction formula

$$\langle \pi^a(q)\beta|\mathcal{O}|\alpha\rangle \;=\; -i\int d^4x\, e^{iqx} q^2 \langle\beta|T\phi^a(x)\mathcal{O}|\alpha\rangle\,. \qquad (3.134)$$

Now integrating by parts the right hand side of the following CAC implied equation:

$$0 = i\int d^4x\; e^{iqx}\langle\beta|T\partial^\mu A^a_\mu O|\alpha\rangle\,,$$

extract the expression of the right hand side of Eq. (3.134). Finally, using

$$A^a_\mu \sim -f_\pi \partial_\mu \phi^a\,,$$

prove the soft pion theorem.

Chapter 4

THE SIGMA MODEL

In this chapter we discuss sigma models in linear and non-linear realizations. In order to construct linear sigma models we use a group theoretical method, which turns out to be powerful and useful. In section 4.1, we construct linear sigma models for two and three flavors. In section 4.2, we introduce non-linear sigma models first by emphasizing geometric structure of the parameter space spanned by the chiral field. A formal method of Callan-Coleman-Wess-Zumino is then described in a concrete way. Finally in section 4.3, we introduce fermions both for linear and non-linear realizations. Through all these discussions we have tried to clarify mathematical (mostly geometric) structure of theories as much as possible by considering some specific examples and by trying to write down all necessary equations explicitly.

4.1 Linear Sigma Model

$SU(2)_R \times SU(2)_L$

In linear sigma models, basic fields are assumed to belong to linear representations of the chiral group. We consider representations of low dimensions, since they play dominant role in the low energy region. The lagrangian is then a chiral invariant combination of the basic fields.

Let us define our notation. We use dimensions to denote representations of isospin symmetries of $SU(2)$. They are 1 (singlet), 2 (fundamental representation), 3 (adjoint representation) and etc. Representations of the chiral group $SU(2)_R \times SU(2)_L$ are then denoted by a set of two integers

(m, n), where m and n are dimensions of the representations of $SU(2)_R$ and $SU(2)_L$, respectively. In what follows, we consider only one fermion. Therefore, we may call the isospin rotation for the right handed fermion as the Right isospin rotation, and the isospin rotation for the left handed fermion as the Left isospin rotation (see the discussion in the subsection *Chiral vs Vector* in section 3.2).

The fundamental representations of the right and left handed fermions are denoted as

$$\begin{aligned}(2,1) \quad &\sim \quad \chi_R = \begin{pmatrix} \chi_R^u \\ \chi_R^d \end{pmatrix},\\ (1,2) \quad &\sim \quad \chi_L = \begin{pmatrix} \chi_L^u \\ \chi_L^d \end{pmatrix},\end{aligned}$$

and are subject to the transformation rule as (see Eq. (3.76))

$$\begin{aligned}(2,1) : \chi_R \quad &\to \quad (\exp(i\boldsymbol{t}\cdot\boldsymbol{r}))_{ab}(\chi_R)_b \equiv g_R\chi_R\,,\\ (1,2) : \chi_L \quad &\to \quad (\exp(i\boldsymbol{t}\cdot\boldsymbol{l}))_{ab}(\chi_L)_b \equiv g_L\chi_L\,.\end{aligned}$$

Let us recall the parity transformation (see Eq. (3.12)):

$$\chi_R \leftrightarrow \eta\chi_L \quad \text{or} \quad (2,1) \leftrightarrow (1,2)\,, \tag{4.1}$$

where $\eta = \pm 1$. Thus, either χ_R or χ_L alone is not an eigenstate of parity. Instead, their linear combinations are:

$$\begin{aligned}\chi_\eta \equiv \chi_R + \chi_L \quad &\to \quad \eta(\chi_L + \chi_R) = \eta\chi_\eta\,,\\ \chi_{-\eta} \equiv \chi_R - \chi_L \quad &\to \quad \eta(\chi_L - \chi_R) = -\eta\chi_\eta\,.\end{aligned}$$

They are nothing but the components of Dirac representations and are denoted as $(2,1) + (1,2)$.

Of particular importance is the (chiral) four-vector representation:

$$(2,2) \sim \chi_L\chi_R^\dagger \equiv \; U \;\; \text{or} \;\; \chi_R\chi_L^\dagger \equiv U^\dagger\,. \tag{4.2}$$

Note that from a group theoretical point of view, there is no distinction between $\chi_L\chi_R^\dagger$ and $\chi_R\chi_L^\dagger$, since $SU(2)$ is real (see section 2.3). Their transformation rules are

$$U \to g_L U g_R^\dagger\,,\;\; U^\dagger \to g_R U^\dagger g_L^\dagger\,. \tag{4.3}$$

The key formula which relates the quark bilinear $\chi_L \chi_R^\dagger$ and a four vector B_α is *

$$B_\alpha = \mathrm{tr}\, \tau_\alpha \chi_L \chi_R^\dagger = \chi_R^\dagger \tau_\alpha \chi_L \,, \tag{4.4}$$

where the (isospin) τ matrices form a four vector, $\tau_\alpha = (1, \tau_1, \tau_2, \tau_3)$. Note that the combination here is a four-vector in the chiral isospin space but a Lorentz scalar in space time; see Eqs. (3.27) and (3.29). The basic trans-formation rule of the four-vector under the $SU(2)_R \times SU(2)_L$ is given by

$$\begin{array}{llcl} \text{Vector tr} & B_0 & \to & B_0 \\ & \boldsymbol{B} & \to & \boldsymbol{B} + \boldsymbol{v} \times \boldsymbol{B} \\ \text{Axial vector tr} & B_0 & \to & B_0 + i\boldsymbol{a} \cdot \boldsymbol{B} \\ & \boldsymbol{B} & \to & \boldsymbol{B} + i\boldsymbol{a} B_0 \end{array} \tag{4.5}$$

Here $\boldsymbol{v}$ and $\boldsymbol{a}$ are the parameters for the vector and axial-vector transformations as defined in Eqs. (3.81) and (3.82). The transformation rule of Eq. (4.5) implies that the group $SU(2)_R \times SU(2)_L$ is isomorphic to $\mathcal{O}(4)$. Their equivalence is also verified by counting their dimensions,

$$\begin{aligned} \dim[SU(2) \times SU(2)] &= 3 + 3 = 6 \,, \\ \dim[\mathcal{O}(4)] &= n(n-1)/2|_{n=4} = 6 \,. \end{aligned} \tag{4.6}$$

The four vector B_α is not an eigenstate of the parity,

$$B_\alpha = \chi_R^\dagger \tau_\alpha \chi_L \stackrel{\text{parity}}{\longrightarrow} \chi_L^\dagger \tau_\alpha \chi_R = B_\alpha^\dagger \,. \tag{4.7}$$

The appropriate combinations with a definite parity are

$$\begin{array}{lll} \text{Positive parity} & \sigma_\alpha & = \chi_L^\dagger \tau_\alpha \chi_R + \chi_R^\dagger \tau_\alpha \chi_L = \bar{\psi} \tau_\alpha \psi \,, \\ \text{Negative parity} & \pi_\alpha & = i(\chi_L^\dagger \tau_\alpha \chi_R - \chi_R^\dagger \tau_\alpha \chi_L) = i\bar{\psi} \tau_\alpha \gamma_5 \psi \,. \end{array} \tag{4.8}$$

Here in the last equations we have introduced the four-component spinor ψ, e.g. in the Weyl representation

$$\psi = \begin{pmatrix} \chi_R \\ \chi_L \end{pmatrix} . \tag{4.9}$$

*Although some equations here look similar to some in section 3.1, they should be distinguished. In the previous section, Lorentz group was discussed, but here we are considering internal chiral group.

Note that $\bar{\psi}\tau_\alpha\gamma_5\psi$ are pure imaginary. Now we have realized that there are two chiral four-vectors of (Lorentz) scalar and pseudoscalar,

$$\sigma_\alpha = \bar{\psi}\tau_\alpha\psi \quad \text{and} \quad \pi_\alpha = i\bar{\psi}\tau_\alpha\gamma_5\psi\,. \tag{4.10}$$

Consider the transformation rule for σ and π

$$\text{Vector} \quad \left\{ \begin{array}{l} \sigma_0 \to \sigma_0 \\ \boldsymbol{\sigma} \to \boldsymbol{\sigma} + \boldsymbol{v} \times \boldsymbol{\sigma} \\ \pi_0 \to \pi_0 \\ \boldsymbol{\pi} \to \boldsymbol{\pi} + \boldsymbol{v} \times \boldsymbol{\pi} \end{array} \right. \tag{4.11}$$

$$\text{Axial vector} \quad \left\{ \begin{array}{l} \sigma_0 \to \sigma_0 + \boldsymbol{a}\cdot\boldsymbol{\pi} \\ \boldsymbol{\sigma} \to \boldsymbol{\sigma} + \boldsymbol{a}\pi_0 \\ \pi_0 \to \pi_0 - \boldsymbol{a}\cdot\boldsymbol{\sigma} \\ \boldsymbol{\pi} \to \boldsymbol{\pi} - \boldsymbol{a}\sigma_0 \end{array} \right. \tag{4.12}$$

From these, we see that there are two invariants in quadratic form:

$$\sigma_0^2 + \boldsymbol{\pi}^2 \quad \text{and} \quad \pi_0^2 + \boldsymbol{\sigma}^2\,, \tag{4.13}$$

which are the lengths of the two independent four-vectors, $(\sigma_0, \boldsymbol{\pi})$ and $(\pi_0, \boldsymbol{\sigma})$. A convenient representation of, for instance, the former is the polar decomposition,

$$\sigma + i\boldsymbol{\tau}\cdot\boldsymbol{\pi} = f\exp(i\boldsymbol{\tau}\cdot\boldsymbol{\theta})\,. \tag{4.14}$$

This form has a direct relevance to the non-linear sigma model, which is discussed in the next section.

Assuming small number of powers in derivatives and fields, we can write down the lagrangian of the linear sigma model that is both chiral and Lorentz invariant. In view of the fact that in nature there is a three-component pseudoscalar pion, we choose the first combination of Eq. (4.13). Thus we have

$$\begin{aligned} L &= \frac{1}{2}\left[(\partial_\mu\sigma)^2 + (\partial_\mu\boldsymbol{\pi})^2\right] - V(\sigma^2 + \boldsymbol{\pi}^2) \\ &= \frac{1}{2}(\partial_\mu\phi)^2 - V(\phi^2)\,, \end{aligned} \tag{4.15}$$

where we have introduced the notation $\phi = (\sigma, \boldsymbol{\pi})$. The potential term of Eq. (4.15) is assumed to be of fourth order in ϕ:

$$V(\phi^2) = \frac{\mu^2}{2}\phi^2 + \frac{\lambda}{4}\phi^4\,. \tag{4.16}$$

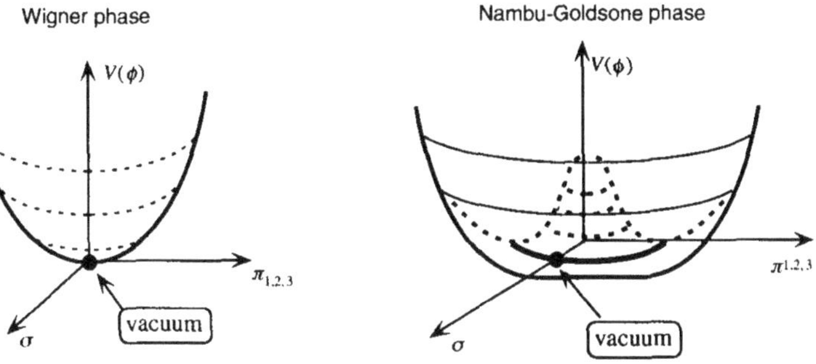

Fig. 4.1 Two potential shapes of the linear sigma model.

The coupling λ must be positive to ensure the stability of the theory, while μ^2 can be either positive or negative.

Let us investigate the structure of the vacuum and discuss the realization of chiral symmetry in the linear sigma model. The hamiltonian derived from Eq. (4.15) takes the form

$$H = \frac{1}{2}p_\alpha^2 + \frac{1}{2}(\partial_i \phi_\alpha)^2 + V(\phi^2)\,, \tag{4.17}$$

where the canonical momentum is defined by $p_\alpha = \partial L/\partial\dot{\phi}_\alpha = \dot{\phi}_\alpha$. Now the vacuum which has the lowest energy is expected to be static ($p_\alpha = 0$) and uniform ($\partial_i\phi_\alpha = 0$) †. Then the points of potential minimum, $\partial V(\phi)/\partial\phi = 0$, determines the vacuum.

When $\mu^2 > 0$, the minimum of the potential occurs at $\phi_\alpha = (\sigma, \boldsymbol{\pi}) = 0$ as shown in Fig. 4.1 (Wigner mode). This point is not affected by the chiral transformation and is chiral invariant (symmetric). Around this point the field can be expanded in small fluctuations which determine the mass of the particles. They have an equal mass: $m_\sigma = m_\pi = \mu$.

When $\mu^2 < 0$, the minimum of the potential occurs at $|\phi| = \sqrt{(}\phi_0^2 + \phi_1^2 + \phi_2^2 + \phi_3^2)^{1/2} = \sqrt{-\mu^2/\lambda}$ (Fig 4.1). The potential minima degenerate infinitely, and therefore, the vacuum can be at any point in the minima. If such a point has a non-vanishing component in a pion direction, the vacuum of that point does not have a definite parity, which contradicts the nature. Such a point, however, can be transformed into ($\sigma_0 \equiv \sqrt{-\mu^2/\lambda}, \pi^i = 0$)

†Nontrivial vacua are also possible when the theory has nontrivial topology, as in the Skyrme model (see the next chapter).

by a suitable chiral transformation. We define this point as the vacuum as shown in Fig 4.1. Around this point the field fluctuations can be introduced as $\phi_\alpha = (\sigma_0+\sigma, \boldsymbol{\pi})$. Then the masses of these fluctuations are $m_\sigma = \sqrt{-2\mu^2}$ and $m_\pi = 0$. As expected there are three massless pions in the Nambu-Goldstone mode.

$SU(3)_R \times SU(3)_L$

The fundamental representations of the $SU(3)$ right and left fermions are denoted as $(3,1)$ and $(1,3)$, and are transformed as

$$\begin{aligned}(3,1) &\sim \chi_R \to (\exp(i\ r))_{ab}(\chi_R)_b\,, \\ (1,3) &\sim \chi_L \to (\exp(i\ l))_{ab}(\chi_L)_b\,, \end{aligned} \tag{4.18}$$

where $r(l) = \sum_{i=1,8}\lambda_i r_i(l_i)$ and

$$\chi_{R,L} = \begin{pmatrix} \chi^u_{R,L} \\ \chi^d_{R,L} \\ \chi^s_{R,L} \end{pmatrix}. \tag{4.19}$$

Unlike $SU(2)$, complex representations of $SU(3)$ are distinguished from the representations of Eq. (4.18):

$$\begin{aligned}(3^*,1) \sim (\chi_R)^* &\to (\exp(i\ r))^*_{ab}(\chi_R)^*_b \\ &= (\chi_R)^*_b(\exp(+i\ r))^\dagger_{ba} \\ &= \chi^\dagger_R g^\dagger_R\,, \end{aligned} \tag{4.20}$$

$$(1,3^*) \to (R \to L)\,. \tag{4.21}$$

As in the previous example, we are interested in representations formed by quark bilinears which transform as,

$$\begin{aligned}(3^*,3)\ \chi_L\chi^\dagger_R &\to g_L\ \chi_L\chi^\dagger_R\ g^\dagger_R\,, \\ (3,3^*)\ \chi_R\chi^\dagger_L &\to g_R\ \chi_R\chi^\dagger_L\ g^\dagger_L\,. \end{aligned} \tag{4.22}$$

The parity eigenstates are their linear combinations $(3^*,3)+(3,3^*)$. We write them as [Gasiorowicz and Geffen 1969]

$$\begin{aligned}\chi_L\chi^\dagger_R + \chi_R\chi^\dagger_L &= \sigma_0 + \sum_{a=1,\ldots,8}\lambda^a\sigma_a \equiv \sigma\,, \\ \chi_L\chi^\dagger_R - \chi_R\chi^\dagger_L &= \pi_0 + \sum_{a=1,\ldots,8}\lambda^a\pi_a \equiv \pi\,. \end{aligned}$$

These fields form scalar and pseudoscalar nonets, whose candidates are $(a_0(1450), f_0(1370), K_0^*(1430))$ and (π, η, η', K).

The transformation rules for the σ and π nonets are

$$\begin{aligned} \delta\sigma &= il\ \chi_L\chi_R^\dagger - \chi_L\chi_R^\dagger\ ir + ir\ \chi_R\chi_L^\dagger - \chi_R\chi_L^\dagger\ il\,, \\ \delta\pi &= il\ \chi_L\chi_R^\dagger - \chi_L\chi_R^\dagger\ ir - ir\ \chi_R\chi_L^\dagger + \chi_R\chi_L^\dagger\ il\,, \end{aligned} \tag{4.23}$$

whose finite transformations are written in the vector and axial-vector form as

$$\begin{aligned} \delta_V(\sigma,\pi) &= e^{iv}(\sigma,\pi)e^{-iv}\,, \\ \delta_A(\sigma,\pi) &= e^{-ia}(\pi,\sigma)e^{-ia}\,. \end{aligned} \tag{4.24}$$

The relevant variable for the $SU(3) \times SU(3)$ linear sigma model is a 3×3 complex valued matrix which are spanned by nine σ's and nine π's,

$$M = \sigma + i\pi\,. \tag{4.25}$$

As in Eq. (4.14), polar decomposition is possible through the matrix decomposition

$$M = HU\,. \tag{4.26}$$

This is the theorem that a complex valued matrix can be always written as a product of a hermitian H and a unitary U matrix. After the hermitian matrix H is diagonalized one finds a useful expression

$$M = \begin{pmatrix} f_u & 0 & 0 \\ 0 & f_d & 0 \\ 0 & 0 & f_s \end{pmatrix} e^{i\lambda\phi}e^{i\phi_0}\,. \tag{4.27}$$

Here we have separated out the $U(1)$ phase explicitly. The octet part $e^{i\lambda\phi}$ $(\lambda\phi = \sum_{i=1,8}\lambda_i\phi_i)$ plays the main role in the non-linear sigma model.

4.2 Non-Linear Sigma Model

In the previous section, we have seen that the construction of linear sigma models is rather straightforward once representations of the chiral group are specified. However, the resulting lagrangian, for instance, Eq. (4.15) faces some problems.

(1) Since there is no small coupling constant, calculation of various amplitudes would require all Feynman diagrams of all orders of perturbation theory, which would make practical calculations formidable. In fact, however, there are examples in which computation of finite numbers of diagrams provides good description of, for instance, meson-meson scatterings. The termination of the diagrams is not,however, justified from a theoretical ground.

(2) In linear representations there are some particles which are not well established experimentally. For instance, the sigma particle in the multiplet $(\sigma, \boldsymbol{\pi})$ is still under controversy [PDG 2000].

In order to avoid, especially the second problem of the linear sigma model, we attempt to eliminate the sigma particle via, e.g.,

$$\sigma^2 + \boldsymbol{\pi}^2 = const. \tag{4.28}$$

If chiral symmetry is spontaneously broken, by setting $const = f_\pi$, the fields can move along the circle of the potential minima where the massless Nambu-Goldstone bosons are the dynamical degrees of freedom, whereas the σ is suppressed as a heavy particle. The potential is then just a constant (minimum value) and does not play a dynamical role. Hence, in the lagrangian, only the kinetic term remains relevant. More generally, the lagrangian is written as polynomials of the Maurer-Cartan one-form (see Eq. (4.65)) in a chiral and Lorentz invariant way.

Under the constraint of (4.28), linear chiral transformations among (σ, π) become non-linear chiral transformations among π's. This is the reason that the model is called the non-linear sigma model [Weinberg 1968; Weinberg 1985].

Features of the non-linear sigma model are:

(1) Since every term in a lagrangian contains some powers of derivatives, it provides an expansion in powers of small momenta at low energies. This enables a field theoretical method including renormalization at each order of momentum, and is the basis of the chiral perturbation theory in the present day [Gasser and Leutwyler 1984; Ecker 1995; Pich 1995].

(2) Although the full chiral transformations $g \in G$ are non-linear, the diagonal transformations $h \in H$ can remain linear, corresponding to the flavor vector symmetry. For $SU(2)_R \times SU(2)_L \to SU(2)_V$, the pions $(\pi^\pm, \pi^0)$ form isotriplet and nucleons (p, n) isodoublet

representations of $SU(2)_V$. For $SU(3)_R \times SU(3)_L \to SU(3)_V$, both mesons and nucleons belong to octet of $SU(3)_V$. The non-linear sigma model is the minimal model that satisfies chiral symmetry in terms of only well established particles while maintaining linear representations for the vector symmetry.

In this section we attempt to describe non-linear sigma models in two different ways. In the first part we emphasize a geometric structure of the variables of the non-linear sigma model, where the relevant angle variables are regarded as those in the curved space imbedded in a flat space of higher dimension. In the second part we explore a rather formal group theoretical method. These methods are explained for mesons first, and then fermions are included.

Construction from linear sigma models

An n-dimensional representation $\phi = (\phi_1, \phi_2, \ldots, \phi_n)$ of a group G is called linear if it is transformed under a group element $g \in G$ linearly, i.e., its infinitesimal variation is given by

$$\delta\phi = \varepsilon_m T^m \phi\,, \tag{4.29}$$

where T_m are $n \times n$ matrix and $m = 1, 2, \ldots M$ with M being the number of real parameters that parameterize the element g. In contrast, the non-linear representation allows $\delta\phi$ to depend on ϕ non-linearly.

A simple example is presented by the rotation group $O(3)$. Consider a three dimensional sphere of radius one as shown in Fig 4.2. Then a point on the sphere $P(x, y, z)$ transforms under an infinitesimal rotation as

$$\begin{pmatrix} \delta x \\ \delta y \\ \delta z \end{pmatrix} = \begin{pmatrix} 0 & \varepsilon_3 & -\varepsilon_2 \\ -\varepsilon_3 & 0 & \varepsilon_1 \\ \varepsilon_2 & -\varepsilon_1 & 0 \end{pmatrix} \begin{pmatrix} x \\ y \\ z \end{pmatrix}. \tag{4.30}$$

Namely, the cartesian coordinate (x, y, z) is a linear representation of the rotation group $O(3)$. Equivalently, we can express the point P using angle variables, $P(\theta, \varphi)$:

$$x = \sin\theta\cos\varphi\,, \quad y = \sin\theta\sin\varphi\,, \quad z = \cos\theta\,. \tag{4.31}$$

As investigated in problem 4.4, the infinitesimal transformation of the angle

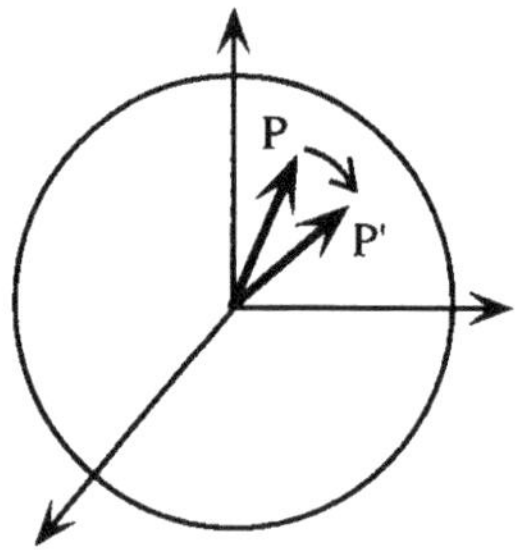

Fig. 4.2 A three dimensional unit sphere and a point P on it.

variables is given by

$$\begin{pmatrix} \delta\theta \\ \delta\varphi \end{pmatrix} = \begin{pmatrix} \sin\varphi\varepsilon_1 - \cos\varphi\varepsilon_2 \\ \cot\theta\cos\varphi\varepsilon_1 + \cot\theta\sin\varphi\varepsilon_2 - \varepsilon_3 \end{pmatrix} . \tag{4.32}$$

Apparently, the variations are non-linear in angle variables θ and ϕ.

Now we consider a sphere of radius f in four dimensional space. A point on the surface is represented by four cartesian coordinates, $(\sigma, \pi_1, \pi_2, \pi_3)$. This point can be parameterized by three angles:

$$\begin{aligned} \sigma &= f\cos\theta_1 , \\ \pi_1 &= f\sin\theta_1\sin\theta_2\cos\theta_3 , \\ \pi_2 &= f\sin\theta_1\sin\theta_2\sin\theta_3 , \\ \pi_3 &= f\sin\theta_1\cos\theta_2 . \end{aligned} \tag{4.33}$$

Generally, there are many equivalent ways to parameterize the cartesian coordinates in terms of three angle variables.

The lagrangian of the linear sigma model written in terms of cartesian coordinates $(\sigma, \boldsymbol{\pi})$ is now transformed into that of the non-linear sigma model written in terms of angle variables $\theta' s$,

$$\begin{aligned} L &= \frac{1}{2}(\partial_\mu\phi^\alpha)^2 \\ &= \frac{1}{2}\left(\frac{\partial\theta_i}{\partial x_\mu}\frac{\partial\phi^\alpha}{\partial\theta_i}\right)^2 = \frac{1}{2}g_{ij}(\theta)\frac{\partial\theta_i}{\partial x_\mu}\frac{\partial\theta_j}{\partial x^\mu} . \end{aligned} \tag{4.34}$$

Here we have introduced the metric tensor by

$$g_{ij}(\theta) = \sum_{\alpha} \frac{\partial \phi^{\alpha}}{\partial \theta_i} \frac{\partial \phi^{\alpha}}{\partial \theta_j} . \tag{4.35}$$

Depending on the choice of the metric tensor we have many equivalent representations of the non-linear sigma model.

(1) Square root representation
The metric is given by

$$g_{ij}(\theta) = \delta_{ij} + \frac{\theta_i \theta_j}{f^2 - \theta^2} . \tag{4.36}$$

Then the lagrangian takes the form

$$L = \frac{f^2}{2} \left[(\partial_\mu \sigma)^2 + (\partial_\mu \boldsymbol{\theta})^2 \right] , \tag{4.37}$$

where $\sigma = \sqrt{1 - \theta^2}$.

(2) Unitary representation
The metric is given by

$$g_{ij}(\theta) = \frac{\theta_i \theta_j}{\theta^2} + \left(\frac{\delta_{ij}}{\theta^2} - \frac{\theta_i \theta_j}{\theta^4} \right) \sin^2 \theta . \tag{4.38}$$

Then the lagrangian can be written in terms of the unitary matrix $U(x) = \exp(i\boldsymbol{\tau} \cdot \boldsymbol{\theta})$ as

$$L = \frac{f^2}{4} \operatorname{tr} \partial_\mu U \partial^\mu U^\dagger , \tag{4.39}$$

where the 2×2 unitary matrix can be written in the form of the quartanion

$$U = \sigma + i\boldsymbol{\tau} \cdot \boldsymbol{\pi} . \tag{4.40}$$

The second representation in terms of the unitary matrix U is often adopted when writing chiral lagrangians. An advantage of the unitary representation is that the form Eq. (4.39) can be used not only for $N_f = 2$ but also for general $SU(N_f)_R \times SU(N_f)_L$ with $N_f \geq 3$.

Finally, we write down the transformation rules and the currents in the unitary representation. Under the chiral $SU(N_f)_R \times SU(N_f)_L$ transformations, the U field transforms as:

$$U \to g_L U g_R^\dagger . \tag{4.41}$$

The equivalent expression is

$$
\begin{aligned}
V:\ U &\rightarrow g_V U g_V^\dagger \\
A:\ U &\rightarrow g_A^\dagger U g_A^\dagger .
\end{aligned}
\tag{4.42}
$$

Then the corresponding vector and axial-vector currents are given by

$$
\begin{aligned}
V_\mu^a &= \frac{if^2}{4}\,\mathrm{tr}\,\tau^a[U^\dagger,\ \partial_\mu U] \ = \ \varepsilon_{abc}\partial_\mu\phi^b\phi^c + \cdots \\
A_\mu^a &= \frac{if^2}{4}\,\mathrm{tr}\,\tau^a\{U^\dagger,\ \partial_\mu U\} \ = \ -f\partial_\mu\phi^a + \cdots ,
\end{aligned}
\tag{4.43}
$$

where $\phi^a = f\theta^a$.

CCWZ construction

In this subsection we formulate a group theoretical method to construct the non-linear sigma model due to Callan-Coleman-Wess-Zumino [Coleman et al. 1969; C. G. Callan et al. 1969]. See also Refs. [Weinberg 1985; Bando et al. 1988]. It is based on the assumption that the symmetry G is spontaneously broken down to $H \subset G$. There the Nambu-Goldstone bosons parameterize matrices in the coset space G/H.

Let us choose a representative, a group element of G, $a_1 \notin H$. This defines the right coset Ha_1 as the set of ha_1 for all $h \in H$. The coset Ha_1 does not have common elements with H. If it did, there would be an element $h_1 a_1 = h_2 \in H$. This implies that $a_1 = h_1^{-1} h_2 \in H$, which contradicts the assumption that $a_1 \notin H$. Now choose another representative $a_2 \notin \{H, a_1 H\}$, and form the second coset Ha_2. Repeating this process, we end up with a decomposition of the group G into a sum of cosets,

$$
G = H + Ha_1 + Ha_2 + \cdots + Ha_N . \tag{4.44}
$$

N becomes infinite for continuous groups which we are interested in.

If $Ha = aH$ for all $a \in G$, H is called an invariant subgroup. In this case the set of cosets $\{H, Ha_1, Ha_2, \ldots\}$ forms a group. This is called the coset group G/H.

Let us now turn to the case of the chiral group $G = SU(N_f)_R \times SU(N_f)_L$, which is spontaneously broken down to the diagonal subgroup $H = SU(N_f)_V$. Denote the number of generators of $SU(N_f)$ by $M \equiv N^2 - 1$. To express representations of the direct-product group $SU(N_f)_R \times SU(N_f)_L$, we place a representation of $SU(N_f)_R$ and that of $SU(N_f)_L$ in the first and second column of a two component vector. For instance, the right and left generators t^a_R and t^a_L are written as

$$\begin{aligned} t^a_R &= (t^a, 0)\,, \\ t^a_L &= (0, t^a)\,. \end{aligned} \tag{4.45}$$

Elements $g \in G$ are expressed similarly: [‡]

$$g = (g_R, g_L)\,, \tag{4.46}$$

where

$$\begin{aligned} g_R &= \exp(it^a r_a) \equiv \exp(i\ r)\,, \\ g_L &= \exp(it^a l_a) \equiv \exp(i\ l)\,. \end{aligned} \tag{4.47}$$

Calculations are then performed for the right and left components separately. In this way, charge commutation relations in Eq. (3.87) are reproduced by identifying

$$\begin{aligned} Q^a_R &\sim t^a_R = (t^a, 0)\,, \\ Q^a_L &\sim t^a_L = (0, t^a)\,. \end{aligned}$$

Now we separate the generators of G into the symmetric (H) and broken parts,

$$\begin{aligned} &S_1, S_2, \cdots, S_M && \text{for subgroup } H \\ &X_1, X_2, \cdots, X_M && \text{other generators} \end{aligned}$$

The commutation relations among S and X are assumed to be

$$\begin{aligned} \left[S^a, S^b\right] &= if_{abc}S^c\,, \\ \left[S^a, X^b\right] &= if_{abc}X^c\,, \\ \left[X^a, X^b\right] &= if_{abc}S^c\,. \end{aligned} \tag{4.48}$$

Although the first two equations are general, the last one is not and is possible only when H is what is called the symmetric group. In this case,

‡The ordering (g_L, g_R) appears to be natural. The present convention of placing the right element on the left component is to match it with the convention of the fermion, in which the right handed component is placed on the upper component in the Weyl representation.

we can introduce the parity. Namely, Eq. (4.48) are invariant under the following parity transformation

$$\text{parity}: \; S \leftrightarrow S, X \leftrightarrow -X\,. \tag{4.49}$$

For the chiral group, S and X are

$$\begin{aligned} S^a &= t^a_R + t^a_L = (t^a, t^a)\,, \\ X^a &= t^a_R - t^a_L = (t^a, -t^a)\,. \end{aligned} \tag{4.50}$$

The Nambu-Goldstone bosons, $\pi_a, a = 1, \cdots M$, parameterize matrices Ξ generated by the broken generators X. Denoting

$$\begin{aligned} \Pi &\equiv \pi_a X^a \\ &= (\pi, -\pi)\,, \end{aligned} \tag{4.51}$$

where $\pi = \sum \pi_a t^a$, we have

$$\begin{aligned} \Xi(\pi) &= \exp(i\Pi) \\ &= (\exp(i\pi), \exp(-i\pi)) \\ &\equiv (\xi(\pi), \xi^\dagger(\pi))\,. \end{aligned} \tag{4.52}$$

For $\pi \neq 0$, Ξ is an element of G which does not belong to H, as corresponding to the representative a_1 of Eq. (4.44).

Consider an element of a symmetric vector group $g_V \in H$,

$$\begin{aligned} g_V &= \exp(iS^a v_a) \\ &= (\exp(it^a v_a), \exp(it^a v_a)) \\ &\equiv (g_V, g_V)\,. \end{aligned} \tag{4.53}$$

Then the right multiplication of $\Xi(\pi)$ by $g_V^\dagger$, since it is a group element of G, can be written as a right coset, a product of $\Xi(\pi')$ and $h^\dagger \in H$,

$$\Xi(\pi) g_V^\dagger = h^\dagger(g_V, \pi)\Xi(\pi')\,, \tag{4.54}$$

where we have indicated that the element $h(g_V, \pi)$ depends generally on g_V and on the original pion field π. We can express this equation in the two-component form

$$(\xi(\pi),\, \xi^\dagger(\pi)) \cdot (g_V^\dagger,\, g_V^\dagger) = (h^\dagger(g_V, \pi),\, h^\dagger(g_V, \pi)) \cdot (\xi(\pi'),\, \xi^\dagger(\pi'))\,, \tag{4.55}$$

or

$$(\xi(\pi)g_V^\dagger,\ \xi^\dagger(\pi)g_V^\dagger) = (h^\dagger(g_V,\pi)\xi(\pi'),\ h(g_V,\pi)\xi^\dagger(\pi')) . \tag{4.56}$$

For this equation to hold, it is sufficient that

$$h(g_V,\pi) = g_V\ ,\quad \xi(\pi') = g_V\xi(\pi)g_V^\dagger \tag{4.57}$$

which is equivalent to

$$\pi' = g_V\pi g_V^\dagger . \tag{4.58}$$

This equation implies that π_a transform as a linear (adjoint) representation under the symmetry group transformation of $g_V \in H$.

Let us summarize what we have demonstrated. The pion field parameterizes representatives of an element of the coset space G/H. From a representative Ξ in one coset, one can generate all the elements of the coset by the symmetry transformation Ξg_V. In fact, the element $(\xi(\pi)g_V, \xi^\dagger(\pi)g_V)$ is equivalent to an element $(\exp(it_a l_a), \exp(it_a r_a))$, and therefore the decomposition g_V times Ξ exhausts all the group elements of G.

One can define chiral transformations of the pion field Ξ by extending the right multiplication of Eq. (4.54) by g_V to that by an arbitrary $g \in G$ (here and in what follows we suppress arguments of group elements, e.g., $\Xi(\pi) \to \Xi$, $\Xi(\pi') \to \Xi'$ and etc.):

$$\Xi g^\dagger = h^\dagger \Xi' . \tag{4.59}$$

Note here that the element $h(g,\pi)$ is a function of g and π, which is determined by the condition below, Eq. (4.63). Then Eq. (4.59) defines the non-linear transformation of the pion field:

$$\Xi \to \Xi' = h\Xi g^\dagger . \tag{4.60}$$

In the two component form,

$$\begin{aligned}(\xi',\ \xi'^{\,\dagger}) &= (h,\ h)\cdot(\xi,\ \xi^\dagger)\cdot(g_R^\dagger,\ g_L^\dagger) \\ &= (h\xi g_R^\dagger,\ h\xi^\dagger g_L^\dagger) ,\end{aligned} \tag{4.61}$$

or

$$\begin{aligned}\xi &\to \xi' = h\xi g_R^\dagger , \\ \xi^\dagger &\to \xi'^{\,\dagger} = h\xi^\dagger g_L^\dagger .\end{aligned} \tag{4.62}$$

The left component must be the adjoint of the right component, which leads to the constraint

$$g_L \xi h^\dagger = h \xi g_R^\dagger \,. \tag{4.63}$$

This equation determines h as a function of g_R . Therefore, we have an alternative expression which is equivalent to Eq. (4.61):

$$(\xi', \xi'^{\,\dagger}) = (g_L \xi h^\dagger, \, g_R \xi^\dagger h^\dagger) \tag{4.64}$$

Now let us consider a non-linear chiral lagrangian. It is constructed from the Nambu-Goldstone bosons Ξ and is invariant under the chiral transformations of Eq. (4.60). Here we consider only the lowest order in derivatives.

The elementary building blocks are chiral right and left currents, or the Maurer-Cartan one-forms:

$$\begin{aligned} R_\mu &= -i(\partial_\mu \Xi^\dagger)\Xi = +i\Xi^\dagger \partial_\mu \Xi \,, \\ L_\mu &= -i(\partial_\mu \Xi)\Xi^\dagger \,, \end{aligned} \tag{4.65}$$

The right chiral transformations on a matrix Ξ is defined by

$$\begin{aligned} \Xi &\to \Xi g_R^\dagger \sim \Xi + \Xi(-i\boldsymbol{t}\cdot\boldsymbol{r}) \\ &\equiv \Xi + \delta_R \Xi \end{aligned} \tag{4.66}$$

and so the right infinitesimal change is

$$\boldsymbol{t}\cdot\boldsymbol{r} = i\Xi^\dagger(\delta_R \Xi) \,. \tag{4.67}$$

Likewise,

$$\boldsymbol{t}\cdot\boldsymbol{l} = -i(\delta_L \Xi)\,\Xi^\dagger \,. \tag{4.68}$$

These relations explain why we have called the right and left currents in Eqs. (4.65).

The two one-forms are elements of the Lie algebra and are expanded in terms of the generators S_i and X_i,

$$\begin{aligned} (\partial_\mu \Xi^\dagger)\Xi &= i\sum (S_i v_\mu^i + X_i a_\mu^i) \,, \\ (\partial_\mu \Xi)\Xi^\dagger &= i\sum (S_i v_\mu^i - X_i a_\mu^i) \,. \end{aligned} \tag{4.69}$$

Coefficients v_μ^i and a_μ^i are given in Eqs. (4.74) and (4.75). One could have included $\Xi\partial_\mu\Xi^\dagger$ and $\Xi^\dagger\partial_\mu\Xi$ in addition to those of Eq. (4.65). The latter are, however, not independent and the former two objects of Eq. (4.65) are sufficient. We write Eq. (4.65) in components,

$$\begin{aligned}(\partial_\mu \Xi^\dagger)\Xi &= (\partial_\mu \xi^\dagger, \, \partial_\mu \xi) \cdot (\xi, \, \xi^\dagger) \\ &= (\partial_\mu \xi^\dagger \, \xi, \, \partial_\mu \xi \, \xi^\dagger) \, . \end{aligned} \tag{4.70}$$

Similarly, we have

$$(\partial_\mu \Xi)\Xi^\dagger = (\partial_\mu \xi \, \xi^\dagger, \, \partial_\mu \xi^\dagger \, \xi) \, . \tag{4.71}$$

Forming the sum and difference of Eq. (4.70) and Eq. (4.71), we obtain

$$\begin{aligned}\partial_\mu \Xi^\dagger \, \Xi + \partial_\mu \Xi \, \Xi^\dagger &= (\partial_\mu \xi^\dagger \, \xi + \partial_\mu \xi \, \xi^\dagger, \, \partial_\mu \xi^\dagger \, \xi + \partial_\mu \xi \, \xi^\dagger) \\ &= 2i(v_\mu, v_\mu) \equiv 2iV_\mu \end{aligned} \tag{4.72}$$

$$\begin{aligned}\partial_\mu \Xi^\dagger \, \Xi - \partial_\mu \Xi \, \Xi^\dagger &= (\partial_\mu \xi^\dagger \, \xi - \partial_\mu \xi \, \xi^\dagger, \, -\partial_\mu \xi^\dagger \, \xi + \partial_\mu \xi \, \xi^\dagger) \\ &= 2i(a_\mu, -a_\mu) \equiv 2iA_\mu \, . \end{aligned} \tag{4.73}$$

Here we have introduced the vector and axial-vector combinations:

$$v_\mu \equiv -\frac{i}{2}(\partial_\mu \xi^\dagger \, \xi + \partial_\mu \xi \, \xi^\dagger) \, , \tag{4.74}$$

$$a_\mu \equiv -\frac{i}{2}(\partial_\mu \xi^\dagger \, \xi - \partial_\mu \xi \, \xi^\dagger) \, . \tag{4.75}$$

From the structure of the right and left handed components, the plus combination of Eq. (4.72) belongs to $\{S\} \sim$ (vector), while the minus combination of Eq. (4.73) to $\{X\} \sim$ (axial).

It is instructive to see small field expansion of v_μ and a_μ. For $SU(2)$, the ξ field is expanded as

$$\xi = \exp\left(\frac{i}{2f_\pi}\pi\right) = 1 + \frac{i}{2f_\pi}\pi - \frac{1}{8f_\pi^2}\pi^2 \cdots \, , \tag{4.76}$$

where $\pi = \boldsymbol{\tau} \cdot \boldsymbol{\pi}$. Thus we find

$$\begin{aligned} v_\mu &= \frac{1}{8f_\pi^2}(-\partial_\mu \pi \; \pi + \pi \; \partial_\mu \pi) + \cdots \, , \\ a_\mu &= -\frac{1}{2f_\pi}\partial_\mu \pi + \cdots \, . \end{aligned} \tag{4.77}$$

Therefore, the leading term of v_μ is proportional to the vector current, and that of a_μ is proportional to the axial-vector current (see Eq. (4.43)).

Let us investigate the transformation properties of the currents under the group transformation of Eq. (4.60). Using Eq. (4.61), we see that

$$\begin{aligned} \partial_\mu \Xi^\dagger \Xi &= (\partial_\mu \xi^\dagger \xi,\ \partial_\mu \xi \xi^\dagger) \\ &\to \Big(\partial_\mu (h\xi^\dagger g_L^\dagger)(g_L \xi h^\dagger),\ \partial_\mu (h\xi g_R^\dagger)(g_R \xi^\dagger h^\dagger)\Big) \\ &= (\partial_\mu h h^\dagger + h\partial_\mu \xi^\dagger \xi h^\dagger,\ \partial_\mu h h^\dagger + h\partial_\mu \xi\xi^\dagger h^\dagger) \\ &= h(-\partial_\mu + \partial_\mu \xi^\dagger \xi,\ -\partial_\mu + \partial_\mu \xi \xi^\dagger) h^\dagger \,. \end{aligned} \tag{4.78}$$

In the last step we have used the unitarity of h, $hh^\dagger = 1$ and

$$\partial_\mu h h^\dagger + h\partial_\mu h^\dagger = 0\,. \tag{4.79}$$

To compute another quantity, we use Eq. (4.64) for an alternative expression of the transformation. We find

$$\begin{aligned} \partial_\mu \Xi\, \Xi^\dagger &= (\partial_\mu \xi \xi^\dagger,\ \partial_\mu \xi^\dagger \xi) \\ &\to h(-\partial_\mu + \partial_\mu \xi\xi^\dagger,\ -\partial_\mu + \partial_\mu \xi^\dagger \xi) h^\dagger \,. \end{aligned} \tag{4.80}$$

Adding or subtracting Eqs. (4.78) and (4.80) we find

$$\begin{aligned} &\partial_\mu \Xi^\dagger \Xi + \partial_\mu \Xi\, \Xi^\dagger \\ &\to h(-2\partial_\mu + \partial_\mu \xi^\dagger \xi + \partial_\mu \xi\xi^\dagger,\ -2\partial_\mu + \partial_\mu \xi^\dagger \xi + \partial_\mu \xi \xi^\dagger) h^\dagger\,, \end{aligned} \tag{4.81}$$

$$\begin{aligned} &\partial_\mu \Xi^\dagger \Xi - \partial_\mu \Xi\, \Xi^\dagger \\ &\to h(\partial_\mu \xi^\dagger \xi - \partial_\mu \xi\xi^\dagger,\ -\partial_\mu \xi^\dagger \xi + \partial_\mu \xi \xi^\dagger) h^\dagger\,. \end{aligned} \tag{4.82}$$

Or, in terms of v_μ and a_μ,

$$\begin{aligned} v_\mu &\to h(i\partial_\mu + v_\mu) h^\dagger\,, \\ a_\mu &\to h a_\mu h^\dagger\,. \end{aligned} \tag{4.83}$$

(Compare the first equation with the gauge transformation of Eq. (1.3).) We observe that the axial (minus) combination a_μ transforms homogeneously under g.

Thus the chiral invariant lagrangian may be formed from a_μ,

$$\begin{aligned} \mathrm{tr}\,(a_\mu a^\mu) &= -\frac{1}{4}\,\mathrm{tr}\,(\partial_\mu \xi^\dagger \xi - \partial_\mu \xi\xi^\dagger)^2 \\ &= -\frac{1}{4}\,\mathrm{tr}\,(\partial_\mu U U^\dagger \partial^\mu U U^\dagger) = +\frac{1}{4}\,\mathrm{tr}\,\partial_\mu U \partial^\mu U^\dagger\,, \end{aligned} \tag{4.84}$$

where we have introduced the chiral field

$$U = \xi^2 = \exp(\frac{i}{f_\pi}\tau \cdot \pi)\,. \tag{4.85}$$

Note that the transformation rule for U here is

$$\begin{aligned} U &= \xi^2 \\ &\to (g_L \xi h^\dagger)(h \xi g_R^\dagger) = g_L U g_R^\dagger\,. \end{aligned} \tag{4.86}$$

This justifies the transformation rule appeared in Eq. (4.41).

From the expansion of Eq. (4.76), the properly normalized kinetic term is

$$\begin{aligned} \mathcal{L}_k &= -\frac{f_\pi^2}{4}\,\mathrm{tr}\,(\partial_\mu U U^\dagger \partial^\mu U U^\dagger) \\ &= \frac{f_\pi^2}{4}\,\mathrm{tr}\,\partial_\mu U \partial^\mu U^\dagger\,. \end{aligned} \tag{4.87}$$

This is the standard form of the non-linear sigma model and is often used in this text book.

4.3 Fermion Field

The construction of chiral invariant lagrangian including fermions (either quarks or baryons) can be performed in a way analogous to the one we discussed in the previous sections for mesons. Introducing a fermion multiplet of the chiral group, we consider possible chiral invariant (singlet) forms. Quarks are fundamental representations of either isospin $SU(2)$ or flavor $SU(3)$, while baryons form either the fundamental representation of isospin $SU(2)$ for (p, n) or octet of flavor $SU(3)$ for $(p, n, \Lambda, \Sigma, \Xi)$.

$SU(2)_R \times SU(2)_L$ *linear model*

Fermions (both quarks and nucleons) are assumed to belong to the (linear) fundamental representation:

$$\psi = \chi_R + \chi_L \sim (2,1) + (1,2)\,. \tag{4.88}$$

Our main concern is to construct interaction lagrangian for fermions and mesons.

Take $(\sigma, \boldsymbol{\pi}) \sim (2,2)$ as introduced in Eq. (4.8). Then the corresponding $(2,2)$ representation of the fermion is formed by

$$\begin{aligned} s &= \chi_L^\dagger \chi_R + \chi_R^\dagger \chi_L \,, \\ p_i &= i(\chi_L^\dagger \tau_i \chi_R - \chi_R^\dagger \tau_i \chi_L) \,. \end{aligned} \tag{4.89}$$

The chiral invariant combination of mesons and fermions is then

$$\begin{aligned} & s\sigma + \boldsymbol{p}\cdot\boldsymbol{\pi} \\ &= \chi_L^\dagger \sigma \chi_R + \chi_R^\dagger \sigma \chi_L + i(\chi_L^\dagger \boldsymbol{\tau}\cdot\boldsymbol{\pi}\chi_R - \chi_R^\dagger \boldsymbol{\tau}\cdot\boldsymbol{\pi}\chi_L) \\ &= (\chi_L^\dagger, \chi_R^\dagger)\begin{pmatrix} 0 & 1 \\ 1 & 0 \end{pmatrix}\sigma\begin{pmatrix} \chi_R \\ \chi_L \end{pmatrix} \\ &\quad + i(\chi_L^\dagger, \chi_R^\dagger)\begin{pmatrix} 0 & 1 \\ 1 & 0 \end{pmatrix}\begin{pmatrix} 1 & 0 \\ 0 & -1 \end{pmatrix}\boldsymbol{\tau}\cdot\boldsymbol{\pi}\begin{pmatrix} \chi_R \\ \chi_L \end{pmatrix} \\ &= \bar{\psi}\sigma\psi + i\bar{\psi}\boldsymbol{\tau}\cdot\boldsymbol{\pi}\gamma_5\psi \,. \end{aligned} \tag{4.90}$$

This is the familiar expression for the interaction term in the linear sigma model.

Including kinetic and mass terms, we have the lagrangian

$$\begin{aligned} L &= \bar{\psi}(i\partial\!\!\!/ - m_0)\psi - g\bar{\psi}(\sigma + i\boldsymbol{\tau}\cdot\boldsymbol{\pi}\gamma_5)\psi \\ &\quad + (\text{meson terms}) \,. \end{aligned} \tag{4.91}$$

Here g is a coupling constant. As discussed in subsection 4.1, if chiral symmetry is spontaneously broken, then σ can take a non-zero vacuum expectation value $\langle 0|\sigma|0\rangle \equiv f_\pi$. Hence the mass of the fermion is

$$M = m_0 + g f_\pi \,. \tag{4.92}$$

When $m_0 = 0$, the mass is entirely generated by the non-zero value of f_π associated with the spontaneous breaking of chiral symmetry.

The last remark on the lagrangian (4.91) is that it implies

$$g_A = 1 \,. \tag{4.93}$$

Then Eq. (4.92) with $m_0 = 0$ recovers the Goldberger-Treiman relation (3.122). Note, however, that the unity of g_A is a consequence of the linear sigma model of the form Eq. (4.91) without derivative couplings. If we allow higher order terms in field variables, one can introduce terms which carry an arbitrary value of g_A (see problem 4.6).

Non-linear model with fundamental representations

Induced by the discussion at the beginning of this chapter, we postulate that the fermion field of non-linear realization is subject to the non-linear chiral transformation (chiral rotation)

$$q \to hq \quad \text{or} \quad \begin{cases} q_R \to hq_R\,, \\ q_L \to hq_L\,. \end{cases} \tag{4.94}$$

Namely, both q_R and q_L transform solely by $h \in H$. The non-linear fermion field q are related to the linear field ψ through the compensator Ξ,

$$q = \Xi\psi\,, \tag{4.95}$$

or in components,

$$(q_R, q_L) = (\xi\psi_R\,, \xi^\dagger\psi_L)\,. \tag{4.96}$$

Global transformations by g_R and g_L,

$$\begin{aligned} \psi_R &\to g_R\psi_R \\ \psi_L &\to g_L\psi_L \\ \xi &\to h\xi g_R^\dagger \\ \xi^\dagger &\to h\xi^\dagger g_L^\dagger \end{aligned}$$

are indeed compensated in the products $\xi\psi_R$ and $\xi^\dagger\psi_L$.

Let us consider the lagrangian

$$L' = \bar{q}(i\partial\!\!\!/ - m_0)q\,. \tag{4.97}$$

This, however, is not chiral invariant, for

$$L' \to \bar{q}(i\partial\!\!\!/ - m_0 + ih^\dagger \partial\!\!\!/ h)q\,. \tag{4.98}$$

But the inhomogeneous term $ih^\dagger\partial\!\!\!/ h$ can be eliminated by adding a term analogous to the gauge coupling term:

$$L = \bar{q}(i\partial\!\!\!/ - m_0 + v\!\!\!/)q\,, \tag{4.99}$$

where v_μ is given in Eq. (4.74). Note also that the coupling term can be written as

$$\bar{q}v\!\!\!/ q \to \bar{q}V\!\!\!\!/\,q\,, \tag{4.100}$$

where V_μ is the two component representation in Eq. (4.72). It is straightforward to check that this lagrangian is invariant under the non-linear chiral transformation

$$\begin{aligned} q &\to hq \\ v_\mu &\to hv_\mu h^\dagger + ih\partial_\mu h^\dagger . \end{aligned} \tag{4.101}$$

There is another term which is chiral invariant in the nonlinear theory:

$$\bar{q}\not{A}q \to \bar{q}\not{a}\gamma_5 q . \tag{4.102}$$

The reason that γ_5 appears here is explained by the two component representation of Eq. (4.73), where the signs of the first and second components are opposite.

Finally, we arrive at the lagrangian [§]

$$L = \bar{q}(i\not{D} - g_A\not{a}\gamma_5 - m_0)q , \tag{4.103}$$

where the covariant derivative is defined by

$$D_\mu = \partial_\mu - iv_\mu , \tag{4.104}$$

and v_μ and a_μ are given in Eqs. (4.74) and (4.75). Note that in Eq. (4.103) there appears a parameter g_A which is arbitrary. In the non-linear representation, the axial coupling constant g_A can not be determined.

Non-linear model with SU(3) octets

The eight-fold way assumes that both meson and baryon fields form octet representations of the diagonal flavor group of $SU(3)$. They can be expressed in 3×3 matrix form. For instance, the baryon field is

$$\begin{aligned} B &= pV_+ + nU_+ + \Sigma^+T_+ + \Sigma^-T_- + \Sigma^0\sqrt{2}t_0 \\ &\quad + \Xi^0U_- + \Xi^-V_- + \Lambda\sqrt{2}t_8 \\ &= \begin{pmatrix} \frac{\Sigma^0}{\sqrt{2}} + \frac{\Lambda}{\sqrt{6}} & \Sigma^+ & p \\ \Sigma^- & -\frac{\Sigma^0}{\sqrt{2}} + \frac{\Lambda}{\sqrt{6}} & n \\ \Xi^- & \Xi^0 & -\frac{2}{\sqrt{6}}\Lambda \end{pmatrix} . \end{aligned} \tag{4.105}$$

Here V, U and T are raising and lowering operators as given in Eq. (2.64). Note that the point reached by, for instance, V_+ from the origin of the

[§]For the sign of the g_A term, see problem 4.6.

$T_3 - Y$ plain carries the quantum numbers of the proton p, explaining the combination pV_+ in Eq. (4.105). Similarly, we can write the pseudoscalar and vector meson octets as

$$\phi = \begin{pmatrix} \frac{\pi^0}{\sqrt{2}} + \frac{\eta}{\sqrt{6}} & \pi^+ & K^+ \\ \pi^- & -\frac{\pi^0}{\sqrt{2}} + \frac{\eta}{\sqrt{6}} & K^0 \\ K^- & \bar{K}^0 & -\frac{2}{\sqrt{6}}\eta \end{pmatrix} \tag{4.106}$$

$$V = \begin{pmatrix} \frac{\rho^0}{\sqrt{2}} + \frac{\omega}{\sqrt{6}} & \rho^+ & K^{*+} \\ \rho^- & -\frac{\rho^0}{\sqrt{2}} + \frac{\omega}{\sqrt{6}} & K^{*0} \\ K^{*-} & \bar{K}^{*0} & -\frac{2}{\sqrt{6}}\omega \end{pmatrix} \tag{4.107}$$

In the non-linear realization the pseudo-scalar meson field ϕ is exponentiated,

$$U = \exp(i\phi/f_\pi)\,, \quad \xi = \exp(i\phi/2f_\pi)\,, \tag{4.108}$$

which can be used as in Eq. (4.103) for constructing lagrangians. What differs here from Eq. (4.103) where the fermion field was introduced as the fundamental representation is that there are two independent terms for the meson-baryon couplings through the axial-vector current. They are singlet components in the products of three octets (one meson and two baryon octets),

$$\begin{aligned} (8 \times 8 \times 8)_{\text{singlet}} &= ((8 \times 8) \times 8)_{\text{singlet}} \\ &= ((1 + 8 + 8 + 10 + 10^* + 27) \times 8)_{\text{singlet}} \\ &\sim (8 \times 8)_{\text{singlet}} + (8 \times 8)_{\text{singlet}}\,. \end{aligned} \tag{4.109}$$

These two terms are often identified with the F and D terms

$$\begin{aligned} &F \text{ type} \qquad \text{tr}\, \bar{B}\gamma_5[\not{A},\, B]\,, \\ &D \text{ type} \qquad \text{tr}\, \bar{B}\gamma_5\{\not{A},\, B\}\,. \end{aligned} \tag{4.110}$$

Therefore, an $SU(3)$ generalized lagrangian is

$$\begin{aligned} L &= \text{tr}\,(\bar{B}i\not{D}B) \\ &+ \frac{F}{2}\,\text{tr}\, \bar{B}\gamma_5[\not{A},\, B] + \frac{D}{2}\,\text{tr}\, \bar{B}\gamma_5\{\not{A},\, B\}\,, \end{aligned} \tag{4.111}$$

where F and D are arbitrary parameters.

Problems

4.1

Verify the chiral transformations (vector and axial-vector transformations) of the four vector B_α in Eq. (4.5).

4.2

Show that the combination $\chi_L^\dagger\tau_\alpha\chi_R+\chi_R^\dagger\tau_\alpha\chi_L$ is purely real, while $\chi_L^\dagger\tau_\alpha\chi_R-\chi_R^\dagger\tau_\alpha\chi_L$ is purely imaginary. This explains why the imaginary unit i is necessary in Eqs. (4.8) and (4.89).

4.3

Suppose that chiral symmetry is spontaneously broken; $\mu^2 < 0$.
(1) Find the chiral transformation that transforms $(0,0,0,\pi_z)$ into $(\sigma,0,0,0)$ (it must be that $\pi_z = \sigma$).
(2) Expand the lagrangian, Eq. (4.15), around the potential minimum $(\sigma_0,0,0,0)$, and find the masses of σ and π.

4.4

Let us outline the derivation of Eq. (4.32). Using Eq. (4.31), we find

$$\begin{pmatrix} \delta x \\ \delta y \\ \delta z \end{pmatrix} = \begin{pmatrix} \cos\theta\cos\varphi\delta\theta - \sin\theta\sin\varphi\delta\varphi \\ \cos\theta\sin\varphi\delta\theta + \sin\theta\cos\varphi\delta\varphi \\ -\sin\theta\delta\theta \end{pmatrix} = \begin{pmatrix} \varepsilon_3 y - \varepsilon_2 z \\ -\varepsilon_3 x + \varepsilon_1 z \\ \varepsilon_2 x - \varepsilon_1 y \end{pmatrix}, \tag{4.112}$$

where x, y and z are expressed by the angle variables θ and ϕ. Solve this equation for $\delta\theta$ and $\delta\varphi$, and verify Eq. (4.32).

4.5

Derive the metric of the non-linear sigma model for

(1) the polar parameterization of Eq. (4.33)
(2) the Weinberg parameterization

$$\sigma = f_\pi \frac{1-\boldsymbol{\theta}^2}{1+\boldsymbol{\theta}^2}, \quad \pi_i = f_\pi \frac{2\theta_i}{1+\boldsymbol{\theta}^2}.$$

4.6
(1) Apply the "phase" transformation of Eq. (4.96) to the linear sigma model of Eq. (4.91). Find the corresponding non-linear lagrangian with $g_A = 1$.
(2) Conversely, apply the inverse phase transformation of Eq. (4.96) to the non-linear lagrangian of Eq. (4.103), find the corresponding linear lagrangian with an arbitrary g_A.

4.7
We have seen that the coupling of the pion to the nucleon appears in two different forms of pseudoscalar and pseudovector (see chiral rotations for ψ and q in Eq. (4.96):

$$\bar{\psi}\gamma_5 \boldsymbol{\tau} \cdot \boldsymbol{\pi}\psi\,, \quad \partial_\mu \boldsymbol{\pi} \cdot \bar{q}\gamma_\mu\gamma_5\boldsymbol{\tau}q\,. \tag{4.113}$$

Verify that they are related in the following points of view:

(1) chiral rotation $\psi \to q = \xi\psi$
(2) non-relativistic reduction

4.8 ['t Hooft 1986]
We would like to consider a linear sigma model of $U(2)_R \times U(2)_L$ symmetry. We use all the field variables of Eq. (4.8) with notations slightly changed:

$$\begin{aligned}(\sigma_0, \boldsymbol{\pi}) &\to (\sigma, \boldsymbol{\pi})\,,\\ (\pi_0, \boldsymbol{\sigma}) &\to (\eta, \boldsymbol{\alpha})\,,\end{aligned}$$

and introduce the matrix representation

$$\phi = (\sigma + i\eta) + (\boldsymbol{\alpha} + i\boldsymbol{\pi}) \cdot \boldsymbol{\tau}\,. \tag{4.114}$$

(1) Derive the transformation rule of the eight fields under the chiral $U(1)_A$ transformation. (Note that under $U(1)_V$ transformation, they are trivially invariant.)
(2) We consider a lagrangian of the form

$$\mathcal{L} = \frac{1}{4}\,\mathrm{tr}\,\partial_\mu\phi\partial^\mu\phi^\dagger - V(\phi)\,, \tag{4.115}$$

(2) We consider a lagrangian of the form

$$\mathcal{L} = \frac{1}{4} \operatorname{tr} \partial_\mu \phi \partial^\mu \phi^\dagger - V(\phi) \,, \tag{4.116}$$

where the potential $V(\phi)$ is given up to the fourth order in ϕ by a sum of three invariants under $U(2) \times U(2)$,

$$V(\phi) = -\frac{\mu^2}{4} \operatorname{tr} \phi\phi^\dagger + \frac{1}{32}(\lambda_1 - \lambda_2)(\operatorname{tr} \phi\phi^\dagger)^2 + \frac{1}{32}\lambda_2 \operatorname{tr} (\phi\phi^\dagger)^2 \,.$$

Write the potential in terms of the eight variables $(\sigma, \boldsymbol{\pi})$ and $(\eta, \boldsymbol{\alpha})$.
(3) When chiral symmetry is spontaneously broken, setting $\sigma \to f_\pi + \sigma$ where $f_\pi = 2\mu^2/\lambda_1$, show that the masses of the four mesons are

$$m_\sigma = 2\mu^2 \,, \quad m_\alpha = \lambda_2 f_\pi^2 \,, \quad m_\eta = m_\pi = 0 \,. \tag{4.117}$$

(4) Consider a less symmetric term. Check the second equation of the following,

$$\begin{aligned} \delta V &= -\frac{\kappa}{4}(\det \phi + \det \phi^\dagger) \\ &= -2\kappa(\sigma^2 + \boldsymbol{\pi}^2 - \eta^2 - \boldsymbol{\alpha}^2) \,, \end{aligned}$$

and verify that this term breaks the $U(1)_A$ symmetry.
(5) When δV is added to the potential, show that

$$f_\pi^2 = 2\mu^2/\lambda_1 + 8\kappa/\lambda_1 \tag{4.118}$$

and the mass of η is given by

$$m_\eta^2 = 8\kappa \,. \tag{4.119}$$

Chapter 5

CHIRAL BAG MODEL

In this chapter we study the chiral bag model as one of successful phenomenological models for the nucleon. This model serves a hybrid picture for the nucleon consisting of the quark bag surrounded by the pion (generally meson) cloud. There is one parameter in the model, the bag radius R, which controls the strength of the quark and pion contributions. By choosing the bag radius appropriately around half femtometer, the model provides a good description of the nucleon.

Although the chiral bag model itself is just an effective model for the nucleon, we can learn many important concepts and techniques in theoretical physics. They are chiral symmetry, soliton dynamics, vacuum polarization (chiral Casimir effect), the cranking method, the large-Nc expansion, and so on. It is a rich source for the study of such interesting subjects.

We introduce the chiral bag model as a hybrid model of the MIT bag model and the Skyrme model. Therefore, we start with the discussion of the latter two models and also of the little bag model for a historical reason. We solve the chiral bag model in a special configuration called the hedgehog ansatz for a solitonic solution for the nucleon. Dynamics of the quark vacuum and the role of the topology will be emphasized.

5.1 The MIT Bag Model

Basic ideas

The idea of the bag model originates in an attempt to model confinement of colors of quarks and gluons inside hadrons [Chodos et al. 1974a;

Bogoliubov 1967; Thomas 1984]. In a recent understanding of QCD based on, for instance, the dual Ginzburg-Landau theory, color electric flux can be confined when color magnetic charges condense [Nambu 1974; Suganuma et al. 1995]. This picture is a dual version of the ordinary superconductivity. When a conductor is sufficiently cooled down, electric charges condense in the form of Cooper pairs. If one applies magnetic fields to the superconducting material, the magnetic flux must be squeezed within a narrow region (the Meissner effect, see Fig. 5.1.). The region where the Cooper pair condensate remains super, while the region where magnetic flux penetrates turns back into a normal conductor.

(a) Squeezing of magnetic field in a superconducting matrial

(b) An intuitive picture of a baryon: Valence quarks and color electric field confined in a spherical bag

Fig. 5.1 Schematic pictures for (a) the Meissner effect and (b) its dual version for a baryon.

In QCD the roles of electric and magnetic fields are interchanged, as in the QCD vacuum the color magnetic charges condense [Nambu 1974; Suganuma et al. 1995]. Indeed, such magnetic charge condensation appears to occur in the specific choice of the gauge fixing of the color $SU(3)$, the maximally abelian gauge. In this gauge, both analytic and numerical simulations show the condensation of color magnetic monopoles. Although this provides an intuitive picture for color confinement, it depends on the choice of gauge. Gauge independent signals for the non-perturbative structure of the vacuum may be characterized by the gluon condensate of $\langle F_{\mu\nu}F^{\mu\nu}\rangle$ [Shuryak 1980] or quark condensate $\langle\bar{\psi}\psi\rangle$.

Now turning to the bag model, let us introduce valence quarks as seeds of hadrons. They are a quark and an antiquark for mesons, and three quarks for baryons. The color electric fields which are produced by the quark sources can now exist only within the hadron interior, where the vacuum is assumed to be brought back to the perturbative vacuum. In reality, the perturbative and non-perturbative regions change gradually from one to the other. However, it would be useful if one divided the spatial region into two phases by a sharp boundary just like bubbles in hot boiling water. This implies that energy density inside hadrons is higher than in the exterior region. The excess of the energy is attributed to the difference of the latent heat in a macroscopic language. The region where the quarks and gluons can exist is called the bag.

Let us consider a single baryon, where the three valence quarks are confined in a bag of volume V. Let the excess of the energy density in the bag interior be denoted by B (bag constant). Then the total energy of the baryon is given by the sum of the kinetic energy and the bag energy. The latter may be characterized, for instance, by the volume and the surface area of the bag. If we assume that the bag energy is dominated by the volume type energy, we find

$$E_{\rm Bag} = \frac{a}{V^{1/3}} + VB\,, \tag{5.1}$$

where the first term is for the kinetic energy and the second term is the bag energy of the volume type. The $V^{-1/3}$-dependence of the kinetic term is determined by the uncertainty principle. Thus the coefficient a must be positive, but its actual value depends on the geometric structure of the bag. Now the important point in Eq. (5.1) is that it takes the minimum value for a certain value of V. At this equilibrium point, the two terms in Eq. (5.1) balance. More precisely, the outward pressure (the V-derivative) in the first term and inward pressure in the second term are equal (but with opposite signs). This point determines the size and the energy of the baryon as shown in Fig. 5.2. In general, to have a stable point, there must be such an equilibrium mechanism from more than two terms.

Quarks in a spherical cavity

In this subsection we solve equations of motion for quarks confined in a spherical cavity. This is achieved by introducing a scalar potential, or equivalently a position dependent mass of the quarks. The scalar nature of the confinement has not been established yet. However, this provides one of the simplest way to realize the permanent confinement of quarks.

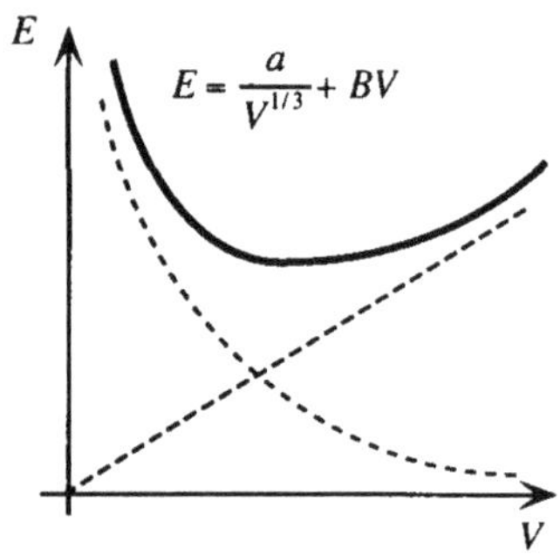

Fig. 5.2 Bag energy as a function of the volume V.

For practical purposes, quarks of u and d flavors are regarded as being massless inside the bag of radius R, as they remain current quarks, while it is postulated that they acquire a non-zero masse M in the bag exterior which are eventually led to infinity. The problem here is, therefore, to find wave functions of relativistic free quarks in a spherical cavity.

Let us start with the lagrangian

$$\mathcal{L} = \bar{\psi}\left(\frac{i}{2}\overleftrightarrow{\not\partial} - M\theta(r-R)\right)\psi\,, \tag{5.2}$$

where $\theta(r-R)$ is the Heaviside's step function. In order to solve the Dirac equation in a spherical cavity, we expand the quark field ψ in the spherical basis, since the hamiltonian

$$h = \psi^\dagger\left(-i\boldsymbol{\alpha}\cdot\boldsymbol{\nabla} + \beta M\theta(r-R)\right)\psi \tag{5.3}$$

satisfies the rotational symmetry and commutes with the total angular momentum $\boldsymbol{j}$:

$$\begin{aligned} &[\boldsymbol{j}, h] = 0\,,\\ &\boldsymbol{j} = \boldsymbol{l} + \boldsymbol{s} = -i\boldsymbol{x}\times\boldsymbol{\nabla} + \frac{\boldsymbol{\sigma}}{2}\,. \end{aligned} \tag{5.4}$$

In Eq. (5.3), the Dirac matrices are

$$\boldsymbol{\alpha} = \begin{pmatrix} 0 & \boldsymbol{\sigma} \\ \boldsymbol{\sigma} & 0 \end{pmatrix}, \quad \beta = \begin{pmatrix} 1 & 0 \\ 0 & -1 \end{pmatrix}. \tag{5.5}$$

Our task now is to find eigenstates of the hamiltonian of Eq. (5.3). The eigenvalue equation reads

$$(-i\boldsymbol{\alpha}\cdot\boldsymbol{\nabla} + \beta M\theta(r-R))\,\psi = E\psi\,. \tag{5.6}$$

Writing the four-component field ψ in terms of two spinor functions as

$$\psi = \begin{pmatrix} u \\ v \end{pmatrix}, \tag{5.7}$$

we find the following coupled eigenvalue equations,

$$\begin{aligned} Mu + i\boldsymbol{\sigma}\cdot\boldsymbol{\nabla}v &= Eu\,, \\ i\boldsymbol{\sigma}\cdot\boldsymbol{\nabla}u - Mv &= Ev\,. \end{aligned} \tag{5.8}$$

Eliminating the lower component from the second equation:

$$v = \frac{i\boldsymbol{\sigma}\cdot\boldsymbol{\nabla}}{E+M}u\,, \tag{5.9}$$

we obtain

$$\left(E^2 - M^2\right)u = -\nabla^2 u\,. \tag{5.10}$$

In deriving this equation, we have pretended that the mass M is a constant. In practice, we can solve Eqs. (5.8) - (5.10) separately for the inside and the outside of the bag and the change in M is taken into account by a boundary condition.

Now we consider the eigenstates of the total angular momentum $\boldsymbol{j}$: $(j, m \equiv j_z)$. For a spin 1/2 particle, they are presented by the spinor harmonics (see also Appendix H):

$$\begin{aligned} \mathcal{Y}^l_{jm}(\hat{\boldsymbol{x}}) &= \sum_\mu (l\mu\ \tfrac{1}{2}\ m-\mu|jm) Y_{l\mu}(\hat{\boldsymbol{x}})\chi_{m-\mu} \\ &\equiv [Y_l, \chi]_{jm} \end{aligned} \tag{5.11}$$

which possess the property

$$\begin{aligned} \boldsymbol{j}^2\mathcal{Y}^l_{jm}(\hat{\boldsymbol{x}}) &= j(j+1)\mathcal{Y}^l_{jm}(\hat{\boldsymbol{x}})\,, \quad \text{total } j\,, \\ j_z\mathcal{Y}^l_{jm}(\hat{\boldsymbol{x}}) &= m\mathcal{Y}^l_{jm}(\hat{\boldsymbol{x}})\,, \quad z \text{ component of } \boldsymbol{j}\,, \\ \mathcal{Y}^l_{jm}(-\hat{\boldsymbol{x}}) &= (-1)^l\mathcal{Y}^l_{jm}(\hat{\boldsymbol{x}})\,, \quad \text{parity}\,. \end{aligned} \tag{5.12}$$

Here we have introduced the standard spherical harmonics $Y_{lm}(\hat{\boldsymbol{x}})$ and two component spinors $\chi_{1/2} = (1, 0)$ and $\chi_{-1/2} = (0, 1)$. In the second line of Eq. (5.11), we have also introduced a notation for coupling of angular momenta: $[l_1, l_2]_{l,m}$.

Using these spinor harmonics, the upper component u is written for a given (l, j, m) as

$$u^l_{jm}(\boldsymbol{x}) = f_l(r)\mathcal{Y}^l_{jm}(\hat{\boldsymbol{x}})\,. \tag{5.13}$$

Substituting this for u in Eq. (5.10), we find the following radial equation for $f_l(r)$

$$-\frac{\partial^2 f_l}{\partial r^2} - \frac{2}{r}\frac{\partial f_l}{\partial r} - \frac{l(l+1)}{r^2} f_l = (E^2 - M^2) f_l\,. \tag{5.14}$$

The solutions to this equation are known and are given by:

For $k^2 \equiv E^2 - M^2 > 0$,

$$j_l(kr) \quad \text{spherical Bessel functions}\,,$$

For $-\kappa^2 \equiv E^2 - M^2 < 0$,

$$k_l(\kappa r) \quad \text{modified spherical Bessel functions}\,. \tag{5.15}$$

We note that we have chosen regular solutions among two independent ones; when $k^2 > 0$, the one which is regular (finite) at the origin, and when $\kappa^2 > 0$ the one which is regular at infinity.

To be specific, we consider the case $k^2 > 0$, which is the most relevant to the following discussion. Thus the eigenstate for the upper component is given by

$$u^l_{jm} = j_l(kr)\mathcal{Y}^l_{jm}(\hat{\boldsymbol{x}})\,. \tag{5.16}$$

The lower component v is now readily obtained by Eq. (5.9):

$$v(\boldsymbol{x}) = \frac{i\,\boldsymbol{\sigma}\cdot\boldsymbol{\nabla}}{E+M}\left(j_l(kr)\mathcal{Y}^l_{jm}(\hat{\boldsymbol{x}})\right)\,. \tag{5.17}$$

To proceed, we use the following convenient formula for the spinor harmonics when applied to by the derivative operator:

$$\boldsymbol{\sigma}\cdot\boldsymbol{\nabla}\left(f(kr)\mathcal{Y}^l_{jm}(\hat{\boldsymbol{x}})\right) = \begin{cases} -\left(\dfrac{df}{dr} - \dfrac{l}{r}f\right)\mathcal{Y}^{l+1}_{jm}(\hat{\boldsymbol{x}}) & j = l + \frac{1}{2} \\ -\left(\dfrac{df}{dr} + \dfrac{l+1}{r}f\right)\mathcal{Y}^{l-1}_{jm}(\hat{\boldsymbol{x}}) & j = l - \frac{1}{2} \end{cases} \tag{5.18}$$

Furthermore, the spinor harmonics satisfy the simple relation

$$\boldsymbol{\sigma}\cdot\hat{\boldsymbol{x}}\,\mathcal{Y}^{l}_{jm}(\hat{\boldsymbol{x}}) = -\mathcal{Y}^{l+\epsilon}_{jm}(\hat{\boldsymbol{x}}) \quad \text{for } j = l + \frac{\epsilon}{2}\,, \tag{5.19}$$

where $\epsilon = \pm 1$. Using this property and the recursion relation of the spherical Bessel functions $j_l(z)$:

$$\begin{aligned} \left(\frac{d}{dz} - \frac{l}{z}\right) j_l &= -j_{l+1}\,, \\ \left(\frac{d}{dz} + \frac{l+1}{z}\right) j_l &= j_{l-1}\,, \end{aligned}$$

we find

$$v(\boldsymbol{x}) = \epsilon \frac{ik}{E+M} j_{l+\epsilon}\,\boldsymbol{\sigma}\cdot\hat{\boldsymbol{x}}\,\mathcal{Y}^{l+\epsilon}_{jm}(\hat{\boldsymbol{x}}) \tag{5.20}$$

To summarize, we have derived the solution to the free Dirac equation in a spherical cavity as

$$\psi^{l}_{j=l+\epsilon/2,\,m}(\boldsymbol{x}) = \mathcal{N} \begin{pmatrix} j_l(kr) \\ \epsilon \dfrac{ik}{E+M} j_{l+\epsilon}(kr)\,\boldsymbol{\sigma}\cdot\hat{\boldsymbol{x}} \end{pmatrix} \mathcal{Y}^{l}_{jm}(\hat{\boldsymbol{x}})\,, \tag{5.21}$$

where $\mathcal{N}$ is the normalization constant.

Eigenvalues

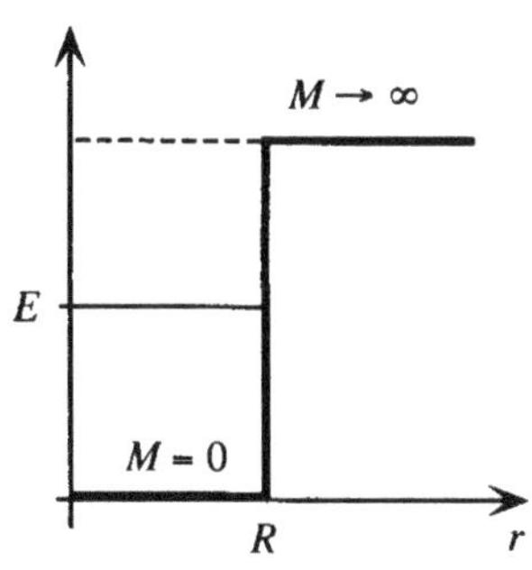

Fig. 5.3 The relation between the energy E and $M\theta(r-R)$.

Now we determine the eigenvalues. The relation between the energy E and the mass M is illustrated in Fig. 5.3, where the mass M for the outside of the bag is taken to be infinity. Let us write the solutions inside and outside of the bag separately:

$$\psi_{j\,m}^{(in)l}(\boldsymbol{x}) = \mathcal{N} \begin{pmatrix} j_l(kr) \\ \epsilon \dfrac{k}{E} j_{l+\epsilon}(kr)\, i\boldsymbol{\sigma}\cdot\hat{\boldsymbol{x}} \end{pmatrix} \mathcal{Y}^l_{jm}(\hat{\boldsymbol{x}}) , \tag{5.22}$$

$$\psi_{j\,m}^{(out)l}(\boldsymbol{x}) = A\mathcal{N} \begin{pmatrix} k_l(\kappa r) \\ \dfrac{\kappa}{E+M} k_{l+\epsilon}(\kappa r)\, i\boldsymbol{\sigma}\cdot\hat{\boldsymbol{x}} \end{pmatrix} \mathcal{Y}^l_{jm}(\hat{\boldsymbol{x}}) . \tag{5.23}$$

Here we have introduced the normalization constant $\mathcal{N}$ and another constant A. Note the following points in these solutions:

(1) We need to work out the solution in the bag exterior when M is finite. This can be done in a rather straightforward manner.
(2) In ψ_{out}, the sign factor ϵ is not present. This is due to the property of the modified Bessel functions k_l.
(3) In ψ_{in}, the factor k/E expresses just the sign of the energy E, since the quarks are massless.
(4) We have introduced the relative normalization factor A. This will be determined by the continuity condition as usual.

Since the Dirac equation is a first order differential equation, we require the continuity of the wave function at $r = R$, which reads

$$\begin{aligned} j_l(kR) &= A\, k_l(\kappa R) \stackrel{M\to\infty}{\longrightarrow} A\frac{e^{-MR}}{MR} , \\ \epsilon\frac{k}{E} j_{l+\epsilon}(kR) &= A\, k_{l+\epsilon}(\kappa R) \stackrel{M\to\infty}{\longrightarrow} A\frac{e^{-MR}}{MR} . \end{aligned} \tag{5.24}$$

In the limit $M \to \infty$, the right hand sides of these equation vanish unless A remains finite. However, it is not possible for both $j_l(kR)$ and $j_{l+\epsilon}(kR)$ to vanish simultaneously. In order for these two equations hold, the coefficient A must be sufficiently large when $M \to \infty$. Then we obtain the boundary condition, or the eigenvalue equation

$$j_l(kR) = \epsilon\frac{k}{E} j_{l+\epsilon}(kR) . \tag{5.25}$$

We note that this equation is equivalently expressed by

$$-i\hat{\boldsymbol{x}}\cdot\boldsymbol{\gamma}\,\psi = \psi \tag{5.26}$$

Finally, the normalization constant $\mathcal{N}$ is determined by (for $\epsilon = 1$)

$$\begin{aligned} \mathcal{N}^{-2} &= \int_0^R d^3x\,\psi^\dagger\psi \\ &= R^3\left(j_l^2 + j_{l+1}^2 + \frac{2l+2}{ER}j_l j_{l+1}\right), \end{aligned} \tag{5.27}$$

where the argument of the Bessel functions on the right hand side is ER. In deriving this equation, we have used the formulae

$$\int z^2 dz (j_l(z))^2 = \frac{z^3}{2}(j_l^2 - j_{l-1}j_{l+1}), \tag{5.28}$$

and

$$j_{l-1} + j_{l+1} = \frac{2l+1}{z}j_n\,. \tag{5.29}$$

Solutions to Eq. (5.25) can be solved numerically. Eigenvalues $y \equiv kR$ for several low lying states are shown in Table 5.1. They are functions of the orbital angular momentum l, total angular momentum (spin) j, parity P and also nodal quantum number n. Several remarks are in order:

(1) Physical energies are given in units of $1/R$. Therefore, the lowest energy is given by

$$E_{1/2^+,0} = \frac{2.04}{R}\,. \tag{5.30}$$

This is about 400 MeV when $R \sim 1$ fm. Therefore, the energy of a meson of two quark system amounts to about 800 MeV, while that of a baryon of three quark system amounts to about 1.2 GeV. These values are of the same order of magnitude as the observed values of, for instance, rho meson mass $m_\rho = 770$ MeV, nucleon mass $M_N = 938$ MeV and delta mass $M_\Delta = 1232$ MeV.

(2) Energies differ for different j but with the same l. A spin-orbit force may be responsible for the splitting. The origin of the spin-orbit force in the MIT bag model is the change in the mass at the bag

Table 5.1 Eigenvalues of the MIT bag model for several low states.

Positive parity				Negative parity			
j^P	l	$n=1$	$n=2$	j^P	l	$n=1$	$n=2$
$1/2^+$	0	2.04	5.40				
$3/2^+$	2	5.12	8.41	$1/2^-$	1	3.81	7.00
$5/2^+$	2	4.33	8.06	$3/2^-$	1	3.20	6.75
$7/2^+$	4	7.58	11.06	$5/2^-$	3	6.37	9.75
$9/2^+$	4	6.52	10.56	$7/2^-$	3	5.43	9.32

boundary, as the spin orbit interaction is proportional to potential derivatives. Here the mass term is regarded as a scalar potential:

$$\frac{\partial}{\partial r}\left(M\theta(r-R)\right) = M\delta(r-R)\,. \tag{5.31}$$

Lagrangian formalism

So far we have considered quark states in a spherical cavity by introducing an infinite mass for the outside region of the bag. We have done so, since our intention was to emphasize a physical picture for quark confinement implemented by the scalar potential

$$U(r) = M\,\theta(r-R)\,. \tag{5.32}$$

In the literature, however, the lagrangian formalism which leads to the boundary condition of Eq. (5.26) is often adopted [Chodos et al. 1974b; Chodos et al. 1974b; DeGrand et al. 1975]. In the limit $M \to \infty$, the quarks are completely confined and can exist only inside the bag. Therefore, it is convenient to formulate the same physics described by Eq. (5.2) only in terms of variables in the bag interior, or at most those at the bag surface.

Let us start with the lagrangian and its action,

$$\begin{aligned} \mathcal{L}' &= \frac{i}{2}\bar{\psi}(\overrightarrow{\not\partial} - \overleftarrow{\not\partial})\psi\,\theta(R-r)\,, \\ S' &= \int_0^R d^3x\,\mathcal{L}'\,, \end{aligned} \tag{5.33}$$

where the integral region is restricted to the inside of the bag and the time integral is ignored since we deal with the static bag and stationary solutions for quark eigenstates. Taking the variation with respect to $\bar{\psi}$, we find

$$\delta_{\bar{\psi}} S = \int_0^R d^3x \, i \, \delta\bar{\psi} \, \overrightarrow{\not{\partial}} \, \psi - \frac{i}{2} \int_{r=R} (\delta\bar{\psi} \, \hat{\boldsymbol{x}} \cdot \boldsymbol{\gamma} \psi) \, , \tag{5.34}$$

The first term gives the ordinary Dirac equation inside the bag, while the second term of the surface integral at the bag surface provides a boundary condition. Apparently, the condition that the second term vanishes alone does not lead to the desired boundary condition of Eq. (5.26). It is sufficient to add another surface term, leading to the lagrangian of the MIT bag model:

$$\mathcal{L}_{MIT} = \frac{i}{2} \bar{\psi} (\overleftarrow{\not{\partial}} - \overrightarrow{\not{\partial}}) \psi \, \theta(R - r) - \frac{1}{2} \bar{\psi} \psi \delta(r - R) \, . \tag{5.35}$$

With this lagrangian, we obtain the desired equations of motion

$$\begin{aligned} \not{\partial} \psi &= 0 \, , \quad \text{for} \quad r < R \\ i \hat{\boldsymbol{x}} \cdot \boldsymbol{\gamma} \psi &= \psi \, , \quad \text{for} \quad r = R \, . \end{aligned} \tag{5.36}$$

5.2 The Little Bag Model

Restoration of chiral symmetry

From the discussions in section 3.2, we know that the surface term of the MIT bag model in Eq. (5.35) breaks chiral symmetry. As we have emphasized in the preceding chapter, chiral symmetry should be satisfied in all respects of strong interaction physics. The standard prescription to restore chiral symmetry is to introduce a chiral field which couples to the quarks at the bag surface in a chiral invariant way [Chodos and Thorn 1975; Inoue and Maskawa 1975]. For this purpose, it is convenient to adopt the non-linear representation for the chiral field of Eq. (4.85), where an $SU(2)$ unitary matrix is parameterized by the isovector pion field $\boldsymbol{\phi} = (\phi_1, \phi_2, \phi_3)$ as

$$U(x) = \exp(i \boldsymbol{\tau} \cdot \boldsymbol{\phi} / f_\pi) \, , \tag{5.37}$$

where $f_\pi = 93$ MeV is the pion decay constant. The U-field is then transformed under the chiral transformations as shown in Eq. (4.86):

$$U \to g_L U g_R^\dagger \, . \tag{5.38}$$

Thus the chiral invariant and Hermite combination for the πqq coupling is

$$\chi_L^\dagger U \chi_R + \chi_R^\dagger U^\dagger \chi_L \,, \tag{5.39}$$

where χ's are two component spinors. In terms of the four component spinor, $\psi = (\psi_R, \psi_L)$, this interaction term is rewritten as (compare with Eq. (4.90)

$$\begin{aligned}
(\chi_L^\dagger, \chi_R^\dagger) \begin{pmatrix} U & 0 \\ 0 & U^\dagger \end{pmatrix} \begin{pmatrix} \chi_R \\ \chi_L \end{pmatrix} \\
&= (\chi_R^\dagger, \chi_L^\dagger) \begin{pmatrix} 0 & 1 \\ 1 & 0 \end{pmatrix} \begin{pmatrix} U & 0 \\ 0 & U^\dagger \end{pmatrix} \begin{pmatrix} \chi_R \\ \chi_L \end{pmatrix} \\
&= \bar{\psi} U_5 \psi \,,
\end{aligned} \tag{5.40}$$

where we have defined

$$U_5 = \exp(i\boldsymbol{\phi} \cdot \boldsymbol{\tau} \gamma_5 / f_\pi) \,. \tag{5.41}$$

Note that in Eq. (5.40) we adopted the Weyl representation for γ-matrices (see also Eq. (3.41) and below),

$$\gamma_0 = \begin{pmatrix} 0 & 1 \\ 1 & 0 \end{pmatrix} , \quad \gamma_5 = \begin{pmatrix} 1 & 0 \\ 0 & -1 \end{pmatrix} . \tag{5.42}$$

Including the kinetic term of the pion field (Eq. (4.87), the chiral invariant extension of the MIT bag model is

$$\begin{aligned}
\mathcal{L}_{\mathrm{CB}} &= (\bar{\psi} i \partial\!\!\!/ \psi - B)\, \theta_V - \frac{1}{2} \bar{\psi} \exp(i\boldsymbol{\tau} \cdot \boldsymbol{\phi} \gamma_5 / f_\pi) \psi \, \delta_V \\
&\quad - \frac{f_\pi^2}{4} \operatorname{tr} (\partial_\mu U U^\dagger \partial^\mu U U^\dagger)\, \theta_{\bar{V}} \,.
\end{aligned} \tag{5.43}$$

This is essentially the form of the chiral bag model proposed in the original works [Inoue and Maskawa 1975; Chodos and Thorn 1975]. The third term of Eq. (5.43) for the kinetic term of the pion is the non-linear sigma model lagrangian, which is uniquely determined up to second order in derivative as discussed in the previous chapter. In Eq. (5.43), the pion is assumed to be excluded from the bag as indicated by the theta function, $\theta_{\bar{V}} = 1 - \theta_V$.

For the lagrangian of Eq. (5.43), we have the vector and axial-vector currents as the sum of the quark and pion contributions,

$$V_\mu^a = V_\mu^a(q)\, \theta_V + V_\mu^a(\pi)\, \theta_{\bar{V}} \,, \tag{5.44}$$

$$A_\mu^a = A_\mu^a(q)\, \theta_V + A_\mu^a(\pi)\, \theta_{\bar{V}} \,, \tag{5.45}$$

where the superscript a is for isospin. The currents are given by

$$V_\mu^a(q) = \frac{1}{2}\bar{\psi}\tau^a\gamma_\mu\psi\,, \tag{5.46}$$

$$A_\mu^a(q) = \frac{1}{2}\bar{\psi}\tau^a\gamma_\mu\gamma_5\psi\,, \tag{5.47}$$

for the quark part, and

$$V_\mu^a(\pi) = -i\frac{f_\pi^2}{4}\,\mathrm{tr}\,\tau^a[U^\dagger, \partial_\mu U]\,, \tag{5.48}$$

$$A_\mu^a(\pi) = +i\frac{f_\pi^2}{4}\,\mathrm{tr}\,\tau^a\{U^\dagger, \partial_\mu U\}\,, \tag{5.49}$$

for the pion part.

Hedgehog solution of the little bag

Historically, the little bag model described by the lagrangian of Eq. (5.43) has been solved in two different ways. In one approach, non-linear effects of the pion fields and those of the pion-quark interaction are assumed to be small and solutions are obtained in perturbation of the pion field starting from the MIT bag model [Thomas 1984]. The pion-quark (πqq) interaction is truncated at the first order, and most calculations can be done analytically. Since the interaction is assumed to be weak, the resulting chiral bag solutions resemble those of the MIT bag model. It would be possible to go to higher order calculations, which, however, does not seem to be practical except for a few calculations [Hulthage et al. 1981].

In the other approach, the pion field and interactions are considered to be strong and the non-linear effects are included to all orders [Brown and Rho 1979; Brown et al. 1979] in the semiclassical method. This reveals a strong relevance of the little (chiral) bag model to the Skyrme model. Here the πqq interaction has a special role such that it transfers the dynamics of the quark into that of the pion, and vice versa. In particular, the pion dynamics outside the bag is transferred into the quark sector in the form of vacuum polarization of the quark bag. Such a transmuting mechanism of the quark and pion dynamics is also related to the underlying topological structure of the model.

A consistent treatment of the model including vacuum polarization effects will be postponed to the next section. Here, we rather investigate in some detail the original work of the so called little bag model [Brown and Rho 1979; Brown et al. 1979; Vento et al. 1980]. Although it suffered from serious problems such as stability, interesting ideas were proposed, which matured and led to the successful chiral bag model of mid 80's [Brown and Rho 1988; Hosaka and Toki 1992; Brown et al. 1984; Hosaka and Toki 1986]. Here, we think it useful to show how calculations without vacuum polarizations break down in order to understand their importance.

Let us first consider the equations of motion derived from Eq. (5.43). They are

$$(r < R) \qquad i\gamma^\mu \partial_\mu \psi = 0\,, \tag{5.50}$$

$$(r = R) \qquad i\hat{\boldsymbol{x}}\cdot\boldsymbol{\gamma}\,\psi = -\exp(i\boldsymbol{\tau}\cdot\boldsymbol{\phi}\,\gamma_5/f_\pi)\psi\,, \tag{5.51}$$

$$(r = R) \qquad \frac{f_\pi^2}{2}\operatorname{tr}\tau^a\,\hat{\boldsymbol{x}}\cdot\boldsymbol{\nabla}UU^\dagger = \frac{i}{2}\bar{\psi}\,\hat{\boldsymbol{x}}\cdot\boldsymbol{\gamma}\,\gamma_5\tau^a\,\psi\,, \tag{5.52}$$

$$(r > R) \qquad \operatorname{tr}\tau^a\partial_\mu^2 UU^\dagger = 0\,. \tag{5.53}$$

The linear boundary condition of Eq. (5.51) is the chiral invariant version of the corresponding one of the MIT bag model and expresses the continuity of the vector current at the bag surface. The third equation in Eqs. (5.52) expresses the continuity of the axial-vector current ((5.45), (5.47) and (5.49)) at the bag surface. This new piece is not present in the MIT bag model and is the most important ingredient of the chiral bag model.

The coupled equations (5.50)-(5.53) are greatly simplified by assuming the maximally symmetric configuration, called the hedgehog ansatz [Skyrme 1962; Vento et al. 1980], where the pion field is assumed to be

$$\phi_a = \hat{\boldsymbol{x}}_a F(r) \quad \text{or} \quad U = \exp(i\boldsymbol{\tau}\cdot\hat{\boldsymbol{x}}F(r))\,, \tag{5.54}$$

with $\hat{\boldsymbol{x}}_a = x_a/|\boldsymbol{x}|$. Here $F(r)$ is called the chiral angle or the chiral profile function. The physical meaning of this ansatz will be discussed in detail in the next chapter. This configuration is invariant under the simultaneous rotations in the coordinate space generated by the total angular momentum operator $\boldsymbol{J} = \boldsymbol{L} + \boldsymbol{\sigma}/2$ and in the isospin space generated by the isospin operator $\boldsymbol{I} = \boldsymbol{\tau}/2$. Here $\boldsymbol{L}$ is the orbital angular momentum, and $\boldsymbol{\sigma}$ and $\boldsymbol{\tau}$ are the Pauli matrices for spin and isospin. The invariance of Eq. (5.54) under the simultaneous transformation implies that the hedgehog configuration has the vanishing grand angular momentum, $\boldsymbol{K} = \boldsymbol{J} + \boldsymbol{I} = 0$ [Adkins et al. 1983; Zahed and Brown 1986]. Therefore, in a quantum theory the hedgehog state may be considered as a superposition of physical states of $J = I$, which are in turn projected out from the hedgehog state.

According to the hedgehog structure of the pion field, the quark eigenstates have good quantum numbers of the grand angular momentum K and the parity P. The quark eigenstates are then labeled by K^P (quark hedgehog). The lowest eigenstate has $\boldsymbol{K} = 0$ with $\boldsymbol{L} = 0$ and $\boldsymbol{S} + \boldsymbol{I} = 0$. The $K^P = 0^+$ quark wave function is written as

$$\begin{aligned} \psi &= N \begin{pmatrix} j_0(kr) \\ i\boldsymbol{\sigma}\cdot\hat{\boldsymbol{x}} j_1(kr) \end{pmatrix} \chi , \\ N^{-2} &= R^3 j_0^2(kR)(1 + w_+^2 - 2w_+/kR) , \\ w_+ &= j_1(kR)/j_0(kR) . \end{aligned} \tag{5.55}$$

The isospinor-spinor function χ for $\boldsymbol{S} + \boldsymbol{I} = 0$ is

$$\chi = \frac{1}{\sqrt{2}}(\uparrow d - \downarrow u) , \tag{5.56}$$

where $\uparrow$ ($\downarrow$) and u (d) are the spin up (down) and isospin up (down) states, respectively. Now we observe that the a-th component of the normal flow of the isospin axial-vector current $\hat{\boldsymbol{x}}\cdot\boldsymbol{A}^a$ is proportional to the a-th component of the radial direction $\hat{\boldsymbol{x}}^a$ and that Eqs. (5.52) reduce to a single equation:

$$-f_\pi F'(R) = \frac{N^2}{2}(j_0^2 + j_1^2) . \tag{5.57}$$

Now the linear boundary condition of Eq. (5.51) for the lowest valence orbit of Eq. (5.55) takes the following form:

$$\frac{1 + \sin F}{\cos F} - \frac{j_1(kR)}{j_0(kR)} = 0 . \tag{5.58}$$

Here we have made the abbreviation $F \equiv F(R)$. Eq. (5.58) determines the valence quark energy as a function of the chiral angle F at the bag surface. In Fig. 5.4, the dimensionless eigenvalue $y \equiv ER$ is plotted as a function of F. $F = 0$ corresponds to the MIT bag solution, where we find $j_0(y) = j_1(y)$ and so $y = 2.04$. As F is increased the eigenvalue y decreases. In particular, for $F > \pi/2$, it becomes negative. This is the origin of the negative energy problem in the little bag model.

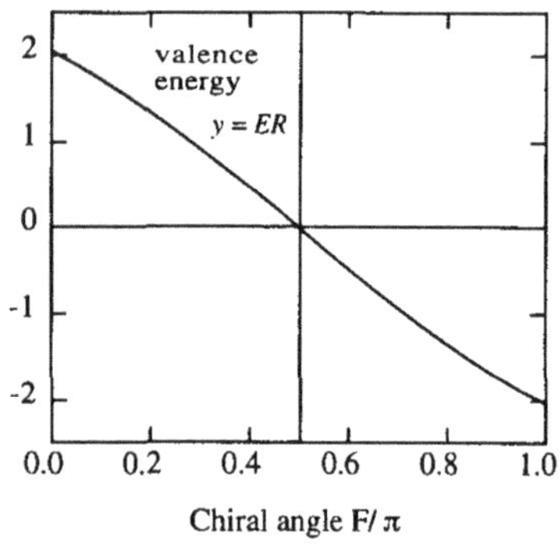

Fig. 5.4 The eigenenergy of the lowest $K^P = 0^+$ quark state as a function of the chiral angle F.

Next we consider the pion field equation (5.53). In the hedgehog ansatz, it reduces to an ordinary differential equation for $r > R$,

$$F'' + 2\frac{F'}{r} - \frac{\sin 2F}{r^2} = 0 \tag{5.59}$$

which is solved numerically subject to the boundary condition (5.58). A solution $F(r)$ at $R = 0.6$ fm is shown in Fig. 5.5, as compared with the skyrmion solution ($R \to 0$) of section 5.3.

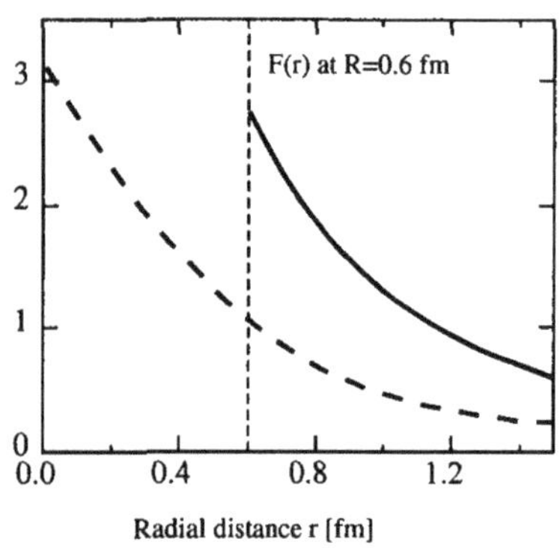

Fig. 5.5 A profile function $F(r)$ of the hedgehog solution in the little bag model. The solution at $R = 0.6$ fm (solid line) is compared with the Skyrme model solution (dashed line) of section 5.3.

The total energy of the system is the sum of the three terms, the quark energy, the pion energy and the volume energy of the bag:

$$E_{\text{tot}} = N_c \frac{y}{R} + E_\pi + \frac{4\pi}{3} R^3 B \,. \tag{5.60}$$

where, the energy of the classical pion field is given by

$$E_\pi = \frac{f_\pi^2}{2} \int_R^\infty d^3x \left(F'^2 + \frac{2}{r^2} \sin^2 F \right) . \tag{5.61}$$

As the bag radius is decreased, the chiral angle at the bag surface F increases and the quark energy y/R decreases. The other two terms in Eq. (5.60) also decrease as R is decreased. The decrease in the pion energy E_π is related to the fact that the non-linear sigma model with only the second order derivative term does not support stable solutions. Hence the total energy, Eq. (5.60), is unbounded from below for small R, leading to the instability of the little bag. We show the total energy as a function of R in Fig. 5.6.

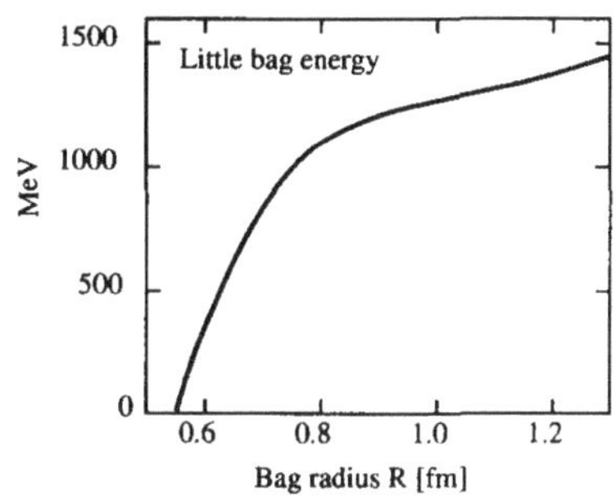

Fig. 5.6 The total energy of the little bag as a function of R.

Fractional baryon number

Another difficulty of the little bag model is the missing baryon number. In order to outline this problem, we examine the case that corresponds to that originally considered by Jackiw and Rebbi. There the baryon number carried by quarks inside the bag becomes 1/2 [Jackiw and Rebbi 1976; Niemi and Semenoff 1986]. They considered the problem of a fermion interacting with a background field, in which the fermion spectrum is symmetric between the positive and negative energy states and there is one zero energy state. In the little bag model, this situation is realized when $F = \pi/2$, where there is symmetry (see Fig. 5.8)

$$E(K^P, n) = -E(K^P, -n)\,. \tag{5.62}$$

Here n is the nodal quantum number.

The quark field ψ is expanded in energy eigenstates as

$$\psi(\boldsymbol{x}) = \sum_{E_n>0} \psi_n(\boldsymbol{x}) e^{-iE_n t} b_n + \sum_{E_m<0} \psi_m(\boldsymbol{x}) e^{iE_m t} d_m^\dagger + \psi_0(\boldsymbol{x}) a_0\,, \tag{5.63}$$

where ψ_n, ψ_m and ψ_0 are the positive, negative and zero energy wave functions, respectively, and b_n, d_m and a_0 are the corresponding annihilation operators. The baryon number operator for the quark field is then defined by

$$\hat{B}_q = \frac{1}{2} \int_0^R d^3x [\bar{\psi}(\boldsymbol{x}), \gamma_0 \psi(\boldsymbol{x})]\,. \tag{5.64}$$

This operator is normal ordered such that it changes the sign under the charge conjugation transformation, $\psi \to i\gamma_2 \psi^*$:

$$j_0 \overset{\text{C.C.}}{\longleftrightarrow} -j_0\,. \tag{5.65}$$

Inserting Eq. (5.63) into Eq. (5.64), we find for the vacuum expectation value of the baryon number

$$\begin{aligned} B_q &= \frac{1}{2}\left(\sum_{E_m<0} \int d^3x\, \psi_m^\dagger(\boldsymbol{x}) \psi_m(\boldsymbol{x}) \right. \\ &- \left. \sum_{E_n>0} \int d^3x\, \psi_n^\dagger(\boldsymbol{x}) \psi_n(\boldsymbol{x}) + \int d^3x\, \psi_0^\dagger(\boldsymbol{x}) \psi_0(\boldsymbol{x}) \right) \\ &= \frac{1}{2}\left(\sum_{E_m<0} 1 - \sum_{E_n>0} 1 + 1 \right)\,. \end{aligned} \tag{5.66}$$

From the symmetry between the positive and negative energy states, the first and the second terms of Eq. (5.66) cancel exactly and only the last term remains. Thus we find

$$B_q = \frac{1}{2} \int d^3x\, \psi_0^\dagger(\boldsymbol{x}) \psi_0(\boldsymbol{x}) = \frac{1}{2}\,. \tag{5.67}$$

Therefore, we conclude that $B_q = 1/2$ when $F = \pi/2$.

Two remarks are in order.

(1) An apparent question is what carries the rest of the baryon number. It is now well understood that the pion field outside the bag does so. The static pion field satisfying the boundary conditions $F(R) = \pi/2$ and $F(\infty) = 0$, carries the baryon number 1/2. Thus the total baryon number, the sum of the quark and the pion contributions, is exactly one. This will be investigated in detail in the next section.

(2) In Eq. (5.66), we have essentially computed a vacuum polarization effect. There, the symmetry of the quark spectrum ensures the cancellation between the positive and negative energy contributions. The remaining contribution from the zero energy state is just 1/2. It can be interpreted as follows: the zero energy state is either in the positive or in the negative energy regions with equal probability. Thus on average the number of valence quarks inside the bag is 1/2.

In order to overcome the problems of the negative energy and the fractional baryon number, we need to consider two aspects of the model; one is the topological property of the pion cloud and the other is the effect of the strong pion-quark coupling at the bag surface as described in Eq. (5.51). The former is accounted for by the Skyrme model and the latter by the chiral Casimir effect. In the following sections we will consider these two aspects.

5.3 The Skyrme Model

Let us briefly discuss the Skyrme model for the nucleons. It is based on an extended version of the non-linear sigma model for the unified description of mesons and nucleons at low energies. Skyrme considered the nucleon as a collective excitation of a meson field, the pion, rather than a single particle state of quarks. The motivation of how he arrived at such an idea is described in Ref. [Sanyuk 1992]. Through the study of the ϕ^4 and the sine-Gordon theory in 1+1 dimension [Skyrme 1958], he realized that the kink solution possesses fermion like particle properties [Coleman 1975; Mandelstam 1975; Rajaraman 1982; Coleman 1985]. From topological reasoning, he was led to the idea that the nucleon in 1+3 dimension would be a kink like solution of a non-linear theory of the three component pion.

We consider stable static solutions to the classical field equation for the pion. A static field configuration of the pion is regarded as a mapping:

$$\{U(\boldsymbol{x})|\boldsymbol{x} \in R^3 \to U \in SU(2) \sim S^3\}\,. \tag{5.68}$$

By imposing the boundary condition $U(|\boldsymbol{x}| \to \infty) = 1$, R^3 may be compactified and the mapping is regarded as $S^3 \to S^3$. The mapping is classified by the winding number:

$$B = \frac{\epsilon_{ijk}}{24\pi^2} \int d^3x \;\mathrm{tr}\, \partial_i U U^\dagger \partial_j U U^\dagger \partial_k U U^\dagger\,. \tag{5.69}$$

The factor $1/(24\pi^2)$ normalizes the spatial integral B to an integer. The number B is mathematically interpreted as the number of times the field configuration $U(\boldsymbol{x})$ covers the functional space $U \in S^3$ while $\boldsymbol{x}$ covers the three dimensional space $R^3 \sim S^3$ once, as shown in Fig. 5.7. Crudely, the winding number is a multiple of the normalized volume of the $SU(2)$ parameter space,

$$B \sim \frac{1}{V_{SU(2)}} \int dV_{SU(2)} = \frac{1}{V_{SU(2)}} \int \left| \frac{\partial V_{SU(2)}}{\partial x} \right| dx\,. \tag{5.70}$$

The integrand of the second equation is the Jacobian for the volume element.

By the definition, the winding number is integer-valued. The integrand of the winding number B is regarded also as the zeroth component of the topological current

$$B_\mu = \frac{\epsilon_{\mu\nu\alpha\beta}}{24\pi^2}\, \mathrm{tr}\, \partial^\nu U U^\dagger \partial^\alpha U U^\dagger \partial^\beta U U^\dagger\,. \tag{5.71}$$

Note that due to the epsilon tensor in Eq. (5.71), the current B_μ conserves trivially:

$$\partial^\mu B_\mu = 0\,. \tag{5.72}$$

Skyrme conjectured that the winding number of Eq. (5.69) should be identified with the baryon number and the current (5.71) with the associated baryonic current [Skyrme 1961]. The baryon number of the form (5.69) can be derived by adiabatically coupling quarks with the pion [Goldstone and Wilczek 1981; Balachandran et al. 1983; Kahana and Ripka 1984]. Transitions between solutions with different baryon numbers are rigorously

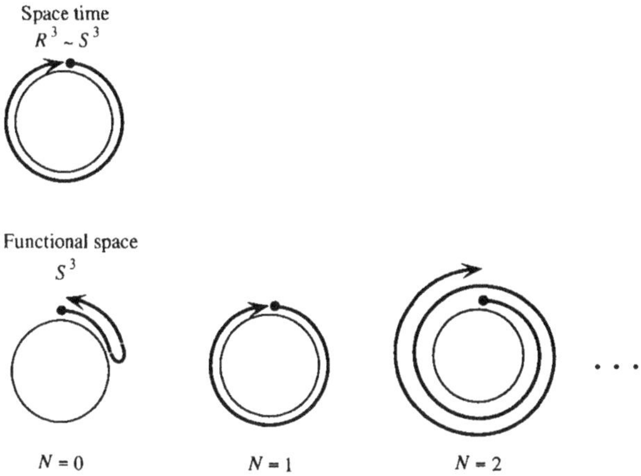

Fig. 5.7 A schematic picture for mappings of the winding number $B = 0$, $B = 1$ and $B = 2$.

forbidden for the topological reason. Therefore, a baryon number one solution, if it exists, does not decay.

The topological argument here is only a necessary condition for the existence of a stable solution. In order to demonstrate the relevant point, we consider the standard non-linear sigma model with two derivatives. Its static energy is given by

$$E_2 = \int d^3x \frac{f_\pi^2}{4} \operatorname{tr} \partial_i U U^\dagger \partial^i U U^\dagger \,. \tag{5.73}$$

Then consider a field configuration of finite size $U(x)$. Perform a scale transformation $x \to ax$, $U(x) \to U(ax)$. Then we see that E_2 changes as

$$E_2 \stackrel{x \to ax}{\longrightarrow} aE_2 \,, \tag{5.74}$$

and so, it collapses as $a \to 0$. In order to prevent this, an n (> 3)derivative term must be added. This provides an energy which behaves as $1/a^{n-3}$. Then the total energy is given by

$$E_{tot}(a) \sim E_2 a + \frac{E_n}{a^{n-3}} \tag{5.75}$$

which has the minimum at some value a. At the minimum point, there is a pressure balance between the two terms, a situation seen in the MIT bag model, where the outward quark pressure and the inward bag pressure balance.

As such a stabilizing term, Skyrme added a fourth order ($n = 4$) derivative term, whose energy behaves as $\sim 1/a$ under the scale transformation $x \to ax$. Skyrme's choice for the lagrangian was

$$\mathcal{L}_{\mathrm{Sk}} = -\frac{f_\pi^2}{4} \operatorname{tr}[\partial_\mu U U^\dagger \partial^\mu U U^\dagger] + \frac{1}{32e^2} \operatorname{tr}[\partial_\mu U U^\dagger, \partial_\nu U U^\dagger]^2 , \qquad (5.76)$$

where the second term is the so called Skyrme term and e the Skyrme parameter. Generally for $U \in \mathrm{SU}(2)$, there are two terms of fourth order derivatives which satisfy the requirements of Lorentz invariance and chiral symmetry. The other term is the anti-commutator $\{\cdots,\cdots\}$. Skyrme's choice of the commutator in Eq. (5.76) has the advantage that the quantization is straightforward since it does not contain fourth order time derivatives. Physically, this term can be interpreted as the kinetic term of the ρ meson in the limit $m_\rho \to \infty$ [Bando et al. 1988]. Also, the Skyrme term appears to be a natural choice from a geometric point of view [Manton 1987].

Solutions of the Skyrme lagrangian for the nucleon was considered by Skyrme himself and later by many authors in detail. Since the method shares many common features with what we will investigate in the chiral bag model, we stop further investigation of the Skyrme model here.

5.4 The Chiral Bag Model

The lagrangian and equations of motion

The original chiral bag model is a hybrid model of the MIT bag model and the Skyrme models:

$$\begin{aligned} \mathcal{L}_{\mathrm{CB}} &= \left(\bar\psi i\partial_\mu\gamma^\mu\psi - B\right)\theta_V - \frac{1}{2}\bar\psi \exp(i\vec\tau\cdot\vec\phi\gamma_5/f_\pi)\psi\,\delta_V \\ &+ \Bigg(-\frac{f_\pi^2}{4}\operatorname{tr}[\partial_\mu U U^\dagger\partial^\mu U U^\dagger] \\ &\qquad + \frac{1}{32e^2}\operatorname{tr}[\partial_\mu U U^\dagger, \partial_\nu U U^\dagger]^2\Bigg)\,\theta_{\bar V} \,. \end{aligned} \qquad (5.77)$$

As in the MIT bag model, the bag radius R is regarded as a parameter of the model. Its meaning, however, differs very much in the chiral bag model from that in the MIT bag model. This will be clarified in the following discussions.

Although the chiral bag lagrangian of Eq. (5.77) resembles very much that of the little bag model of Eq. (5.43), its physical content is quite different once topological aspects are taken into account. A crucial point is the fact that the quark vacuum is polarized by the surface πqq interaction in Eq. (5.77). From the setup of the model as a hybrid of the Skyrme model and the MIT bag model, one would expect that the chiral bag model should reduce to the Skyrme model when $R \to 0$ and to the MIT bag model when $R \to \infty$. Even such trivially looking conditions can be realized only when the vacuum polarization effects are appropriately taken into account.

Since the chiral bag lagrangian is non-linear, a full quantum mechanical approach is very difficult. We adopt here the semiclassical method starting from a mean field solution. In the mean field approximation, the pion field may be replaced by a classical (c-number) field. As in the little bag and the Skyrme model, the pion field takes the hedgehog ansatz:

$$U_h(\boldsymbol{x}) = \exp(i\boldsymbol{\tau}\cdot\hat{\boldsymbol{x}}F(r)) \tag{5.78}$$

with the spherically symmetric profile function $F(r)$. That the hedgehog ansatz is energetically favored over other field configurations must be proven. In the Skyrme model, it has been verified numerically [Braaten et al. 1990]. We assume that it holds also in the chiral bag model.

Using the ansatz of Eq. (5.78) in Eq. (5.77), we find the equations of motion for the chiral bag model:

$r < R$

$$i\gamma^\mu \partial_\mu \psi = 0\,, \tag{5.79}$$

$r = R$

$$i\hat{\boldsymbol{x}}\cdot\boldsymbol{\gamma}\,\psi = -\exp(i\boldsymbol{\tau}\cdot\hat{\boldsymbol{x}}F(R)\,\gamma_5/f_\pi)\psi\,, \tag{5.80}$$

$r = R$

$$\left(f_\pi^2 + \frac{2\sin^2 F}{e^2 r^2}\right) F' = \frac{i}{2}\langle\bar{\psi}\boldsymbol{\tau}\cdot\hat{\boldsymbol{x}}\,\gamma_5\, e^{i\boldsymbol{\tau}\cdot\hat{\boldsymbol{x}}\gamma_5 F}\psi\rangle\,, \tag{5.81}$$

$r > R$

$$F'' + \frac{2}{r}F' - \frac{\sin 2F}{r^2}$$
$$+ \frac{2}{e^2 f_\pi^2 r^2}\left(\sin^2 F\, F'' + \frac{\sin 2F}{2}F'^2 - \frac{\sin^2 F \sin 2F}{2r^2}\right) = 0\,. \quad (5.82)$$

In Eq. (5.81), $\langle\cdots\rangle$ is an expectation value in the hedgehog state. Its precise meaning is explained in the following section.

Quark hedgehogs

Let us consider the Dirac equation for quarks, Eq. (5.79) subject to the boundary condition (5.80). All the quark eigenstates including the positive and negative energy states are considered. They will be needed when computing vacuum expectation values. As in the little bag model, the conserved quantity is the grand spin $\boldsymbol{K} = \boldsymbol{J} + \boldsymbol{I}$ (hedgehog ansatz). Eigenstates of $\boldsymbol{K}$ are constructed as direct products of the three eigenstates of $\boldsymbol{L}$, $\boldsymbol{S}$ and $\boldsymbol{I}$. We adopt the following coupling scheme [Mulders 1984]:

$$|KM\rangle = |[L\,[S\,I]^G]^K_M\rangle\,. \quad (5.83)$$

First we form the eigenstates of $\boldsymbol{G} = \boldsymbol{S} + \boldsymbol{I} = \boldsymbol{\sigma}/2 + \boldsymbol{\tau}/2$:

$$\begin{aligned} \chi^0_0 &= \frac{1}{\sqrt{2}}(\uparrow d - \downarrow u)\,, \\ \chi^1_1 &= \uparrow u\,, \quad \chi^1_0 = \frac{1}{\sqrt{2}}(\uparrow d + \downarrow u)\,, \quad \chi^1_{-1} = \downarrow d\,, \end{aligned} \quad (5.84)$$

which satisfy

$$\begin{aligned} \boldsymbol{G}^2\chi^G_\mu &= G(G+1)\chi^G_\mu \quad G = 0,\, 1\,, \\ G_3\chi^G_\mu &= \mu\chi^G_\mu\,. \end{aligned} \quad (5.85)$$

There is a useful relation that spin matrices σ_i generate the $G = 1$ states from the $G = 0$ state:

$$\chi^1_\mu = \sigma_\mu \chi^0_0\,, \quad (5.86)$$

where the spherical basis of the Pauli matrices are

$$\begin{aligned} \sigma_{+1} &= -\frac{1}{\sqrt{2}}(\sigma_x + \sigma_y)\,, \\ \sigma_{-1} &= +\frac{1}{\sqrt{2}}(\sigma_x - \sigma_y)\,, \end{aligned} \quad (5.87)$$

$$\sigma_0 = \sigma_z . \tag{5.88}$$

Now the eigenstates of K and $K_z = M$ are given by

$$|KM\rangle = \sum_{m\mu} (lmG\mu|KM)\, Y_{lm}(\hat{\boldsymbol{x}})\chi_\mu^G , \tag{5.89}$$

where $(lmG\mu|KM)$ are the standard Clebsch-Gordan coefficients of $SU(2)$, and $Y_{lm}(\hat{\boldsymbol{x}})$ are the spherical harmonics. Since $S = I = 1/2$ for quarks, there are four combinations for a given (K, M):

$$\begin{aligned}
|0\rangle &= [K\, [\tfrac{1}{2}\, \tfrac{1}{2}]^0]^K_M = Y_{KM}(\hat{\boldsymbol{x}})\chi_0^0 , \\
|1\rangle &= [K\, [\tfrac{1}{2}\, \tfrac{1}{2}]^1]^K_M = \sum_{m\mu} (Km1\mu|KM) Y_{Km}(\hat{\boldsymbol{x}})\chi_\mu^1 , \\
|2\rangle &= [K{+}1\, [\tfrac{1}{2}\, \tfrac{1}{2}]^1]^K_M \\
&\qquad = \sum_{m\mu} (K+1\, m\; 1\mu|KM) Y_{K+1\, m}(\hat{\boldsymbol{x}})\chi_\mu^1 , \\
|3\rangle &= [K{-}1\, [\tfrac{1}{2}\, \tfrac{1}{2}]^1]^K_M = \sum_{m\mu} (K-1\, m\; 1\mu|KM) Y_{K-1\, m}(\hat{\boldsymbol{x}})\chi_\mu^1 .
\end{aligned} \tag{5.90}$$

Note that the parity of $(|0\rangle, |1\rangle)$ differs from the parity of $(|2\rangle, |3\rangle)$.

In order to solve the Dirac equation and evaluate various matrix elements, we need matrix elements represented in the four bases, $|0\rangle$, $|1\rangle$, $|2\rangle$ and $|3\rangle$, for a given K. Here we present the results for various operators:

$$i\boldsymbol{\sigma}\cdot\boldsymbol{\nabla} = \begin{pmatrix} 0 & 0 & -iPD_{K+2} & iQD_{-K+1} \\ 0 & 0 & -iQD_{K+2} & -iPD_{-K+1} \\ -iPD_{-K} & -iQD_{-K} & 0 & 0 \\ iQD_{K+1} & -iPD_{K+1} & 0 & 0 \end{pmatrix}, \tag{5.91}$$

$$\boldsymbol{\sigma}\cdot\hat{x} = \begin{pmatrix} 0 & 0 & -P & Q \\ 0 & 0 & -Q & -P \\ -P & -Q & 0 & 0 \\ Q & -P & 0 & 0 \end{pmatrix}, \tag{5.92}$$

$$\boldsymbol{\tau}\cdot\hat{x} = \begin{pmatrix} 0 & 0 & P & -Q \\ 0 & 0 & -Q & -P \\ P & -Q & 0 & 0 \\ -Q & -P & 0 & 0 \end{pmatrix}, \tag{5.93}$$

$$\boldsymbol{\sigma}\cdot\hat{x}\,\boldsymbol{\tau}\cdot\hat{x} = \begin{pmatrix} -1 & 0 & 0 & 0 \\ 0 & 1 & 0 & 0 \\ 0 & 0 & -Q^2/K & 2PQ \\ 0 & 0 & 2PQ & Q^2/K \end{pmatrix}, \tag{5.94}$$

$$\boldsymbol{\sigma}\cdot\boldsymbol{\tau} = \begin{pmatrix} -3 & 0 & 0 & 0 \\ 0 & 1 & 0 & 0 \\ 0 & 0 & 1 & 0 \\ 0 & 0 & 0 & 1 \end{pmatrix}, \tag{5.95}$$

$$\boldsymbol{\sigma}\times\boldsymbol{\tau}\cdot\hat{x} = 2i\begin{pmatrix} 0 & 0 & P & -Q \\ 0 & 0 & 0 & 0 \\ -P & 0 & 0 & 0 \\ Q & 0 & 0 & 0 \end{pmatrix}. \tag{5.96}$$

In these equations, D_N, P and Q are defined by

$$D_N = \frac{d}{dr} + \frac{N}{r}, \tag{5.97}$$

$$P = \sqrt{\frac{K+1}{2K+1}}, \quad Q = \sqrt{\frac{K}{2K+1}}. \tag{5.98}$$

Solutions to the Dirac equation

The free Dirac equation with the eigenenergy E,

$$\begin{pmatrix} E & i\boldsymbol{\sigma}\cdot\boldsymbol{\nabla} \\ i\boldsymbol{\sigma}\cdot\boldsymbol{\nabla} & E \end{pmatrix}\begin{pmatrix} u \\ v \end{pmatrix} = 0 \tag{5.99}$$

can be expanded in terms of the four basis functions of Eq. (5.90). As usual, the radial part of the wave function is given by a spherical Bessel function, the order of which is the orbital angular momentum l. Thus we have two functions for quark eigenstates with definite parity:

$$\begin{aligned} \psi^+ &= \begin{pmatrix} a_0\, j_K(Er)|0\rangle + a_1\, j_K(Er)|1\rangle \\ a_2\, j_{K+1}(Er)|2\rangle + a_3\, j_{K-1}(Er)|3\rangle \end{pmatrix} \\ &\equiv \begin{pmatrix} u_0|0\rangle + u_1|1\rangle \\ v_2|2\rangle + v_3|3\rangle \end{pmatrix}, \\ \psi^- &= \begin{pmatrix} b_2\, j_{K+1}(Er)|2\rangle + b_3\, j_{K-1}(Er)|3\rangle \\ b_0\, j_K(Er)|0\rangle + b_1\, j_K(Er)|1\rangle \end{pmatrix} \end{aligned} \tag{5.100}$$

$$\equiv \begin{pmatrix} u_2|2\rangle + u_3|3\rangle \\ v_0|0\rangle + v_1|1\rangle \end{pmatrix} . \tag{5.101}$$

These states transform under the parity transformation $\boldsymbol{x} \to -\boldsymbol{x}$:

$$\psi^+(\boldsymbol{x}) \;\to\; \gamma^0 \psi^+(-\boldsymbol{x}) = (-1)^K \psi^+(\boldsymbol{x}) , \tag{5.102}$$

$$\psi^-(\boldsymbol{x}) \;\to\; \gamma^0 \psi^-(-\boldsymbol{x}) = (-1)^{K+1} \psi^-(\boldsymbol{x}) . \tag{5.103}$$

We note that the argument of the spherical Bessel functions in Eqs. (5.100) and (5.101) is Er rather than the standard one kr. For massless quarks, this notation simplifies the expressions of the wave functions, without introducing extra sign factors for negative energy solutions.

The coefficients a_i or b_i in Eqs. (5.100) and (5.101) are determined as follows. For definiteness, we consider a_i for the natural parity state. The coefficients b_i for the unnatural parity states are obtained similarly. Substituting the wave function of Eq. (5.100) in the Dirac equation (5.99), one finds

$$\begin{pmatrix} E & 0 & -iPE & -iQE \\ 0 & E & -iQE & iQE \\ iPE & iQE & E & 0 \\ iQE & -iPE & 0 & E \end{pmatrix} \begin{pmatrix} a_0 \\ a_1 \\ a_2 \\ a_3 \end{pmatrix} = 0 , \tag{5.104}$$

where

$$P \equiv \left(\frac{K+1}{2K+1} \right)^{1/2} , \quad Q \equiv \left(\frac{K}{2K+1} \right)^{1/2} . \tag{5.105}$$

In deriving these equations, the matrix elements of $\boldsymbol{\sigma} \cdot \boldsymbol{\nabla}$ of Eq. (5.91) have been used. Since the rank of the 4×4 matrix in Eq. (5.104) is two, the matrix equation has two independent solutions. A convenient choice for them is

$$\begin{pmatrix} a_0 \\ a_1 \\ a_2 \\ a_3 \end{pmatrix} = N_1 \begin{pmatrix} iP \\ iQ \\ 1 \\ 0 \end{pmatrix} \equiv A_1 \quad \text{and} \quad N_2 \begin{pmatrix} iQ \\ -iP \\ 0 \\ 1 \end{pmatrix} \equiv A_2 , \tag{5.106}$$

where N_i are the normalization constants given by

$$\begin{aligned} N_1^{-2} &= R^3 j_K^2(ER) \left(ER(1+w_+^2) - 2(K+1)w_+ \right) / ER , \\ N_2^{-2} &= R^3 j_K^2(ER) \left(ER(1+w_-^2) - 2Kw_- \right) / ER , \end{aligned} \tag{5.107}$$

$$w_+ = j_{K+1}(ER)/j_K(ER)\,,$$
$$w_- = j_{K-1}(ER)/j_K(ER)\,.$$

For $K = 0$, the spherical part of the wave function reduces to

$$\begin{aligned} |0\rangle &= \frac{1}{\sqrt{4\pi}}\chi_0^0\,, \\ |2\rangle &= \sum_{m\mu}(1m1\mu|00)Y_{1m}\chi_\mu^1 = -\frac{1}{\sqrt{4\pi}}\boldsymbol{\sigma}\cdot\hat{\boldsymbol{x}}\chi_0^0\,, \qquad (5.108) \\ |1\rangle &= |3\rangle = 0\,. \qquad (5.109) \end{aligned}$$

and only the A_1 term in Eq. (5.106) survives. Thus, one recovers the $K = 0$ wave function of Eq. (5.55). For $K > 0$, the coefficients a_i are the linear combinations of A_1 and A_2:

$$(a_i) = c_1 A_1 + c_2 A_2 = c_1 N_1 \begin{pmatrix} iP \\ iQ \\ 1 \\ 0 \end{pmatrix} + c_2 N_2 \begin{pmatrix} iQ \\ -iP \\ 0 \\ 1 \end{pmatrix}, \qquad (5.110)$$

where the constants c_1 and c_2 are determined by the boundary condition (5.80), which is explicitly given in terms of $\psi = (u, v)$:

$$\begin{aligned} v &= i\cos F\,\boldsymbol{\sigma}\cdot\hat{\boldsymbol{x}}\,u - \sin F\,\boldsymbol{\sigma}\cdot\hat{\boldsymbol{x}}\,\boldsymbol{\tau}\cdot\hat{\boldsymbol{x}}\,v\,, \\ u &= i\sin F\,\boldsymbol{\sigma}\cdot\hat{\boldsymbol{x}}\,\boldsymbol{\tau}\cdot\hat{\boldsymbol{x}}\,u - i\cos F\,\boldsymbol{\sigma}\cdot\hat{\boldsymbol{x}}\,v\,. \qquad (5.111) \end{aligned}$$

Using the matrix elements of $\boldsymbol{\sigma}\cdot\hat{\boldsymbol{x}}$ and $\boldsymbol{\sigma}\cdot\hat{\boldsymbol{x}}\,\boldsymbol{\tau}\cdot\hat{\boldsymbol{x}}$, Eqs. (5.92) and (5.94) these equations reduce to

$$\begin{pmatrix} P(z^\kappa - \kappa w_+) & Q(z^\kappa + \kappa\omega_-) \\ Q(z^{-\kappa} - \kappa w_+) & -P(z^{-\kappa} + \kappa\omega_-) \end{pmatrix} \begin{pmatrix} c_1 N_1 \\ c_2 N_2 \end{pmatrix} = 0\,, \qquad (5.112)$$

where $\kappa = 1$ and -1 are for the natural and unnatural parity states, respectively, and

$$z = \frac{1+\sin F}{\cos F}\,, \qquad z^{-1} = \frac{1-\sin F}{\cos F}\,. \qquad (5.113)$$

Nontrivial solutions to Eq. (5.112) are allowed when the determinant vanishes:

$$w_+ w_- + \kappa\left[(P^2 z^{-\kappa} + Q^2 z^\kappa)w_+ - (P^2 z^\kappa + Q^2 z^{-\kappa})w_-\right] - 1 = 0\,, \quad (5.114)$$

or equivalently,

$$\cos F(j_K^2 - j_{K+1}j_{K-1}) - \kappa j_K(j_{K+1} - j_{K-1}) + \frac{\sin F}{ER} j_K^2 = 0\,, \qquad (5.115)$$

which is the eigenvalue equation for an arbitrary $K > 0$. The equation for $K = 0$ is obtained simply by setting $j_{K-1} = 0$. Several eigenvalues ($\omega = ER$) of low lying states are shown in Fig. 5.8 as functions of the chiral angle F . For general F, the quark spectrum is asymmetric with respect to $\omega = 0$. Exceptions are when $F = n\pi$ and when $F = \pi/2 + n\pi$ with ($n = 0, \pm1, \pm2, \cdots$). The case of $F = n\pi$ corresponds to the MIT bag model and all the quark states have their symmetric partners on opposite energy sides. Furthermore, K^κ and $(K+1)^\kappa$ states are degenerate. The other case of $F = \pi/2 + n\pi$ has been studied in the previous section as the Jackiw and Rebbi's case [Jackiw and Rebbi 1976], where the spectrum is symmetric except for the zero energy state. The quark spectrum of Fig. 5.8 has also a discrete symmetry $\kappa \to -\kappa$ under $F \to F + \pi$, which is expected from Eq. (5.80) as well.

5.5 Chiral Casimir Effects

When there are interactions, not only particle (valence) states but also the vacuum (sea) structure are modified. Such a change in the vacuum can be observed as originally investigated for the electromagnetic field between two parallel plates of conductors. The boundary condition for the electromagnetic field is modified in the presence of the plates, which produce a tiny attractive force between them. In the chiral bag model, the πqq interaction at the bag surface in Eq. (5.81) induces a chiral Casimir effect and quantities in the quark sector are strongly affected.

Vacuum polarization

Let us consider the quark field interacting, for instance, with a background potential with strength characterized by one parameter a. The potential is then assumed to disappear when $a = 0$. Consider a bilinear operator $\bar\psi\mathcal{M}\psi$. In the mean field approximation, the vacuum expectation value of $\bar\psi\mathcal{M}\psi$ is defined by

$$\mathcal{M}_{\rm vac} = \sum_m \mathcal{M}_m(a) - \sum_m \mathcal{M}_m(0)\,, \qquad (5.116)$$

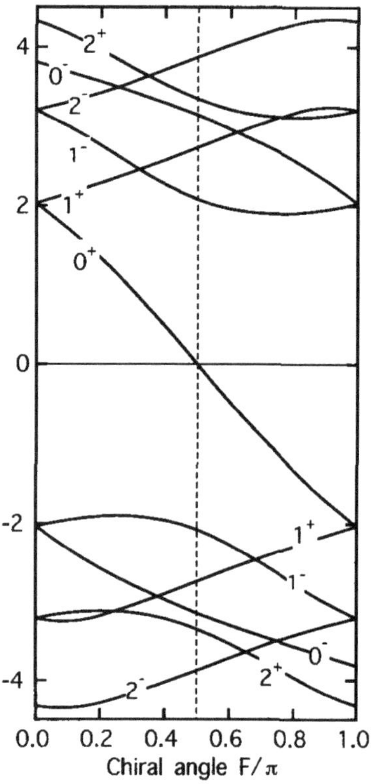

Fig. 5.8 Eigenvalues of the chiral bag model as functions of the chiral angle F.

where $\mathcal{M}_m(a)$ is the matrix element $\bar{\psi}_m \mathcal{M} \psi_m$ for a negative energy quark state ψ_m. The subtraction of the second term is to ensure that $\mathcal{M}_{\text{vac}}$ in the trivial vacuum ($a = 0$) vanishes. For $a \neq 0$, however, $\mathcal{M}_{\text{vac}}$ can be non-zero.

In the chiral bag model, there are two different types of vacuum polarization. One is related to quark confinement by the scalar boundary condition (5.36), and the other is induced by the πqq interaction at the bag surface. The first one has been extensively studied as the Casimir effect of the MIT bag model [Milton 1983; Milton 1980b; Milton 1980a; Milton et al. 1978]. For example, the vacuum expectation value of the energy of the MIT bag measured from the normal vacuum is computed by

$$E_{\rm vac}({\rm MIT}) = \sum_m E_m(R) - \sum_m E_m(R \to \infty) . \tag{5.117}$$

The divergent behavior of this quantity and other related properties are discussed in Refs. [Milton 1980a; Bender and Hays 1976; Vepstas et al. 1984; Vepstas and Jackson 1990]. The second vacuum polarization is the one we consider here and is important in the chiral bag model. The vacuum polarization induces non-zero expectation values for various physical quantities depending on the strength of the chiral field at the bag surface $F \equiv F(R)$.

Since Casimir effects involve the sum over all the states, it often happens that vacuum expectation values diverge. Therefore, we need careful studies either in analytic or numerical methods to extract physically meaningful finite quantities. This is our main purpose in the following part of this section.

In order to handle divergent quantities, we introduce a regularization scheme. First we normal-order the operator $\bar{\psi}\mathcal{M}\psi$ by

$$\bar{\psi}\mathcal{M}\psi \to \frac{1}{2}[\bar{\psi}, \mathcal{M}\psi] . \tag{5.118}$$

This provides the desired transformation properties that are consistent with charge conjugation and G-parity transformations including the vacuum contributions [Bjorken and Drell 1965]. Expanding the quark field in terms of the energy eigenstates as given in the previous subsection,

$$\psi(x) = \sum_{E_n>0} u_n(\boldsymbol{x}) e^{-iE_n t} b_n + \sum_{E_m<0} v_m(\boldsymbol{x}) e^{iE_m t} d_m^\dagger , \tag{5.119}$$

the vacuum expectation value of $\mathcal{M}$ as in Eq. (5.118) is written as a mode sum over the positive and negative energy states. If there is a zero energy mode, it can be included either in positive or negative energy states.

Then the mode sum of Eq. (5.119) can be regularized (made finite) by introducing a cutoff. One convenient choice is the point splitting method, in which a singular operator such as a fermion bilinear is regularized by separating the space-time positions of ψ and $\bar{\psi}$ along the imaginary time direction, *i.e.*,

$$\bar{\psi}(t, \boldsymbol{x})\mathcal{M}\psi(t, \boldsymbol{x}) \to \bar{\psi}(t + i\eta, \boldsymbol{x})\mathcal{M}\psi(t, \boldsymbol{x}) . \tag{5.120}$$

In this prescription the vacuum expectation value is expressed as

$$\begin{aligned}\mathcal{M}_0 &= \langle 0|\frac{1}{2}[\bar{\psi}, \mathcal{M}\psi]|0\rangle \\ &= +\frac{1}{2}\lim_{\eta\to 0}\sum_{E_m<0} \bar{v}_m \mathcal{M} v_m e^{-\eta|E_m|} \\ &\qquad -\frac{1}{2}\lim_{\eta\to 0}\sum_{E_n>0} \bar{u}_n \mathcal{M} u_n e^{-\eta|E_n|} \\ &\equiv -\frac{1}{2}\lim_{\eta\to 0}\sum_{\text{all } n} \text{sgn}(E_n)\bar{\psi}_n \mathcal{M}\psi_n e^{-\eta|E_n|}\,. \end{aligned} \tag{5.121}$$

The sum here is essentially a one-loop quantum correction.

In addition to the linear regulator of Eq. (5.121), we can use the gaussian type regulator in order to achieve a rapid convergence in numerical computations:

$$\mathcal{M}_0 = -\frac{1}{2}\lim_{\eta\to 0}\sum_{\text{all } n} \text{sgn}(E_n)\bar{\psi}_n \mathcal{M}\psi_n e^{-(\eta E_n)^2}\,. \tag{5.122}$$

We will adopt this regularization scheme in Chapter 6 when computing double mode sums that appear in the nucleon moment of inertia and magnetic moments.

Baryon Number

This is the classic example indicating the importance of the vacuum polarization. The baryon number carried by a quark is $1/N_c$ and the mode sum of (5.121) is just the spectral asymmetry:

$$B_0(F) = -\frac{1}{N_c}\lim_{\eta\to 0}\sum_{n} \text{sgn}(E_n) e^{-\eta|E_n|}\,. \tag{5.123}$$

The regulator $\exp(-\eta|E_n|)$ is introduced to regularize the infinite mode sum. Note that the sum also includes color indices over N_c quarks. The special case $F = \pi/2$ has been studied in section 5.2 as the Jackiw-Rebbi's case, where the analytic computation was possible due to the symmetry of the quark spectrum between positive and negative energy states.

A toy model in $1+1$ *dimension*

Before we evaluate the baryon number in the realistic case, let us examine a simple case in which eigenvalues, and therefore, the mode sum for the baryon number in Eq. (5.123) are calculated analytically. A toy example is a $1+1$ dimensional model of a confined quark interacting with the pion at the boundary as shown in Fig. 5.9. The relevant lagrangian is given by

$$L = \bar{\psi} i \partial\!\!\!/ \psi \theta(|\boldsymbol{x}| < R) - \frac{1}{2} \bar{\psi} \exp(iF\gamma_5) \psi \delta(|\boldsymbol{x}| - R) . \tag{5.124}$$

with the boundary condition at the surface:

$$i\gamma_1 \psi = -\exp(iF\gamma_5)\psi . \tag{5.125}$$

In $1+1$ dimension, the fermion ψ is two component spinor and the Dirac matrices are given by

$$\begin{aligned} \gamma^0 &= \sigma_1 = \begin{pmatrix} 0 & 1 \\ 1 & 0 \end{pmatrix} , \\ \gamma^1 &= i\sigma_2 = \begin{pmatrix} 0 & -1 \\ 1 & 0 \end{pmatrix} , \\ \gamma_5 &= -\gamma^0\gamma^1 = \begin{pmatrix} 1 & 0 \\ 0 & -1 \end{pmatrix} . \end{aligned} \tag{5.126}$$

They are the chiral (Weyl) representations of the Dirac matrices in $1+1$ dimension, since the γ_5 matrix is diagonalized. In this basis, the upper component of ψ represents the right-moving quark and the lower component the left-moving one.

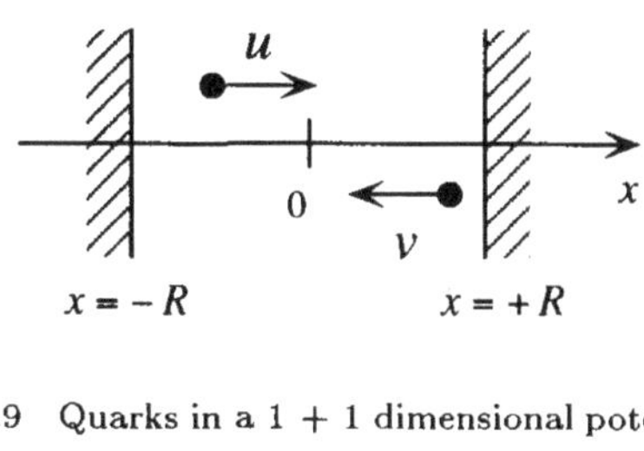

Fig. 5.9 Quarks in a 1 + 1 dimensional potential.

Setting $\psi = \exp(-iEt)(u(x), v(x))$, the free Dirac equation is written as

$$\begin{aligned} Eu + iu' &= 0\,, \\ Ev - iv' &= 0\,, \end{aligned} \tag{5.127}$$

where $u' \equiv du/dx$. The stationary solution is then

$$\psi^{(+)} = \begin{pmatrix} \exp(iEx) \\ \exp(-iEx) \end{pmatrix} . \tag{5.128}$$

This two component form represents the stationary solution as it is the superposition of the left- and right-moving components. Furthermore, this has positive parity, since

$$\begin{aligned} \psi^{(+)}(x) &\overset{parity}{\longrightarrow} \gamma_0 \psi^{(+)}(-x) \\ &= \begin{pmatrix} 0 & 1 \\ 1 & 0 \end{pmatrix} \begin{pmatrix} \exp(iE(-x)) \\ \exp(-iE(-x)) \end{pmatrix} \\ &= \begin{pmatrix} \exp(iEx) \\ \exp(-iEx) \end{pmatrix} . \end{aligned} \tag{5.129}$$

Now the boundary condition of Eq. (5.125) at $x = +R$ reduces to the single eigenvalue equation

$$\exp(i(F + 2ER)) = i\,, \tag{5.130}$$

whose solutions are

$$E_n^{(+)} = \frac{1}{2R}\left(\frac{\pi}{2} + 2n\pi - F\right) . \tag{5.131}$$

Similarly, we find the eigenvalues of negative parity states

$$\psi^{(-)} = \begin{pmatrix} \exp(iEx) \\ -\exp(-iEx)) \end{pmatrix} \tag{5.132}$$

and

$$E_n^{(-)} = \frac{1}{2R}\left(-\frac{\pi}{2} + 2n\pi - F\right) . \tag{5.133}$$

These eigenvalues are shown in Fig. 5.10 (a) as functions of F. When the positive and negative parity energy levels are included, the whole spectrum has the discrete translational invariance in F modulo π. Under the displacement by π, the positive and negative parity levels are interchanged. Similar structure of the spectrum is also seen in the actual chiral bag model in Fig. 5.8. Fig. 5.10 (b) shows the spectrum of $K = 0$ states of the chiral bag model, which indeed resembles Fig. 5.10 (a).

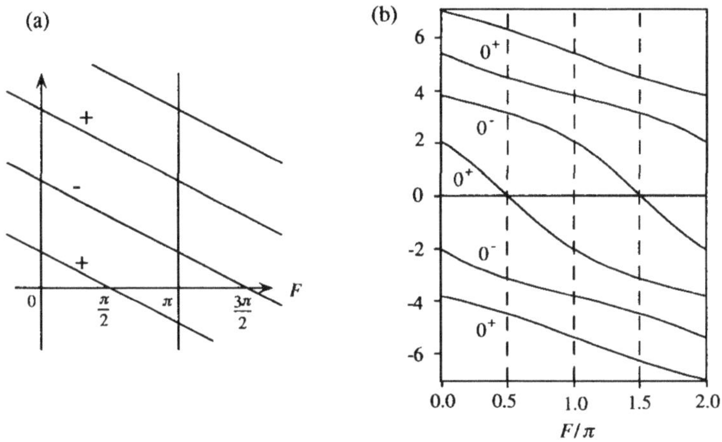

Fig. 5.10 (a) Eigenvalues of the 1 + 1 dimensional model as functions of F, and (b) eigenvalues of $K = 0$ states in the chiral bag model.

Once the analytic expressions for the eigenvalues in Eqs. (5.131) and (5.131) are known, the sum of Eq. (5.123) is easily computed. The result is given by

$$B_q(F) \quad = \quad -\frac{F}{\pi}, \quad \text{for} \quad (n-1/2)\,\pi < F < (n+1/2)\,\pi\,. \tag{5.134}$$

At $F = (n + 1/2)\pi$ when there is a zero mode, we need to be slightly careful. If we put valence quarks in the zero energy level, we can count their baryon number half of the time and the net baryon number should be 1/2, the result obtained by Jackiw and Rebbi. If there is no valence quark in the zero energy level, we should count the anti-baryon number of the quark-holes, which results once again in $-1/2$. This is the reason that there is the discontinuity of one in Eq. (5.134) at $F = \pi/2$.

Realistic case

For the actual chiral bag model in 1 + 3 dimensions, computation becomes complicated, but still analytic method is possible. Goldstone and Jaffe [Goldstone and Jaffe 1983] were the first who performed the mode sum of Eq. (5.123) analytically by exploring the multiple reflection method for the quark propagator [Hansson and Jaffe 1983]. They found the famous expression for $|F| < \pi/2$.

$$B_0(F) = -\frac{1}{\pi}(F - \sin F \cos F) . \tag{5.135}$$

It is not difficult to confirm this result numerically both in the linear and gaussian regularization schemes. With today's computer power, the calculation is easily done even on a personal computer.

Surprisingly, this expression is exactly the negative of the baryon number carried by the pion field outside the bag:

$$\begin{aligned} B_\pi &= \frac{\epsilon_{ijk}}{24\pi^2} \int_R^\infty d^3x \ \mathrm{tr}\, \partial_i U U^\dagger \partial_j U U^\dagger \partial_k U U^\dagger \\ &= \frac{1}{\pi}(F - \sin F \cos F) . \end{aligned} \tag{5.136}$$

Therefore, for a single nucleon, B_0 and B_π cancel exactly, and the total baryon number is entirely carried by the valence quarks:

$$B_{\text{tot}} = B_{\text{val}} + \underbrace{B_0 + B_\pi}_{0} = B_{\text{val}} = 1 . \tag{5.137}$$

As the chiral angle F passes $\pi/2$, the result of Eq. (5.135) for the vacuum contribution has to be considered carefully. Consider the G-parity and DC (discrete chiral) transformations:

$$\mathrm{G}: \ \psi \ \rightarrow \ -\tau_2 \gamma^2 \gamma^0 \bar{\psi}^T , \ \ F \rightarrow -F , \tag{5.138}$$

$$\mathrm{DC}: \ \psi \ \rightarrow \ \gamma_5 \psi , \ \ F \rightarrow F + \pi , \tag{5.139}$$

Using the symmetry of the baryon number operator $\frac{1}{2}\int d^3x[\bar{\psi}^\dagger, \psi]$, under G and DC, as shown in Eqs. (5.138) and (5.139), one finds

$$B_0(F) = -B_0(-F) = B_0(F + \pi) . \tag{5.140}$$

The first equation shows that the baryon number of the vacuum $B_0(F)$ is an odd function of F as expected from the G-parity of the baryon current. In general, G-parity even (or odd) quantities are even (or odd) functions of F. The second equation of (5.140) indicates that $B_0(F)$ is a periodic function with periodicity π as expected from the symmetry of the quark spectrum under $F \rightarrow F + \pi$. Finally, it also implies that

$$B_0(\tfrac{\pi}{2} + F) = -B_0(\tfrac{\pi}{2} - F) . \tag{5.141}$$

For a single nucleon, the chiral angle varies within $0 \leq F \leq \pi$. The result of Eq. (5.135) is valid only for $0 \leq F \leq \pi/2$. As F passes $\pi/2$ and the lowest 0^+ level crosses the zero energy, the vacuum contribution B_0 changes by one. Taking this change into account, we find

$$B_0(F) = \theta(F - \tfrac{\pi}{2}) - \frac{1}{\pi}(F - \sin F \cos F) \tag{5.142}$$

for $0 \leq F \leq \pi$. The baryon number of the quark bag B_q including the valence contribution for $0 \leq F \leq \pi/2$ is therefore

$$\begin{aligned} B_q(F) &= \theta(\tfrac{\pi}{2} - F) + B_0(F) \\ &= 1 - \frac{1}{\pi}(F - \sin F \cos F) . \end{aligned} \tag{5.143}$$

Finally, adding the pion contribution of Eq. (5.136) we find once again the unit total baryon number as in Eq. (5.137) independent of F.

Axial flux

Another example of the mode sum is the quark axial flux which is the normal flow of the axial-vector current of the quark at the bag surface

$$A_q(F) \equiv \hat{x}_i \hat{x}_a \bar{\psi}\gamma^i t^a \psi|_{r=R} \, . \tag{5.144}$$

The normal ordered expression is

$$A_q(F) = \frac{1}{4}[\bar{\psi}, \boldsymbol{\tau}\cdot\hat{\boldsymbol{x}}\,\boldsymbol{\gamma}\cdot\hat{\boldsymbol{x}}\gamma_5\psi] \, , \tag{5.145}$$

Using the linear boundary condition (5.80), the axial flux is proved to be the right hand side of Eq. (5.81). The vacuum expectation value of the nucleon matrix element is then given by

$$A_0(F;\eta) = -\frac{1}{4}\lim_{\eta\to 0}\sum_n \mathrm{sgn}(E_n)\bar{\psi}_n \boldsymbol{\tau}\cdot\hat{\boldsymbol{x}}\,\boldsymbol{\gamma}\cdot\hat{\boldsymbol{x}}\gamma_5\psi_n \, e^{-\eta|E_n|} \, . \tag{5.146}$$

Due to spherical symmetry of the vacuum, it is convenient to define the angle average:

$$\begin{aligned} A_0(F;\eta) &\equiv \frac{1}{4\pi}\int d\Omega\, A_0(F;\eta) \\ &= -\frac{1}{2}\lim_{\eta\to 0}\sum_n \mathrm{sgn}(E_n)\alpha_n e^{-\eta|E_n|} \, . \end{aligned} \tag{5.147}$$

Angle integration of the matrix element can be performed analytically and we find

$$\begin{aligned}\alpha_n &\equiv \frac{1}{8\pi}\int d\Omega\bar{\psi}_n\boldsymbol{\tau}\cdot\hat{\boldsymbol{x}}\,\boldsymbol{\gamma}\cdot\hat{\boldsymbol{x}}\gamma_5\psi_n \\ &= \frac{1}{2}\left[u_0^\dagger u_0 - u_1^\dagger u_1 + C_- v_2^\dagger v_2 - C_+ v_3^\dagger v_3\right]_{r=R}, \qquad (5.148)\end{aligned}$$

where

$$C_\pm = \frac{2[K(K+1)]^{1/2}}{2K+1} \pm \frac{1}{2K+1}, \qquad (5.149)$$

and u_i and v_i are the spherical Bessel functions as defined in Eqs. (5.100) and (5.101). When computing α_n the matrix elements of $\boldsymbol{\sigma}\cdot\boldsymbol{\tau}$ in the basis $|0\rangle$, $|1\rangle$, $|2\rangle$ and $|3\rangle$ of Eqs. (5.90) are used. For the unnatural parity state of $\kappa = -1$, the spinors u and v are interchanged.

It turns out that the mode sum of Eq. (5.147) is linearly divergent as $1/\eta$. Analytic method is available to extract the coefficient of the leading divergent term [Mulders 1984; Zahed et al. 1985]. It is noted that the linearly divergent term is special for the bag model with a sharp boundary. In fact, in ordinary field theories in infinitely extended space, the axial-vector current at the one-loop level, Fig. 5.11, is only logarithmically divergent. The latter divergence may be handled by the standard renormalization, but the bag-originating divergence can not [Zahed et al. 1985].

Fig. 5.11 Axial flux at the one-loop level.

For practical purposes, numerical computation is inevitable. To obtain sufficiently accurate results, the summation is taken up to the maximum energy around $E_{\max}R \sim 100$, where the angular momentum reaches around $K \sim 100$. With these maximum values, the mode sum is calculated for small η, $0.1 \leq \eta/R \leq 0.5$ *.

*These numerical computations were done almost more than fifteen years ago, when computer power was much smaller than it is today. It is possible now to extend the range of summation further and obtain more accurate results.

It is convenient to define the dimensionless quantity per quark

$$a_0(F;\eta) \equiv \frac{R^3}{N_c} A_0(F;\eta)\,, \tag{5.150}$$

and redefine $\eta/R \to \eta$. The function $a_0(F;\eta)$ is expanded in terms of η:

$$\begin{aligned} a_0(F;\eta) &= \frac{c_{-1}(F)}{\eta} + c_{\ln}(F)\ln\eta \\ &\quad + c_0(F) + c_1(F)\eta + c_2(F)\eta^2 + \cdots \\ &\equiv \frac{c_{-1}(F)}{\eta} + c_{\ln}(F)\ln\eta + c_{\text{fin}}(F;\eta)\,, \end{aligned} \tag{5.151}$$

where we have defined the finite term as

$$c_{\text{fin}} = c_0(F) + c_1(F)\eta + c_2(F)\eta^2 + \cdots\,. \tag{5.152}$$

Our task is to determine the coefficients $c_i(F)$'s. This was done using both analytic and numerical methods [Vepstas et al. 1984; Mulders 1984; Hosaka and Toki 1986; Vepstas and Jackson 1990].

For the leading coefficient $c_{-1}(F)$, analytic result is:

$$c_{-1}(F) = \frac{1}{16\pi}\sin 2F\,. \tag{5.153}$$

In order to determine the coefficient of the logarithmic divergence, $c_{\ln}(F)$, the following identity was used:

$$\frac{d\epsilon_0(F;\eta)}{dF} = 4\pi(1+\eta\frac{d}{d\eta})a_0(F;\eta)\,, \tag{5.154}$$

where the dimensionless vacuum energy (the Casimir energy) is defined by

$$\begin{aligned} \epsilon_0(F) &\equiv RE_0(F) \\ &= -\frac{1}{2}\lim_{\eta\to 0}\sum_n \operatorname{sgn}(E_n)\, RE_n e^{-\eta|E_n|} \\ &= \frac{1}{2}\lim_{\eta\to 0}\sum_n |RE_n| e^{-\eta|E_n|}\,. \end{aligned} \tag{5.155}$$

The identity (5.154) is verified by using the relation:

$$\frac{dE_n}{dF} = \frac{R^2}{2}\int d\Omega \bar{\psi}_n \boldsymbol{\tau}\cdot\hat{\boldsymbol{x}}\,\boldsymbol{\gamma}\cdot\hat{\boldsymbol{x}}\,\gamma_5\psi_n\,. \tag{5.156}$$

In the vacuum energy of Eq. (5.155), the F independent term $\epsilon_0(F=0;\eta)$ (the Casimir energy of the MIT bag model) is understood to be subtracted:

$$\epsilon_0(F;\eta) \to \epsilon_0(F;\eta) - \epsilon_0(F=0;\eta)\,. \tag{5.157}$$

The F dependence of the divergent term of the energy (5.155) was analytically worked out to leading order by Zahed, Wirzba and Meissner [Zahed et al. 1985]. They found

$$\epsilon_0(F;\eta) = -\frac{2}{15\pi}\sin^2 F \ln\eta + \epsilon_{\rm fin}(F;\eta)\,, \tag{5.158}$$

where $\epsilon_{\rm fin}(F;\eta)$ is the finite term. Substituting Eqs. (5.151) and (5.158) in Eq. (5.154), one finds that

$$c_{\ln} = -\frac{\sin^2 F}{30\pi^2}\,, \tag{5.159}$$

$$\frac{d\epsilon_{\rm fin}}{dF} = 4\pi(c_{\rm fin}+c_{\ln})\,. \tag{5.160}$$

In this way all divergences in the axial flux have been extracted analytically, and the finite terms are separated as

$$a_0(F) = \left(\frac{1}{16\pi\eta} - \frac{1}{30\pi^2}\ln\eta\right)\sin 2F + (\text{finite terms})\,. \tag{5.161}$$

The presence of the divergences raises a question of how to solve the problem. In particular, the linearly divergent term implies that the ordinary renormalization method does not work. There are several prescriptions:

(1) It was found that the divergences in Eq. (5.161) can be canceled by a similar mode sum of quarks in the bag exterior [Vepstas and Jackson 1990]. At first sight, it does not seem to make sense to consider quarks in the bag exterior, since it appears to contradict quark confinement. However, the divergences are of ultraviolet nature and at sufficiently high energy, quarks would replace the hadrons. Since this cancellation mechanism is interesting, it is briefly discussed in the following subsection.

(2) The above framework for the divergence cancellation is not sufficient, since it can not control finite terms. In a usual renormalization procedure, the ambiguity in finite terms can be fixed by imposing physical conditions. Such conditions are inspired by symmetries of the theory, since the divergent terms (quantum effects)

often break symmetries satisfied at the classical level. We may use the identity between the axial flux and energy which is inferred by the boundary condition and hence the chiral symmetry of the model (see Eq. (5.189)).

(3) Another prescription is to rely on a rather trivial requirement; when $R \to 0$, the chiral bag should reduce to the skyrmion. It turns out that this can not be achieved if the divergent as well as a part of finite terms are present. As discussed in the next section, such terms cause mismatching in the power counting of the chiral angle ($F(R)$) in various quantities between the inside and outside of the bag. By removing the mismatching, all divergent terms disappear.

Cancellation of divergences

We briefly outline the mechanism of divergence cancellation in the quark mode sum by summing the contributions from the chiral bag interior and exterior. In the bag interior ($r \leq R$), quarks obey the free Dirac equation subject to the boundary condition at $r = R$:

$$-i\hat{x}\cdot\gamma\psi = e^{i\boldsymbol{\tau}\cdot\hat{x}F(R)\gamma_5}\psi\,. \tag{5.162}$$

The Casimir energy in the bag interior is the sum over all the eigenenergies. As shown in Eq. (5.158), its divergent behavior is known; when the linear regulator is adopted, it is

$$\epsilon_0(F;\eta) = -\frac{2}{15\pi}\sin^2 F\,\ln\eta + (\text{finite terms})\,. \tag{5.163}$$

Next in the spirit of Ref. [Vepstas and Jackson 1990], we introduce free quarks that exist outside the bag, which is supposed to mimic the ultraviolet nature of QCD in the exterior region. They are continuum states subject to the boundary condition analogous to Eq. (5.162):

$$+i\hat{x}\cdot\gamma\psi = e^{i\boldsymbol{\tau}\cdot\hat{x}F(R)\gamma_5}\psi\,. \tag{5.164}$$

Scattering states of quarks are characterized by phase shifts. For example, the phase shift of the $K = 0$ with parity P state is given by

$$\delta^P_{K=0}(\omega) = \tan^{-1}\left(\frac{\cos F j_0(\omega R) - (P+\sin F)j_1(\omega R)}{\cos F n_0(\omega R) - (P+\sin F)n_1(\omega R)}\right), \tag{5.165}$$

where j_l and n_l are the spherical Bessel and Neumann functions. Phase shifts for general K are expressed in a rather complicated form as given in Ref. [Vepstas and Jackson 1990].

Let us define the total phase shift, the sum over all partial waves:

$$\delta_F(\omega) = \sum_{K,P} (2K+1)\delta_K^P(\omega)\,, \tag{5.166}$$

where the factor $(2K+1)$ reflects the $(2K+1)$-fold degeneracy of the K-state. The suffix F on the left hand side shows the F dependence of the phase shift. In order to see purely the pionic effects of F dependence, we subtract the F-independent part:

$$\tilde{\delta}_F(\omega) = \delta_F(\omega) - \delta_{F=0}(\omega)\,. \tag{5.167}$$

We are then interested in the asymptotic behavior of the phase shift in the limit $|\omega| \to \infty$, which can be found as

$$\begin{aligned}\lim_{|\omega|\to\infty} \tilde{\delta}_F(\omega) &= -(F - \sin F \cos F) \\ &+ \operatorname{sgn}(\omega)\frac{\pi}{4}\sin^2 F - \frac{2}{15\omega}\sin^2 F + \mathcal{O}(1/\omega^2)\,. \end{aligned} \tag{5.168}$$

Now, the total phase shift $\tilde{\delta}_F(\omega)$ is related to the density of states $n(\omega)$ by

$$\frac{dn}{d\omega} = \frac{1}{\pi}\frac{d\tilde{\delta}_F}{d\omega}\,, \tag{5.169}$$

which is one form of the Levinson's theorem [Levinson 1949; Goldberger and Watson 1965]. This relation is useful in performing the following computations.

First we calculate the baryon number outside the bag. The spectral asymmetry for the continuum states is given in the integral form:

$$B_{0,out}(F) = -\frac{1}{2}\lim_{\eta\to 0}\left(\int_{-\infty}^{\infty} \operatorname{sgn}(\omega) e^{-|\eta\omega|}\frac{1}{\pi}\frac{d\tilde{\delta}_F}{d\omega} d\omega\right)\,. \tag{5.170}$$

The integral can be performed in the limit $\omega \to \infty$ to yield

$$B_{0,out}(F) = \frac{1}{\pi}\left(F - \tfrac{1}{2}\sin F \cos F\right)\,. \tag{5.171}$$

Thus, as expected, the baryon number outside the bag cancels precisely the contribution from the bag interior, Eq. (5.135).

Next we calculate the Casimir energy in the bag exterior. It is given as

$$\epsilon_{0,out}(F;\eta) = -\frac{1}{2}\int_{-\infty}^{\infty} |\omega| e^{-|\eta\omega|} \frac{1}{\pi}\frac{d\tilde{\delta}_F}{d\omega} d\omega \,. \tag{5.172}$$

Inserting Eq. (5.168) here, we obtain for small η

$$\epsilon_{0,out}(F;\eta) = \frac{2}{15\pi}\sin^2 F \ln\eta + \text{finite terms} \,. \tag{5.173}$$

Here we observe that the divergence from the bag exterior is exactly the negative of that from the bag interior, so they cancel exactly. Such a cancellation mechanism is a natural consequence of the fact that we adopted the same theories of free quarks both inside and outside the bag, which communicate through the boundary condition.

5.6 The Hedgehog Solution

Removing divergences

In order to solve the equations of motion (5.79) - (5.82), we need to handle the divergent piece of Eq. (5.81). To do so, let us look closely at the axial flux continuity in the limit $R \to 0$. Let $F_R(r)$ be the profile function (the chiral angle) for the chiral bag of radius R, see Fig. 5.12 (a), and define for simplicity

$$F \equiv F_R(r = R) \,. \tag{5.174}$$

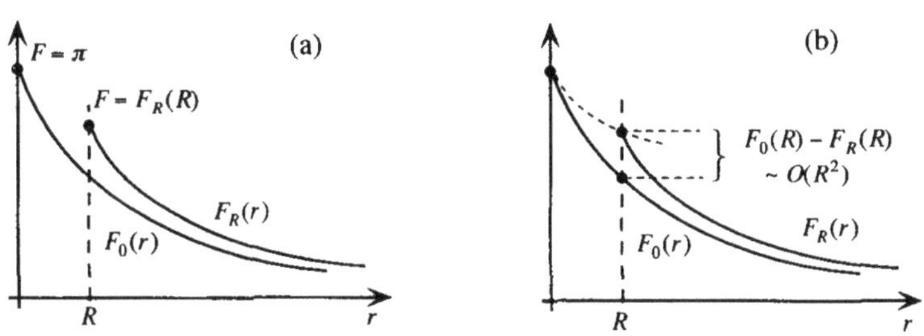

Fig. 5.12 Profiles of chiral angles $F_R(r)$.

In the vicinity $R \sim 0$ where $F \sim \pi$, or $\pi - F \to 0$, the F dependence of $a_0(F)$ has been numerically worked out [Hosaka and Toki 1986; Mulders 1984]:

$$\begin{aligned} a_0(F) &\sim \left(\frac{1}{16\pi\eta} - \frac{1}{30\pi^2}\ln\eta + \frac{3.75}{30\pi^2}\right)\sin 2(\pi - F) \\ &\qquad +1.14\left(\frac{\pi - F}{\pi}\right)^3 \\ &\equiv a_1(\pi - F)^1 + a_3(\pi - F)^3 . \end{aligned} \tag{5.175}$$

Expressing the chiral angle at the bag surface F as

$$F \xrightarrow{R\to 0} \pi\,(1 - bR)\,, \tag{5.176}$$

the axial flux continuity at the bag surface in Eq. (5.81) becomes

$$N_c\left(a_1\frac{b}{R^2} + a_3 b^3 + \mathcal{O}(R^2)\right) = \left(f_\pi^2 + \frac{2b^2}{e^2}\right) b + \mathcal{O}(R^2)\,. \tag{5.177}$$

It is apparent that the a_1 term on the left hand side diverges and Eq. (5.177) no longer holds in the limit $R \to 0$. The a_1 term comes from the $\sin 2F$ dependent terms in Eq. (5.175), which include all the divergent terms as well as a part of finite terms. Thus, it is sufficient to make the following subtraction and define the finite contribution: $\tilde{a}_0(F)$, $\sin 2F$ term must be entirely subtracted:

$$\tilde{a}_0(F) \equiv a_0(F;\eta) - \frac{1}{2}\left[\frac{\partial a_0(F;\eta)}{\partial F}\right]_{F=0} \sin 2F\,. \tag{5.178}$$

The F dependence of $\tilde{a}_0(F)$ for $0 \le F \le \pi$ has been worked out numerically. For $\pi/2 < F < \pi$ or $R \lesssim 0.5$ fm there is only the vacuum contribution, while for $0 < F < \pi/2$ or $R \gtrsim 0.5$ fm there are both the vacuum and the valence contributions. The crossing of the lowest 0^+ valence level at $E = 0$ when $F = \pi/2$ yields a discontinuity in $\tilde{a}_0(F)$, which, however, is canceled by the valence quark contribution which is also discontinuous:

$$\begin{cases} \text{when} & E_{val} > 0, \quad \text{finite} \\ \text{when} & E_{val} \le 0, \quad \text{zero} \end{cases} \tag{5.179}$$

The resulting quark axial flux including both the valence and the vacuum terms is a smooth function of F in the entire region of $0 \le F \le \pi$. In

Fig. 5.13 (a) the total quark axial flux

$$a_q(F) = \tilde{a}_0(F) + a_{\rm val}(F)\theta(\pi/2 - F) \tag{5.180}$$

per quark is plotted as a function of F, where $a_{\rm val}(F)$ is the valence contribution.

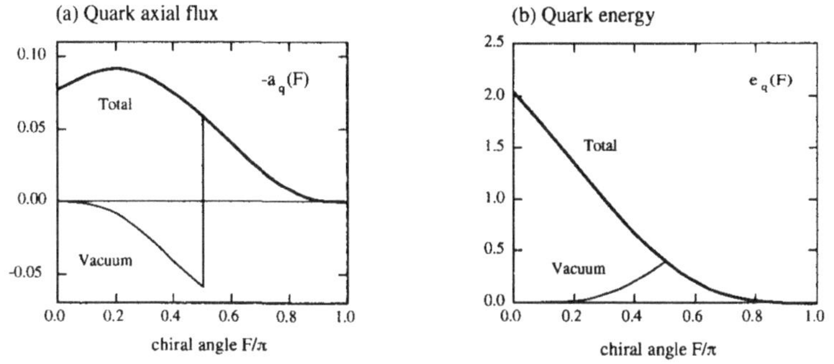

Fig. 5.13 (a) Finite part of the quark axial flux $a_q(F)$, and (b) quark energy $\epsilon_q(F)$.

It is convenient to parameterize the axial flux $a(F)$ for further numerical computations. We perform this for the vacuum and valence parts separately:

$$\begin{aligned} \tilde{a}_0(F) &= \sum_{m=1}^{5} \alpha_m \sin 2mF + \frac{3}{8\pi^2}F\,, \\ &\sum_m 2m\alpha_m + \frac{3}{8\pi^2} = 0\,, \end{aligned} \tag{5.181}$$

$$a_{\rm val}(F) = \sum_{m=0}^{4} \beta_m \left(\frac{\pi}{2} + F\right)^{2m}, \tag{5.182}$$

where the numerical coefficients are

$$(\alpha_m) = \begin{pmatrix} -0.15126 \times 10^{-1} \\ -0.21516 \times 10^{-2} \\ 0.18851 \times 10^{-3} \\ -0.54827 \times 10^{-4} \\ 0.16992 \times 10^{-4} \end{pmatrix},$$

$$(\beta_m) = \begin{pmatrix} 0.11937 \times 10^0 \\ -0.23875 \times 10^{-1} \\ 0.36975 \times 10^{-2} \\ -0.38894 \times 10^{-3} \\ 0.22078 \times 10^{-4} \end{pmatrix} . \tag{5.183}$$

This parameter set provides $\tilde{a}_0(F) \to 1.18(F/\pi)^3$ at small F, which is slightly different from the number given in Eq. (5.175). This indicates a small numerical ambiguity within a few percent. We have also parameterized other quantities such as magnetic moments, though not shown here.

Let us turn to Eq. (5.177) with the $1/R^2$ term removed. It is the boundary condition to determine F at $(r = R)$, or in the vicinity of $R \sim 0$ to determine b. It is also possible to use the same equation to fix the parameter e if we identify b with that of the skyrmion. In other words, we impose

$$F_R(R) \sim F_0(r = R)', \quad R \to 0\,, \tag{5.184}$$

where the right hand side is the skyrmion solution. Alternatively, we can write this condition as

$$F_R(R) - F_0(r = R) \sim \mathcal{O}(R^2)\,, \tag{5.185}$$

as shown in Fig. 5.12 (b). We call this condition the smoothness condition in the limit $R \to 0$ [Hosaka and Toki 1986; Mulders 1984]. Under this assumption, from Eq. (5.177) with the singular term removed, we obtain a condition which determines e,

$$e = \sqrt{\frac{1}{a_3 N_c}\left(2 + \frac{e^2 f_\pi^2}{\pi^2 b^2}\right)}\,, \tag{5.186}$$

We find the expected N_c dependence for $e \sim 1/\sqrt{N_c}$. Using the skyrmion solution for the slope parameter of the chiral angle at the origin, $b/ef_\pi = 0.639$ [Adkins et al. 1983; Zahed and Brown 1986] and $N_c = 3$, we obtain

$$e \sim 4.5\,. \tag{5.187}$$

This number is remarkably close to those widely used in the Skyrme model calculations: $e \sim 5$.

Quark energy

Using Eqs. (5.159) and (5.160), the condition imposed on the quark axial flux $a_0(F)$ can remove the logarithmic divergences in the quark energy. In other words, the subtraction of the $\sin 2F$ terms from the quark axial flux is equivalent to the subtraction of the $\sin^2 F$ terms from the quark energy:

$$\tilde{\epsilon}_0(F) = \epsilon_0(F) - \frac{1}{2}\left[\frac{\partial^2 \epsilon_0(F)}{\partial F^2}\right]_{F=0} \sin^2 F\,. \tag{5.188}$$

This completely establishes a finite quark energy. Equivalently, the subtraction conditions (5.178) and (5.188) amounts to the following identity:

$$\frac{\partial \epsilon_q(F)}{\partial F} = 4\pi a_q(F)\,. \tag{5.189}$$

This identity is satisfied classically (or is satisfied by each single quark state), but is broken quantum mechanically when mode sum is performed. Therefore, the subtraction schemes of Eqs. (5.178) and (5.188) are imposed so as to recover the identity. Eq. (5.189) can be used to calculate the energy $\epsilon_q(F)$ from the axial flux $a_q(F)$. Integrating over F, one finds

$$\epsilon_q(F) = 4\pi \int_\pi^F a_q(F)\,. \tag{5.190}$$

The resulting function $\epsilon_q(F)$ is plotted in Fig. 5.13 (b), where the valence and vacuum contributions are shown separately.

Hedgehog profile and energy

Having made the quark axial flux and energy finite, it is straightforward to obtain the hedgehog solution. Differential equation for $F_R(r)$ for various bag radius R can be solved by using the computer code COLSYS [U. Ascher and Russell 1981]. This is a program for solving coupled non-linear equations, and is particularly useful when the chiral bag model is extended to that with vector mesons. Profile functions $F(r)$ $(r \geq R)$ are plotted in Fig. 5.14 (a) for bag radii $R = 0$ and 0.6 fm. At small bag radii $R \lesssim 0.4$ fm, the chiral bag solution is almost identical to the skyrmion solution (dotted line), while for $R > 0.4$ fm, it begin to deviate and is slightly enhanced.

The energy of the hedgehog solution is given as

$$E_h(R) = E_q(R) + E_\pi(R) + E_V(R)\,. \tag{5.191}$$

Here the quark energy $E_q = N_c\,\epsilon_q(F)/R$ is computed by Eq. (5.190), while the pion energy $E_\pi(R)$ is given by

$$\begin{aligned} E_\pi(F) &= 4\pi\int_R^\infty r^2\,dr\left[\frac{f_\pi^2}{2}\left(F'^2+\frac{2}{r^2}\sin^2 F\right)\right. \\ &\qquad\left.+\frac{\sin^2 F}{2e^2r^2}\left(2F'^2+\frac{\sin^2 F}{r^2}\right)\right]\,. \end{aligned} \tag{5.192}$$

Finally the volume energy E_V is given by

$$E_V = \frac{4\pi}{3}R^3 B\,, \tag{5.193}$$

where the bag pressure B is taken to be $B^{1/4} = 150$ MeV as a reasonable choice.

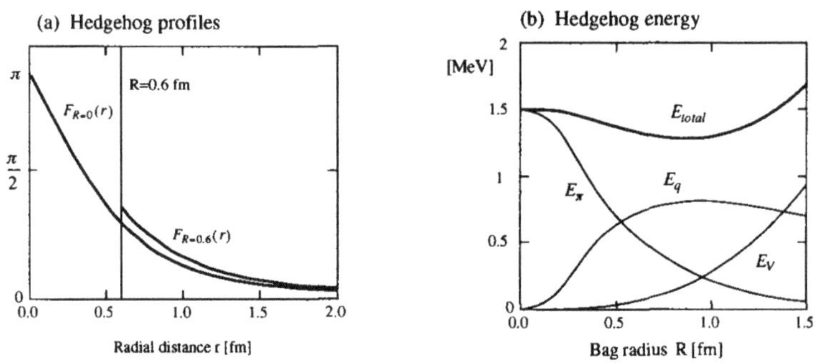

Fig. 5.14 (a) Chiral angle $F_R(r)$, and (b) the energy of the hedgehog solution.

In Fig. 5.14 (b) the total energy E_h is plotted as a function of R with various components shown separately. Unlike in the little bag model, the total energy is finite and is only weakly dependent on R for $0 \le R \le 1$ fm. As anticipated, the skyrmion limit is reproduced smoothly in the limit $R \to 0$. In the region of $R \lesssim 1$ fm, the energy is only weakly dependent on R, although the absolute value looks somewhat too large as compared with the nucleon mass, $M_N = 938$ MeV. Quantum mechanically, the hedgehog state can be interpreted as a superposition of the nucleon and delta (generally baryons of $I = J$). More details will be discussed in the next chapter.

The weak R dependence is a characteristic feature of the chiral bag model, not only for the energy but also for other physical quantities. In particular, the baryon (isoscalar) and isovector charges are exactly independent of R. This can happen if physics outside and inside the bag are connected through the pion-quark interaction at the bag surface. If such a mechanism were exact, one would expect that all physical quantities would become completely independent of R. Then the bag radius R would be a redundant parameter; no matter where it was located the physics would not change.

An exact equivalence is possible when the two descriptions of physics, inside by quarks and outside by the pions, are equivalent, and the interaction can transfer precisely one physics into the other. An example of this ideal case was demonstrated only in a 1+1 dimensional model [Nadkarni et al. 1985]. In 1 + 3 dimensional models, such equivalence is realized only approximately. In actual model calculations, depending on the model set up, we would expect an optimal radius R, where physical observables are best described [Hosaka et al. 1990; Hosaka and Toki 1992; Brown et al. 1986]. This will be studied in the next chapter in detail.

Problems

5.1
Find the energy minimum using the uncertainty principle for the harmonic oscillator and the Coulomb problem:

$$H_{HO} = \frac{p^2}{2m} + \frac{k}{2}x^2\,, \quad H_C = \frac{p^2}{2m} - \frac{k}{r^2}\,.$$

Notice that there is always a balance between the potential and kinetic terms.

5.2
Show that the hamiltonian h in Eq. (5.3):

$$h = \psi^\dagger\left(-i\boldsymbol{\alpha}\cdot\boldsymbol{\nabla} + \beta M\theta(r-R)\right)\psi \tag{5.194}$$

commutes with the total angular momentum operator $\boldsymbol{j} = -i\boldsymbol{x}\times\boldsymbol{\nabla}$.

5.3
Verify the relation Eqs. (5.18) and (5.19) explicitly for an S wave state

$$\psi^0_{1/2,m} = \begin{pmatrix} j_0 \\ i\sigma\cdot\hat{\boldsymbol{x}}j_1 \end{pmatrix}\chi_m$$

5.4
Work out eigenstates of quarks confined by a scalar-vector potential [Tegen et al. 1982]

$$U(r) = (1+\gamma_0)M\theta(r-R)\,.$$

5.5
For the hedgehog configuration, it is convenient to work with the expression:

$$U = \exp(i\boldsymbol{\tau}\cdot\hat{\boldsymbol{x}}F) = c + i\boldsymbol{\tau}\cdot\hat{\boldsymbol{x}}s\,,$$

where $c = \cos F$ and $s = \sin F$. Then it follows, for instance,

$$\partial_i U = -sF'\hat{x}_i + i\tau_j\left(\frac{P_{ij}}{r}s + \hat{x}_i\hat{x}_j cF'\right)\,,$$

where $P_{ij} = \delta_{ij} - \hat{x}_i\hat{x}_j$ is the orthogonal projection operator satisfying $x_i P_{ij} = 0$. Verify that

$$\partial_i^2 U = -F'' - \frac{2s^2}{r^2} + i\boldsymbol{\tau}\cdot\hat{\boldsymbol{x}}\left(F'' + \frac{2}{r}F' - \frac{2cs}{r^2}\right)$$

and the equation of motion

$$F'' + 2\frac{F'}{r} - \frac{\sin 2F}{r^2} = 0\,.$$

5.6

Derive the expressions for baryon number and energy densities in the hedgehog ansatz, $U_H = \exp(i\boldsymbol{\tau}\cdot\boldsymbol{x}F(r))$,

$$B = \frac{\epsilon_{ijk}}{24\pi^2}\int d^3x \; \mathrm{tr}\,\partial_i U U^\dagger \partial_j U U^\dagger \partial_k U U^\dagger$$

and

$$E = -\frac{f_\pi^2}{4}\,\mathrm{tr}\,[\nabla_i U U^\dagger \nabla_i U U^\dagger] - \frac{1}{32e^2}\,\mathrm{tr}\,[\nabla_i U U^\dagger, \nabla_j U U^\dagger]^2$$

5.7

One can get some feeling for the spectral asymmetry defined in Eq. (5.123) by simple computations.

(1) Compute the sum

$$S = 1 - 1 + 1 - 1 + 1 - \cdots$$

by including the regulator $\sim e^{-n\cdot t}$:

$$S = \lim_{t\to 0}(e^{-0\cdot t} - e^{-1\cdot t} + e^{-2\cdot t} - e^{-3\cdot t}) + \cdots\,.$$

(2) Verify Eq. (5.134).

Chapter 6

NUCLEON PROPERTIES

The hedgehog solution obtained in the previous chapter can not be identified with the physical nucleon. It is merely a classical solution to the non-linear field equation and does not carry quantum numbers of the nucleon. In order to discuss physical properties, the hedgehog solution must be quantized so that appropriate quantum numbers such as spin and isospin can be assigned. To do this we apply the semiclassical cranking method. At first sight the application of the method to the chiral bag model appears to be a very specific problem of physics. However, ideas and techniques used there are rather general. It is a good subject to learn important methods and ideas in theoretical physics.

Section 6.1 is an introductory and pedagogic part, where we describe elementary aspects of the cranking method with an emphasis on the quantization of collective motion. There we present some simple examples of rotational motion of a point like particle, which shares common features with the hedgehog quantization. The spin and isospin rotation of the hedgehog solution is then quantized in section 6.2. We derive the mass formula for rotating hedgehog with a detailed computation of the quark moment of inertia.

As applications, we first study in section 6.3 axial properties of the nucleon. The computation of axial couplings is a good exercise to verify the importance of chiral symmetry. Then in section 6.5 electromagnetic properties are discussed in detail. The inclusion of vector mesons, though rather specific, will be discussed in section 6.6 since it can provide one of the best descriptions of the nucleon.

6.1 Semiclassical Method

General aspects

We consider a theory of a set of boson field $\phi(x) = (\phi_1(x), \phi_2(x), \ldots, \phi_N(x))$. The inclusion of fermions will be discussed later. We denote the lagrangian by $\mathcal{L}(\phi)$, where the argument of the lagrangian expresses symbolically both ϕ and its derivatives $\partial\phi$. To be specific, we consider a lagrangian of the form

$$\mathcal{L} = \frac{1}{2}(\partial_t \phi)^2 - \frac{1}{2}(\nabla\phi)^2 - V(\phi)\,, \tag{6.1}$$

where the potential $V(\phi)$ is assumed to be a function of only ϕ.

As usual the equation of motion is given by

$$\partial_\mu \frac{\partial \mathcal{L}}{\partial_\mu \phi} = \frac{\partial \mathcal{L}}{\partial \phi}\,. \tag{6.2}$$

We assume that there is a nontrivial static (time independent) solution ϕ_c to the field equation (6.2). By the word "nontrivial", it is meant that the solution ϕ_c is not just a constant but is an x-dependent function. Note that at this point the field ϕ_c is just a c-number.

The semiclassical cranking method provides a systematic expansion scheme when quantizing the field ϕ in the presence of the classical solution ϕ_c [Rajaraman 1982; Coleman 1985]. The expansion parameter will be clarified soon. To start with, we separate the field ϕ into the (large) classical solution ϕ_c and a (small) fluctuation φ:

$$\phi(t, \boldsymbol{x}) = \phi_c(\boldsymbol{x}) + \varphi(t, \boldsymbol{x})\,. \tag{6.3}$$

Here we have shown explicitly the space time coordinates which the fields depend on. Let us perform the Taylor expansion of the lagrangian around the classical solution ϕ_c:

$$\mathcal{L}(\phi_c + \varphi) = \mathcal{L}(\phi_c) + \frac{\partial \mathcal{L}}{\partial \phi}(\phi_c)\varphi + \frac{1}{2}\frac{\partial^2 \mathcal{L}}{\partial \phi^2}(\phi_c)\varphi^2 + \cdots\,. \tag{6.4}$$

This expression is symbolic in that the coefficients such as $\partial\mathcal{L}/\partial\phi$ involve derivatives acting on φ.

The quantum fluctuation φ is regarded as small and can be treated at each order of φ, separately. So we first perform the variation with respect to φ in the second term of Eq. (6.4). This leads to *

$$\frac{\partial \mathcal{L}}{\partial \phi}(\phi_c) = 0 \tag{6.5}$$

which is nothing but the equation of motion for the classical solution ϕ_c and is automatically satisfied. Next we take variation of the third term in Eq. (6.4), which leads to

$$\frac{\partial^2 \mathcal{L}}{\partial \phi^2}(\phi_c)\varphi = 0 \,. \tag{6.6}$$

This is the equation of motion for the fluctuation φ. It takes the form of the eigenvalue equation

$$-\partial_t^2 \varphi = \mathcal{H}\varphi \,. \tag{6.7}$$

Here $\mathcal{H}$ is a linear differential operator which is assumed to contain only spatial coordinates (including spatial derivatives) but no time coordinate. Hence the general solution can be expanded in eigenmodes:

$$\varphi(t, \boldsymbol{x}) = \sum_n a_n \exp(-i\omega_n t) f_n(\boldsymbol{x}) \,, \tag{6.8}$$

where $f_n(\boldsymbol{x})$ satisfies

$$\mathcal{H} f_n(\boldsymbol{x}) = \omega_n^2 f_n(\boldsymbol{x}) \,. \tag{6.9}$$

What we would like to discuss here is the case when there are symmetries in the system. Apparently there are translational and rotational symmetries. Also, the system may have internal symmetries. We assume that $\phi(x) = (\phi_1(x), \phi_2(x), \ldots)$ form an irreducible representation of a symmetry group G,

$$\phi_i \to \sum_{j=1,..,N} g_{ij}\phi_j \,. \tag{6.10}$$

where $g \sim g_{ij}$ is a representation matrix of a group G.

The symmetry requires that if ϕ_c is a solution, then $g\phi_c$ is also a solution. Hence we have

$$\frac{\partial \mathcal{L}}{\partial \phi_i}(g\phi_c) = \frac{\partial \mathcal{L}}{\partial \phi_i}(\phi_c) = 0 \,. \tag{6.11}$$

*Once again it should be understood that this expression as well as Eq. (6.6) contains derivatives on φ, for instance, the left hand side of Eq. (6.2).

As in Chapter 2, we introduce the generators of the symmetry group and denote them as $t_m (m = 1, \ldots, M)$. Under an infinitesimal transformation, we have

$$\begin{aligned} 0 &= \frac{\partial \mathcal{L}}{\partial \phi_i}(g\phi_c) \\ &\sim \frac{\partial \mathcal{L}}{\partial \phi_i}((1 + i\epsilon t)\phi_c) \\ &= \frac{\partial \mathcal{L}}{\partial \phi_i}(\phi_c) + \frac{\partial^2 \mathcal{L}}{\partial \phi_i \phi_j}(\phi_c)\epsilon(it\phi_c)_j \, , \end{aligned} \tag{6.12}$$

where

$$\epsilon(it\phi_c)_j = \sum_{m=1,..,M} \sum_{k=1,..,N} \epsilon_m i(t^m)_{jk}(\phi_c)_k \, . \tag{6.13}$$

Using the classical equation of motion of Eq. (6.5), we observe

$$\frac{\partial^2 \mathcal{L}}{\partial \phi_i \partial \phi_j}(\phi_c) \; (it^m \phi_c)_j = 0 \; \text{ for } \; m = 1, \ldots M \, . \tag{6.14}$$

Since ϕ is an irreducible representation, $it^m \phi_c \neq 0$ for at least one of m's. By comparing Eq. (6.14) with Eq. (6.7), we conclude that $it^m \phi_c$ is a solution to the equation for the fluctuations of Eq. (6.7) with zero energy (zero mode) $\omega = 0$. This follows from the time independence of ϕ_c.

In this demonstration, we have seen that if there is a non-trivial classical solution ϕ_c, there appear (at most) M zero energy modes associated with the symmetry, where M is the number of generators. Since the symmetry is global, the zero energy modes are interpreted as collective modes. The word "collective" means that the configuration $\phi_c(\boldsymbol{x})$ moves uniformly at all spatial points. If we are interested only in low energy dynamics, these collective zero energy modes play a dominant role.

In many cases, some of the modes become irrelevant, due to residual symmetries of the classical solution ϕ_c, which is a part of the full symmetry of the system. The "relevant" dynamical variables are the parameters of the symmetry transformations which really change the classical solution ϕ_c.

A simple example

Perhaps the simplest and most familiar example is found in the quantum mechanics of a particle in the three dimensional space subject to a spheri-

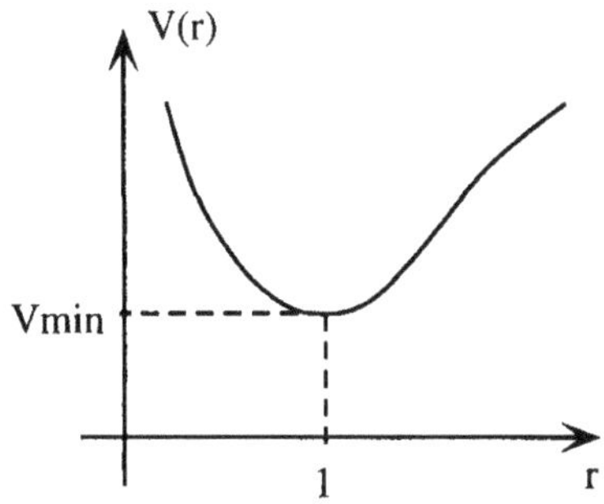

Fig. 6.1 Radial dependence of the potential $V(r)$.

cally symmetric potential. The lagrangian is given by

$$L(\boldsymbol{x}, \dot{\boldsymbol{x}}) = \frac{1}{2}\dot{\boldsymbol{x}}^2 - V(r), \tag{6.15}$$

where $r = |\boldsymbol{x}|$. We assume that the potential $V(r)$ has the minimum at $r = 1$, as shown in Fig. 6.1. A classical solution is given by $r = 1$ but it can be anywhere on the unit sphere. Thus we choose it on the z-axis as

$$\boldsymbol{x}_c = (0, 0, 1). \tag{6.16}$$

The system has rotational symmetry, under which $\boldsymbol{x}$ is transformed as

$$x_i \to \sum_j R_{ij} x_j, \tag{6.17}$$

where R is a familiar 3×3 rotation matrix. Its explicit form as a function of the Euler angles α, β, γ is

$$\begin{aligned} R &= \exp(i\alpha J_z)\exp(i\beta J_y)\exp(i\gamma J_z) \\ &= \begin{pmatrix} c_\alpha & s_\alpha & 0 \\ -s_\alpha & c_\alpha & 0 \\ 0 & 0 & 1 \end{pmatrix} \begin{pmatrix} c_\beta & 0 & -s_\beta \\ 0 & 1 & 0 \\ s_\beta & 0 & c_\beta \end{pmatrix} \begin{pmatrix} c_\gamma & s_\gamma & 0 \\ -s_\gamma & c_\gamma & 0 \\ 0 & 0 & 1 \end{pmatrix} \\ &= \begin{pmatrix} c_\alpha c_\beta c_\gamma - s_\alpha s_\gamma & c_\alpha c_\beta s_\gamma + s_\alpha c_\gamma & -c_\alpha s_\beta \\ -s_\alpha c_\beta c_\gamma - c_\alpha s_\gamma & -s_\alpha c_\beta s_\gamma + c_\alpha c_\gamma & s_\alpha s_\beta \\ s_\beta c_\gamma & s_\beta s_\gamma & c_\beta \end{pmatrix}. \end{aligned} \tag{6.18}$$

Here $J_{x,y,z}$ are the angular momentum operators and $c_\alpha = \cos\alpha$, $s_\alpha = \sin\alpha$ etc. Note that the rotation matrix R is a product of the rotation around the

z-axis by γ, the rotation around the y-axis by β and the rotation around the z-axis by α. Apparently, the first rotation around the z-axis by γ does not change the point $\boldsymbol{x}_c = (0,0,1)$. Therefore, the relevant dynamical variables are β and α, which are the polar angles and may be denoted by θ and ϕ. This situation is shown in Fig. 6.2.

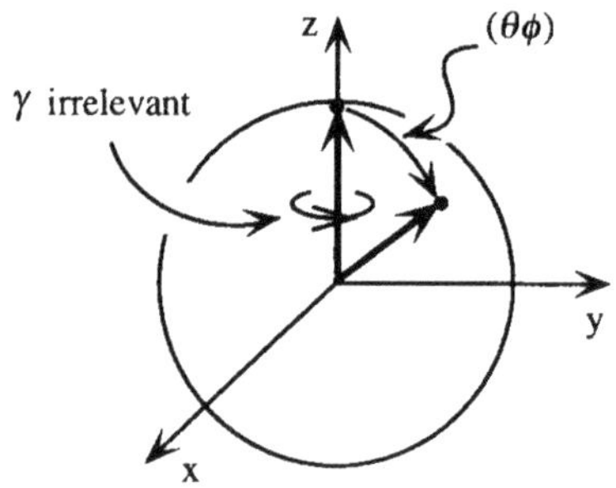

Fig. 6.2 A particle moving on a unit sphere.

Regarding R_{ij} as time dependent variables, the lagrangian of Eq. (6.15) is now written in terms of the angle variables α and β:

$$\begin{aligned} L(\alpha,\beta) &= \frac{1}{2}\,{}^t(\dot{R}\boldsymbol{x}_c)(\dot{R}\boldsymbol{x}_c) - V(r=1) \\ &= \frac{1}{2}(\dot{\beta}^2 + \sin^2\beta\dot{\alpha}^2)\,. \end{aligned} \tag{6.19}$$

This is the lagrangian for the motion of a particle on the unit sphere. Although the quantization is straightforward, we make a remark on the relation between the rotational D-functions and the spherical harmonics. The wave functions of the free particle are the representations of the rotation operator $g(\alpha,\beta,\gamma)$, in which the last variable γ is inactive due to the symmetry of $\boldsymbol{x}_c$. Since the representations of the rotational group ($SU(2)$) are labeled by J and M (see section 2.3), the wave functions are given by

$$\begin{aligned} \langle JM|g(\alpha\beta\gamma)|JM'\rangle &= D^L_{MM'}(\alpha\beta\gamma) \\ &\to D^L_{MM'}(\alpha\beta 0) \sim Y_{LM}(\theta\phi)\,. \end{aligned} \tag{6.20}$$

More details are presented in Appendix B.

Collective hamiltonian

Turning back to the expansion for φ as in Eq. (6.8), we would separate the zero modes from other non-zero modes:

$$\varphi(t,\boldsymbol{x}) = \sum_m a_{0,m}(it^m\phi_c(\boldsymbol{x})) + \sum_{n>0} a_n f_n(\boldsymbol{x})\exp(-i\omega_n t)\,. \tag{6.21}$$

The zero modes need to be treated carefully, since they can no longer be small oscillations as non-zero modes f_n, $(n>0)$. The field $it^m\phi_c$ is tangent to the direction of the small displacement generated by t^m. A finite change in the field is then generated by a finite group transformation $g\phi_c$. Formally, this can be achieved by combining the classical solution and the zero mode fluctuation:

$$\phi_c(\boldsymbol{x}) + \sum_m a_{0,m}(it^m\phi_c(\boldsymbol{x})) \to g(t)\phi_c(\boldsymbol{x})\,, \tag{6.22}$$

where g is a group element $g \in G$.

Thus an appropriate "semiclassical" ansatz for the fluctuation would be

$$\begin{aligned}\phi(t,\boldsymbol{x}) &= g(t)\phi_c(\boldsymbol{x}) + \sum_{n>0} a_n f_n(\boldsymbol{x})\exp(-i\omega_n t)\\ &\equiv g(t)\phi_c(\boldsymbol{x}) + \varphi(t,\boldsymbol{x})\,.\end{aligned} \tag{6.23}$$

Note that in the fluctuation φ on the right hand side the zero modes have been subtracted and included in the first term. This ansatz of Eq. (6.23) may be regarded as the one in the "laboratory frame". Another equivalent ansatz may be considered in the "body-fixed frame" [Ring and Shuck 1980],

$$\begin{aligned}\phi^{\rm bf}(t,\boldsymbol{x}) &= g^{-1}(t)\phi^{\rm lab}(t,\boldsymbol{x})\\ &= \phi_c(\boldsymbol{x}) + g^{-1}(t)\varphi(t,\boldsymbol{x})\,.\end{aligned} \tag{6.24}$$

We will discuss the descriptions in these two reference frames in section 6.2 (see for instance Fig. 6.3). Here we consider the ansatz in the laboratory frame. Inserting the ansatz of Eq. (6.23) into the original lagrangian of Eq. (6.1), we find

$$\begin{aligned}\mathcal{L} &= \frac{1}{2}(\partial_t\phi^{\rm bf})^2 - \frac{1}{2}(\boldsymbol{\nabla}\phi^{\rm bf})^2 - V(\phi^{\rm bf})\\ &= \frac{1}{2}(\phi_c^\dagger \dot{g}^\dagger\,\dot{g}\phi_c) - \frac{1}{2}(\boldsymbol{\nabla}\phi_c)^2 - V(\phi_c)\end{aligned}$$

$$
\begin{aligned}
&+ \quad \frac{1}{2}\phi_c^\dagger \dot{g}^\dagger \dot{\varphi} + \frac{1}{2}\dot{\varphi}^\dagger \dot{g}\phi_c \\
&+ \quad \frac{1}{2}\dot{\varphi}^\dagger\dot{\varphi} - \frac{1}{2}(\nabla\varphi)^2 - V'(\phi_c, \varphi)
\end{aligned}
\tag{6.25}
$$

where

$$
V'(\phi_c, \varphi) = V(\phi) - V(\phi_c) \,. \tag{6.26}
$$

In Eq. (6.25), the first line is for the classical solution and its rigid kinetic motion (the first term), the second line corresponds to the Coriolis coupling (the coupling between the collective velocity $\dot{g}$ and the velocity of the fluctuation $\dot{\phi}$) and the third term is for the fluctuation φ. (See problem 6.1 for the physical meaning.) In the semiclassical quantization the collective motion of g is first quantized. The Coriolis coupling will be considered separately in perturbation.

For the time being we consider the rigid motion of the classical solution. We rewrite the kinetic term to find

$$
\begin{aligned}
\mathcal{L}_{\text{rigid}} &= \frac{1}{2}(\phi_c^\dagger \dot{g}^\dagger \dot{g}\phi_c) - \frac{1}{2}(\nabla\phi_c)^2 - V(\phi_c) \\
&= \frac{1}{2}(ig^\dagger\dot{g}\phi_c)^\dagger(ig^\dagger\dot{g}\phi_c) - \frac{1}{2}(\nabla\phi_c)^2 - V(\phi_c) \,,
\end{aligned}
\tag{6.27}
$$

where we have used the identity $\dot{g}g^\dagger + g\dot{g}^\dagger = 0$ which is derived from the unitarity $gg^\dagger = 1$. The lagrangian for the collective rigid motion is obtained by integrating $\mathcal{L}_{\text{rigid}}$ over three dimensional space:

$$
L_{\text{rigid}} = \frac{1}{2}\,\text{tr}\,\dot{\alpha}\mathcal{I}\dot{\alpha} - M \,, \tag{6.28}
$$

where the collective velocity $\dot{\alpha}$, the moment of inertial $\mathcal{I}$ and the classical energy M are given by

$$
\begin{aligned}
\dot{\alpha} &= ig^\dagger\dot{g} \,, \\
\mathcal{I} &= \int d^3x \phi_c \phi_c^\dagger \,, \\
M &= \int d^3x \left(\frac{1}{2}(\nabla\phi_c)^2 + V(\phi)\right) .
\end{aligned}
\tag{6.29}
$$

Note that the second equation for the moment of inertia is in the matrix form as the combination $\phi_c\phi_c^\dagger$ indicates. Upon quantization of the collective motion, we define the canonical momentum which is conjugate to α by

$$\pi = \frac{\partial L}{\partial \dot{\alpha}} = \mathcal{I}\dot{\alpha}, \tag{6.30}$$

and so

$$\dot{\alpha} = \mathcal{I}^{-1}\pi. \tag{6.31}$$

Then we find the collective hamiltonian as

$$H_{\text{rigid}} = \frac{1}{2}\,\text{tr}\,\pi\mathcal{I}^{-1}\pi + M. \tag{6.32}$$

The hamiltonian (6.32) describes free motion of the collective coordinates α in the parameter space of the symmetry transformation g. Therefore, the wave functions are given by the group representations of g:

$$\psi^J_{rr'}(\alpha) = \langle Jr|g(\alpha)|Jr'\rangle. \tag{6.33}$$

Here $|Jr\rangle$ is a representation of the group G; J specifies the representation and r labels each component of the representation. Details how we can write the wave function in this way are discussed in Appendix B.

Matrix elements and power series in velocity $\dot{\alpha}$

To leading order in the semiclassical method, we have reduced a field theory of an infinite number of degrees of freedom into a quantum mechanics of a finite number of collective degrees of freedom. When there is a symmetry in the system, dynamical variables are taken as symmetry transformations, α. Consequently, we deal with a free hamiltonian as given by Eq. (6.32), and the wave functions are just the harmonics in the collective space spanned by α. At low energies, such collective motion dominance is often postulated, but actually it must be justified in each case.

In computing physical quantities, say for example, an operator $A(\phi)$, we substitute the ansatz (see Eq. (6.23)) for ϕ,

$$\phi(t,\boldsymbol{x}) = g(t)\phi_c(\boldsymbol{x}). \tag{6.34}$$

If the operator A contains time derivatives (on ϕ), and/or the Coriolis coupling is treated in perturbation with respect to the velocity, A may be expanded in terms of the small velocity $\dot{\alpha}$:

$$A(\phi(t,\boldsymbol{x})) = A_0(\alpha) + A_1(\alpha)\dot{\alpha} + A_2(\alpha)\dot{\alpha}^2 + \cdots$$

Table 6.1 Order of various quantities in the small parameter $\lambda \sim 1/N_c$.

$\dot{\alpha}$	π	$\mathcal{I}$	M	K	ω_n
λ	1	λ^{-1}	λ^{-1}	λ	1

$$\equiv \quad A(\alpha, \dot{\alpha}) . \tag{6.35}$$

In this expression $\boldsymbol{x}$ dependence is not shown explicitly. The collective velocity $\dot{\alpha}$ may be regarded as a small parameter if the kinetic energy

$$K = \frac{1}{2} \operatorname{tr} \dot{\alpha} \mathcal{I} \dot{\alpha} = \frac{1}{2} \operatorname{tr} \pi \mathcal{I}^{-1} \pi . \tag{6.36}$$

is smaller than the energy of non-zero fluctuations φ_n,

$$K << \omega_{n>0} . \tag{6.37}$$

In the chiral bag model the small $\dot{\alpha} = \mathcal{I}^{-1}\pi$ can be justified in the large-N_c limit as will be discussed in Chapter 7. Although details will be presented there, it is useful to mention that the velocity is of order $1/N_c \sim \lambda$, where N_c is the number of colors. The λ dependence of other quantities is presented in Table 6.1.

Finally we show that matrix elements are computed by

$$\int d[\alpha] \; \psi^{J\,*}_{rr'}(\alpha) A(\alpha, \dot{\alpha}) \psi^{J}_{rr'}(\alpha) . \tag{6.38}$$

When evaluating Eq. (6.38), the collective momentum $\pi = \mathcal{I}\dot{\alpha}$ is replaced by the derivative $\pi \to i\partial/\partial\alpha$. Note that contrary to the ordinary quantization procedure, there is no minus sign here; see the discussion in Appendix B.

6.2 Isospin Rotation of the Hedgehog Solution

Hedgehog ansatz

Let us now apply the semiclassical method developed in the previous section for the quantization of the spin and isospin for the nucleon. The relevant symmetries are associated with spatial rotation (spin) and isospace rotation (isospin). Under these symmetry transformations, the general pion field $U(t, \boldsymbol{x}) = \exp(i\boldsymbol{\tau} \cdot \boldsymbol{\phi}(t, \boldsymbol{x}))$ is transformed as

$$U(t, \boldsymbol{x}) \quad \to \quad TU(t, D\boldsymbol{x})T^{\dagger} , \tag{6.39}$$

where T is a 2×2 isospin matrix and D is a 3×3 rotation matrix of Eq. (6.18) (see also Eq. (6.44)). For general configurations of U, the spin and isospin transformations are independent zero modes. For the hedgehog configuration, however, due to the symmetry of the ansatz, not all of them are independent.

In the hedgehog ansatz the isospin of the pion field is oriented to the radial direction $\hat{\boldsymbol{x}}$;

$$\boldsymbol{\phi} = \hat{\boldsymbol{x}} F(r)\,, \quad U_h = \exp(i\boldsymbol{\tau} \cdot \hat{\boldsymbol{x}} F(r))\,. \tag{6.40}$$

The key observation is that the hedgehog ansatz of Eq. (6.40) is spherically symmetric in the sense that;

$$\boldsymbol{\phi} \sim [Y_1(\hat{\boldsymbol{x}}), \boldsymbol{e}]^0_0\,, \tag{6.41}$$

where $Y_1(\hat{\boldsymbol{x}})$ is the spherical harmonics of rank one (see Eq. (H.3)) and $\boldsymbol{e}_a$ ($a = 1, 2, 3$) are the three unit vectors (base) in the isospin space. Explicitly,

$$\boldsymbol{\phi} \sim \hat{x}_1 \boldsymbol{e}_1 + \hat{x}_2 \boldsymbol{e}_2 + \hat{x}_3 \boldsymbol{e}_3\,. \tag{6.42}$$

Note that $\hat{x}_i$ are components of the vector $\hat{\boldsymbol{x}} = (\hat{x}_1, \hat{x}_2, \hat{x}_3)$, while $\boldsymbol{e}_i$ are basis vectors.

Under the spatial (S) and isospin (T) rotations, they transform as

$$\begin{aligned} x_i &\to x'_i = D_{ij}(S) x_j\,, \\ \boldsymbol{e}_i &\to \boldsymbol{e}'_i = D_{ij}(T) \boldsymbol{e}_j\,. \end{aligned} \tag{6.43}$$

Here the 3×3 rotation matrix $D_{ij}(B)$ ($B = S$ or T) is related to an elementary 2×2 matrix B by

$$D_{ij}(B) = \frac{1}{2} \operatorname{tr} \left[\tau_i B \tau_j B^\dagger\right]\,. \tag{6.44}$$

Hence the simultaneous rotation in space and isospace transforms the hedgehog ansatz into

$$\boldsymbol{\phi} \sim \hat{x}_i \boldsymbol{e}_i \quad \to \quad D_{ij}(S) D_{ik}(T) \hat{x}_j \boldsymbol{e}_k\,. \tag{6.45}$$

Apparently, if $S = T$, $\boldsymbol{\phi} \sim \hat{x}_i \boldsymbol{e}_i$ is left invariant.

Sometimes the invariance of the hedgehog under the K-transformation is interpreted as that it has the grand spin $\boldsymbol{K} = \boldsymbol{I} + \boldsymbol{J} = 0$, where $\boldsymbol{I}$ and $\boldsymbol{J}$ are the isospin and spin operators (generators). However, this statement is somewhat misleading, since the hedgehog solution is a classical object and does not possess quantum numbers yet. It will make sense when zero modes of the collective motions are quantized; they have quantum numbers (I, I_z) and (J, J_z) which are subject to $I = J$ and $I_z + J_z = 0$, the constraints following from the condition $\boldsymbol{K} = 0$. A detailed procedure for the semiclassical quantization of the hedgehog ansatz is described in Appendix C.

Cranking

Let us now investigate the quantization of the chiral bag. In the semiclassical picture, the hedgehog solution rotates slowly and spontaneously:

$$\begin{aligned} U(t, \boldsymbol{x}) &= A(t) U_h(\boldsymbol{x}) A(t)^\dagger \,, \\ \psi(t, \boldsymbol{x}) &= A(t) \psi_h'(t, \boldsymbol{x}) \,, \end{aligned} \tag{6.46}$$

where $A(t)$ is now regarded as a time dependent dynamical variable for isospin rotation. It is a 2×2 matrix in the isospin space.

Let us introduce two reference frames; one is the laboratory frame at rest and the other is the body-fixed frame which is rotating together with the hedgehog. The relation between the two reference frames is shown in Fig. 6.3. The quantities on the left hand sides of Eq. (6.46) are those in the laboratory frame, while U_h and ψ_h' on the right hand sides are in the body-fixed frame. In the ansatz of Eq. (6.46), the pion (boson) field is just the rigidly rotating hedgehog. Namely, the coupling between the rotation and fluctuations (Coriolis term) is ignored. Contrary, for the quark field, the Coriolis term should be considered [Kahana et al. 1986; Jackson et al. 1987], which modifies the wave function from the hedgehog solution derived in the previous chapter.

Substituting the ansatz of Eq. (6.46) in the chiral bag lagrangian of Eq. (5.77) and integrating over $\int d^3x$, one finds:

$$\begin{aligned} L_{CB} &= \int_0^R d^3x \bar{\psi} i \partial_\mu \gamma^\mu \psi + \frac{1}{2} \int_{r=R} d^3x \bar{\psi} e^{i\boldsymbol{\tau}\cdot\boldsymbol{\phi}\gamma_5/f_\pi} \psi \\ &+ \int_R^\infty d^3x \left(-\frac{f_\pi^2}{4} \operatorname{tr}[\partial_\mu U U^\dagger \partial^\mu U U^\dagger] \right. \\ &\qquad \left. + \frac{1}{32e^2} \operatorname{tr}[\partial_\mu U U^\dagger, \partial_\nu U U^\dagger]^2 \right) \\ &= \int_0^R d^3x \left(i\psi'^\dagger \partial_t \psi' + i\psi'^\dagger \boldsymbol{\alpha}\cdot\boldsymbol{\nabla}\psi' \right) - \frac{1}{2} \boldsymbol{\Omega}\cdot\boldsymbol{Q}' \end{aligned}$$

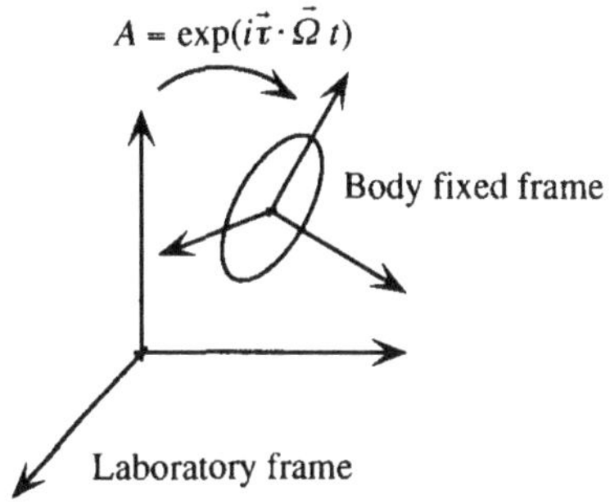

Fig. 6.3 A schematic relation between the laboratory and body-fixed frames.

$$+\frac{1}{2}\Lambda_\pi \boldsymbol{\Omega}^2 - E_\pi \,. \tag{6.47}$$

Here the quark isospin (in the body-fixed frame) $\boldsymbol{Q}'$, the angular velocity $\boldsymbol{\Omega}$ and the moment of inertia of the pion field Λ_π are given by

$$\boldsymbol{Q}' = \int_0^R d^3x\, \psi'^\dagger \boldsymbol{\tau} \psi' \,, \tag{6.48}$$

$$\boldsymbol{\Omega} = i\,\mathrm{tr}\,(\boldsymbol{\tau} \dot{A}^\dagger A) \,, \tag{6.49}$$

$$\Lambda_\pi = \frac{8\pi}{3} f_\pi^2 \int_R^\infty r^2 dr\, \sin^2 F \left(1 + \frac{1}{e^2 f_\pi^2}\left(F'^2 + \frac{\sin^2 F}{r^2}\right)\right) . \tag{6.50}$$

Also E_π in Eq. (6.47) is the hedgehog pion energy as given in Eq. (5.192). In Eq. (6.47), the term proportional to $\boldsymbol{\Omega}^2$ is the rotational energy of the hedgehog pion. The moment of inertia of the pion, Eq. (6.50), is derived from the term involving time derivatives in the Skyrme lagrangian of Eq. (5.76). The actual computation is slightly involved and is left for problem 6.2. The second term $\frac{1}{2}\boldsymbol{\Omega}\cdot\boldsymbol{Q}'$ is the Coriolis interaction for quarks in the rotating body-fixed frame. Since the Coriolis term modifies the quark wave function only, let us investigate the quark part in the following.

Coriolis coupling and moment of inertia

We have two equivalent expressions for the lagrangian in the two different reference frames related through the coordinate transformation of Eq. (6.46) as shown in Eq. (6.47),

$$\begin{aligned} L_{CB}^{quark} &= \int_0^R d^3x \left(\psi^\dagger i\partial_t\psi + \psi^\dagger i\boldsymbol{\alpha}\cdot\boldsymbol{\nabla}\psi \cdots\right) \\ &= \int_0^R d^3x \left(\psi'^\dagger i\partial_t\psi' + \psi'^\dagger i\boldsymbol{\alpha}\cdot\boldsymbol{\nabla}\psi' - \frac{1}{2}\boldsymbol{\Omega}\cdot\psi'^\dagger i\boldsymbol{\tau}\psi'\right). \quad (6.51) \end{aligned}$$

From the first and second terms of Eq. (6.51), one can write the equations of motion in the laboratory and body-fixed frames as

$$i\partial_t\psi = -i\boldsymbol{\alpha}\cdot\boldsymbol{\nabla}\psi\,, \quad (6.52)$$

$$i\partial_t\psi' = -i\boldsymbol{\alpha}\cdot\boldsymbol{\nabla}\psi' + \frac{1}{2}\boldsymbol{\Omega}\cdot\boldsymbol{\tau}\psi'\,. \quad (6.53)$$

These equations are subject to the boundary conditions

$$i\hat{\boldsymbol{x}}\cdot\boldsymbol{\gamma}\,\psi = -A(t)\exp(i\boldsymbol{\tau}\cdot\hat{\boldsymbol{r}}F(R)\gamma_5/f_\pi)A(t)^\dagger\psi\,, \quad (6.54)$$

$$i\hat{\boldsymbol{x}}\cdot\boldsymbol{\gamma}\,\psi' = -\exp(i\boldsymbol{\tau}\cdot\hat{\boldsymbol{r}}F(R)\gamma_5/f_\pi)\psi'\,. \quad (6.55)$$

Furthermore they imply the relation

$$H^{lab} = H^{bf} - \frac{1}{2}\boldsymbol{\Omega}\cdot\boldsymbol{Q}'\,. \quad (6.56)$$

The point is that in the laboratory frame, the equation of motion in Eq. (6.52) is simple (for free field) but the boundary condition (6.54) is complicated (time dependent), while in the body-fixed frame the Coriolis interaction appears as in Eq. (6.53) but the boundary condition (6.55) is simple. Practically, the Coriolis interaction is easier to handle in perturbation theory than dealing with the time dependent boundary condition (6.54). Therefore, we work first in the rotating body-fixed frame and then go back to the laboratory frame.

In the body-fixed frame, first we ignore the Coriolis interaction. Then the equations of motion for the quark field reduce to those in the previous chapter. The solution was obtained for the $B = 1$ ground state including valence quarks and vacuum polarization effects. As depicted in Fig. 6.4, in the ground state all quark levels are occupied up to the valence level. We denote this state as $|h_q\rangle$. It satisfies the equation of motion

$$H_q|h_q\rangle = E_{h_q}|h_q\rangle\,, \quad (6.57)$$

where $E_{h_q} \equiv \epsilon_q/R$ with ϵ_q being given in Eq. (5.190).

The excited states emerge as particle-hole excitations $|n\bar{m}\rangle$ of $|h_q\rangle$ as shown in Fig. 6.4 and satisfy

$$H_q|n\bar{m}\rangle = (E_n - E_m + E_{h_q})|n\bar{m}\rangle\,, \tag{6.58}$$

where E_n and E_m are single particle energies. The ground state $|h_q\rangle$ and the excited states $|n\bar{m}\rangle$ are the basis states of the perturbation method for the Coriolis interaction.

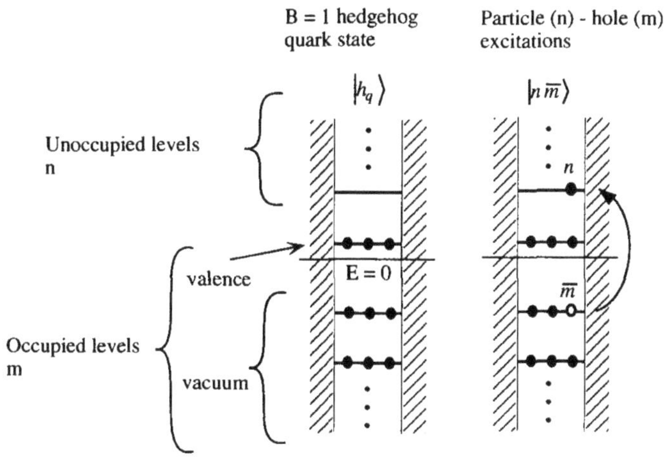

Fig. 6.4 A schematic picture for the $B = 1$ hedgehog quark state $|h_q\rangle$ and its particle-hole excitations.

To leading order in perturbation theory the wave function is given by

$$|h_q'\rangle = |h_q\rangle - \frac{1}{2}\sum_{m,n} \frac{|n\bar{m}\rangle\langle n\bar{m}|\boldsymbol{\Omega}\cdot\boldsymbol{Q}|h_q\rangle}{E_n - E_m}\,. \tag{6.59}$$

Here the sum runs over the occupied (m) and unoccupied (n) states. In actual computations, we can replace the particle-hole matrix elements as

$$\langle n\bar{m}|\boldsymbol{\Omega}\cdot\boldsymbol{Q}|h_q\rangle = \langle n|\boldsymbol{\Omega}\cdot\boldsymbol{Q}|m\rangle\,. \tag{6.60}$$

Then the second order energy in the body-fixed frame is given by

$$E_q^{bf} = E_q - \frac{1}{4}\sum_{m,n} \frac{|\langle n|Q_3|m\rangle|^2}{E_n - E_m}\Omega^2\,. \tag{6.61}$$

In the second term, only Q_3 appears by choosing the axis of rotation along the z-axis, where $\boldsymbol{\Omega} = (0,0,\Omega)$. Note that the energy correction in the rotating frame is negative. This is just a general consequence of the second order perturbation theory.

Now let us go back to the laboratory frame using Eq. (6.46). The quark wave function is given by $\psi' = A\psi$:

$$\begin{aligned} |\psi\rangle &= A|h_q'\rangle \\ &= A|h_q\rangle - \frac{1}{2}\sum_{m,n}\frac{A|n\rangle\langle n|\boldsymbol{\Omega}\cdot\boldsymbol{Q}|m\rangle}{E_n - E_m}\,. \end{aligned} \tag{6.62}$$

The energy in the laboratory frame is computed from Eqs. (6.56) and (6.61):

$$\begin{aligned} E_q^{lab} &= E^{bf} - \frac{1}{2}\langle h_q'|\boldsymbol{\Omega}\cdot\boldsymbol{Q}|h_q'\rangle \\ &= E_q + \frac{1}{4}\sum_{m,n}\frac{|\langle n|Q_3|m\rangle|^2}{E_n - E_m}\Omega^2\,. \end{aligned} \tag{6.63}$$

Now the second term appears as positive contrary to the second term of Eq. (6.61). This can be interpreted as the rotational kinetic energy of the hedgehog quark by writing it as

$$\begin{aligned} E_q^{lab} &= E_q + \frac{1}{2}\Lambda_q\Omega^2\,, \\ \Lambda_q &= \frac{1}{2}\sum_{m,n}\frac{|\langle n|Q_3|m\rangle|^2}{E_n - E_m}\,, \end{aligned} \tag{6.64}$$

where Λ_q is the moment of inertia of the rotating quark hedgehog. This is nothing but the Inglis formula in the nuclear physics [Inglis 1954; Bohr and Mottelson 1975; Ring and Shuck 1980]. The sum over n and m is schematically shown in Fig. 6.5.

$$\sum_{mn}\frac{|\langle n|\vec{\Omega}\cdot\vec{Q}|m\rangle|^2}{E_n - E_m} \sim \sum_{mn}$$

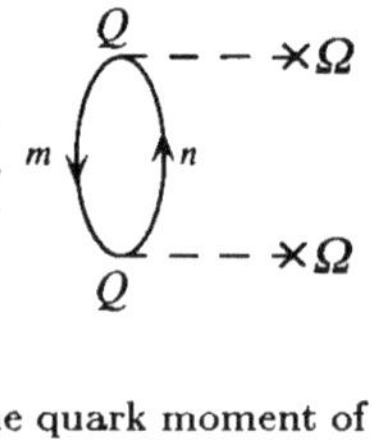

Fig. 6.5 Particle-hole excitations for the quark moment of inertia Λ_q.

Rotational energy

Adding the quark energy of Eq. (6.64) and the meson energy in Eq. (6.47), we find the total energy of the rotating hedgehog in the laboratory frame

$$E_{\text{tot}} = E_h + \frac{1}{2}\Lambda_{\text{tot}}\boldsymbol{\Omega}^2 , \quad \Lambda_{\text{tot}} = \Lambda_q + \Lambda_\pi \tag{6.65}$$

where E_h is given in Eq. (5.191). After the quantization of Eq. (6.65) as described in Appendix C, the energies of physical baryons with $I = J$ are given by,

$$E_J = E_h + \frac{I(I+1)}{2\Lambda_{\text{tot}}} . \tag{6.66}$$

We note that there is a simple rule of replacement upon quantization (see Appendix C),

$$\boldsymbol{\Omega} \to \boldsymbol{\sigma}/(2\Lambda_{\text{tot}}) . \tag{6.67}$$

Let us make a general remark on the energy formula of Eq. (6.66). Since the rotational energy is positive, we find $E_I > E_h$ for all I, in contrast with a general consequence from the projection theory. In a quantum theory, the hedgehog state could be written as a superposition of states with different spin and isospin. Let us write it in a schematic way:

$$|h\rangle = \sum_I a_I |I\rangle . \tag{6.68}$$

Then it follows that

$$E_h = \langle h|H|h\rangle = \sum_I |a_I|^2 E_I . \tag{6.69}$$

Since $E_{1/2} < E_{3/2} < \cdots$, this implies that $E_h > E_{I=1/2}$.

In the $1/N_c$ expansion, the hedgehog energy E_h is of order N_c, while the rotational energy is of order $1/N_c$. Thus, the term of order N_c^0 is missing in Eq. (6.66). Such a term is known as the ground state correlation or the zero point energy in quantum theory, and is not included in the cranking method [Coleman 1985; Rajaraman 1982; Ring and Shuck 1980]. In the projection theory, the missing term appears as the subtraction of the fluctuation energy associated with rotational motion:

$$E_J = E_h - \frac{\langle h|\hat{I}^2|h\rangle}{2\Lambda_{\text{tot}}} + \frac{I(I+1)}{2\Lambda_{\text{tot}}} . \tag{6.70}$$

Note that this equation is consistent with Eq. (6.69), since

$$\begin{aligned}\sum_I |a_I|^2 E_I &= \sum_I |a_I|^2 \left(E_h - \frac{\langle h|\hat{I}^2|h\rangle}{2\Lambda_{\rm tot}} + \frac{I(I+1)}{2\Lambda_{\rm tot}} \right) \\ &= E_h - \frac{\langle h|\hat{I}^2|h\rangle}{2\Lambda_{\rm tot}} + \sum_I |a_I|^2 \frac{I(I+1)}{2\Lambda_{\rm tot}} \\ &= E_h\,, \end{aligned} \tag{6.71}$$

where we have used the normalization condition $\sum_I |a_I|^2 = 1$ and the relation $\langle h|I^2|h\rangle = \sum_I |a_I|^2 I(I+1)$. Since $\langle h|I^2|h\rangle$ is of order N_c, the subtraction term is of order N_c^0.

The absence of the subtraction energy of order N_c^0 is considered to be the origin of the large baryon masses. Although reliable computations of such a term have not been done†, it is possible to estimate it in the limit of large deformation. Expanding in power series analogous to the Kamlah expansion [Kamlah 1968; Ring and Shuck 1980], one can establish a relation between the energy subtraction and the nucleon-delta mass splitting. Such an analysis implies that they are of the same order of magnitude when $N_c = 3$ [Hosaka 1988]. Therefore, it is likely that the masses of physical baryons obtained in the semiclassical method are significantly reduced by the quantum corrections.

Moment of inertia of quarks

In this subsection, we show how to compute the moment of inertia, in particular the quark moment of inertia Λ_q in some detail. It is a double mode sum over two indices for all the particle (n) and hole ($\tilde{m}$) states, as depicted in Fig. 6.4. Analytic calculations are no longer available, and therefore numerical methods have been explored. In the literature, two approaches were developed. One is a straightforward extension of the previous single mode sum, where the sum is computed with a suitable regulator [Park and Rho 1988]. In the other approach, the Strutinsky's smearing method was applied in which a discrete mode sum was replaced by a continuous integral over energy [Wuest et al. 1986; Jackson et al. 1987]. This method is useful, since the integral is dominated by a low energy region after divergent

†The zero point energy was calculated in the chiral quark soliton model in Refs. [Weigel et al. 1995].

terms were suitably subtracted. In the latter method, they computed not only the quark moment of inertia but also other physical quantities such as magnetic moments using this method. Here we will adopt the former method of the discrete sum with a cutoff function. This is convenient when studying divergent behavior in a way analogous to the single mode sum in the previous section.

For actual numerical computation, let us introduce the gaussian regulator:

$$\begin{aligned}\Lambda_q &= \frac{1}{2}\sum_{m,n}\frac{|\langle n|Q_3|m\rangle|^2}{E_n - E_m} \\ &\to \frac{1}{2}\sum_{m,n}\frac{|\langle n|Q_3|m\rangle|^2}{E_n - E_m}e^{-\eta^2(E_n-E_m)^2}. \end{aligned} \tag{6.72}$$

This ensures much faster convergence than the linear regulator, and allows one to extrapolate η down to a smaller value around $\eta \sim 0.075$ when $E_{\rm max} \sim 100$. After performing angular integrals for the isospin matrix elements $\langle n|Q_3|m\rangle$, we obtain [‡]:

$$\Lambda_q(F;\eta) = \frac{N_c}{6}\sum_{K\geq 0}\frac{|\langle K||\tau||\tilde{K}'\rangle|^2}{\tilde{E}_{K'} - E_K}e^{-\eta^2(\tilde{E}_{K'}-E_K)^2}, \tag{6.73}$$

where the tilded quantities are those of the unoccupied states. For a given K, there are three terms with K' satisfying the selection rule $|K' - K| \leq 1$, since the isospin operator τ is rank one. The reduced matrix elements $\langle K||\tau||K'\rangle$ with $K' = K, K \pm 1$ are given in Appendix D.

A straightforward extension of the previous analysis of the single mode sum is first to separate the double mode sum into the vacuum and valence parts, Λ_0 and $\Lambda_{\rm val}$, respectively.

$$\Lambda_q(F;\eta) = \Lambda_0(F;\eta) + \theta(\tfrac{\pi}{2} - F)\Lambda_{\rm val}(F;\eta). \tag{6.74}$$

As shown in Fig. 6.6, the vacuum part includes transitions from negative energy to positive energy levels, whereas the valence part is defined as the transitions from the lowest 0^+ valence level to positive energy excited levels subtracted by the transitions from negative to the 0^+ valence level. For $F < \pi/2$, the lowest 0^+ valence level is in the positive energy region,

[‡]Computation of such matrix elements requires the methods in the spherical basis which are presented in Appendix H.

and both the vacuum and valence parts appear, while for $F > \pi/2$, the 0^+ level dives into the negative energy region and the valence part disappears. This explains why the theta function $\theta(\pi/2 - F)$ appears in Eqs. (6.74).

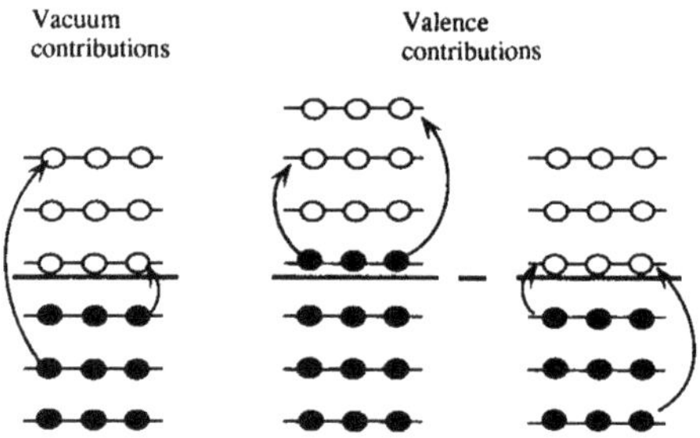

Fig. 6.6 Separation of particle hole sum into the valence and vacuum parts.

For numerical study, it is convenient to define a dimensionless quantity

$$\lambda(F;\eta) \equiv R\Lambda_q(F;\eta) \tag{6.75}$$

for both the vacuum and valence parts. As shown by Jackson et al. [Jackson et al. 1987], and by Park and Rho [Park and Rho 1988], the vacuum part $\lambda_0(F;\eta)$ contains a linearly divergent term which is proportional to $1/\eta$ and depends on $\sin^2 F$. This $\sin^2 F$ dependence may be used when extracting the finite term that behaves as $\mathcal{O}(F^4)$ for small F, or equivalently $\mathcal{O}(\pi - F)^4$ for $F \to \pi$. Therefore, we have the structure

$$\begin{aligned} \lambda(F;\eta) \quad &\sim \quad (\text{divergent coefficient}) \cdot \sin^2 F \\ &+ \quad (\text{finite coefficient}) \cdot (\mathcal{O}(F^4) \text{ or } \mathcal{O}((\pi - F)^4)\,. \end{aligned} \tag{6.76}$$

In fact, the divergent terms are completely eliminated by imposing the subtraction condition analogous to Eqs. (5.178) and (5.188),

$$\lambda_{\text{vac}}(F) \equiv \lim_{\eta \to 0} \left[\lambda_0(F;\eta) - \frac{1}{2} \frac{\partial^2 \lambda_0}{\partial F^2}(0;\eta) \sin^2 F \right]. \tag{6.77}$$

In order to see the divergent behavior of the double mode sum, we show in Fig. 6.7 the divergent coefficient of $\sin^2 F$ in Eq. (6.77), $c_{\text{div}}(\eta)$, as a function of η. The calculated values are fitted numerically by the function

$$c_{\text{div}}(\eta) = \frac{c_{-1}}{\eta^1} + c_{\ln} \ln \eta + c_0 + \eta c_1 + \cdots\,. \tag{6.78}$$

Quality of the fitting is excellent as shown in Fig. 6.7.

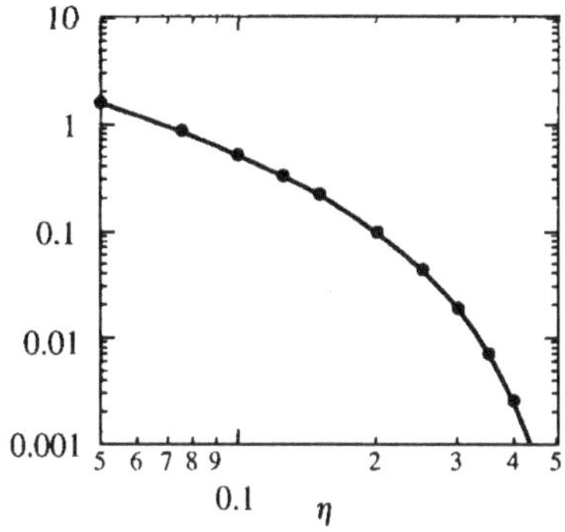

Fig. 6.7 Divergent behavior of the coefficient of the $\sin^2 F$ in Eq. (6.76).

The $\mathcal{O}((\pi - F)^4)$ behavior of the moment of inertia as $F \to \pi$ is expected from the Skyrme model and is consistent with the smoothness condition in the limit $R \to 0$. In the Skyrme model, the moment of inertia in the region $0 \le r \le R$

$$\begin{aligned}\lambda_{\rm Sk} &\equiv R\Lambda_{\rm Sk} \\ &= \frac{8\pi R}{3} f_\pi^2 \int_0^R r^2 dr \sin^2 F \left(1 + \frac{1}{e^2 f_\pi^2}\left(F'^2 + \frac{\sin^2 F}{r^2}\right)\right), \quad (6.79)\end{aligned}$$

is easily computed for small R. Employing the small r behavior of the skyrmion solution, $F(r) \sim \pi(1 - br)$, where b is a numerical constant, one can verify that

$$\lambda_{\rm Sk} \sim (\pi - F)^4 . \quad (6.80)$$

Now adding the valence contribution $\lambda_{\rm val}$ when $0 \le F \le \pi/2$, we get the total quark moment of inertia

$$\lambda_q(F) = \lambda_{\rm vac}(F) + \theta(\tfrac{\pi}{2} - F)\lambda_{\rm val}(F) . \quad (6.81)$$

In Fig. 6.8, λ_q is plotted as a function of F, where the vacuum and the valence contributions are shown separately. The valence part $\lambda_{\rm val}(F)$ grows indefinitely when $F \to 0$, because the 0^+ valence level and the lowest 1^+ level are degenerate in this limit, where the transition $0^+ \to 1^+$ is the

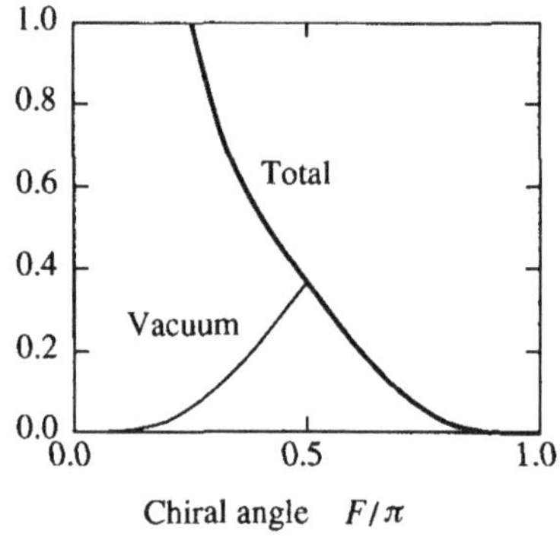

Fig. 6.8 Normalized quark moment of inertia λ_q as a function of the chiral angle F. For $F \leq \pi/2$ the vacuum contribution is separately shown.

relevant one. It should be noted that this divergence is not of ultraviolet nature but is rather related to the symmetry property of the quark states.

Nucleon mass

The total moment of inertia is the sum of the quark and the pion contributions:

$$\begin{aligned} \Lambda_{\text{tot}} &= \Lambda_\pi + \Lambda_q \\ &= \Lambda_\pi + R(\lambda_{\text{vac}} + \lambda_{\text{val}}) . \end{aligned} \tag{6.82}$$

Various terms of the moment of inertia are shown in Fig. 6.9 as functions of the bag radius R. We observe a weak R dependence for small bag radii $R \lesssim 0.4$ fm, as expected from the smoothness condition in the limit $R \to 0$. In this region, calculated values are $\Lambda_{\text{tot}}(R) \sim 1.2$ and are slightly overestimated as compared with the experimental value deduced from the nucleon-delta mass splitting: $\Lambda_{\text{exp}} = 1.01 \text{ fm}^{-1}$. This means that the calculated $N\Delta$ mass splitting in this model is slightly underestimated. In Fig. 6.9 we also show the masses of the nucleon and delta, M_N and M_Δ together with their mass difference.

As we go to larger bag radii, Λ_{tot} increases rapidly and the calculated nucleon-delta mass splitting becomes smaller[§]. In fact, the underestimate of

[§]It is interesting to note that the large R limit is correctly reproduced where M_N and M_Δ are degenerate due to the divergent moment of inertia $\Lambda_q \to \infty$. This seems to oppose

the $N\Delta$ mass splitting at large bag radii is an expected result. At small bag radii, the $N\Delta$ mass splitting is produced by the collective rotation of the pion. As the bag radius is increased, such a pionic effect is replaced by the quark-gluon effect of the spin-spin interaction. The latter is not included in the present calculation. Kahana and Millana [Kahana and Milana 1987], and Kusaka and Toki [Kusaka and Toki 1988] considered the effect of one gluon-exchanges in the chiral bag model and showed that it contributed significantly to the $N\Delta$ mass splitting at large bag radii. The resulting $N\Delta$ mass splitting is only weakly dependent on R within $0 \leq R \lesssim 1$ fm.

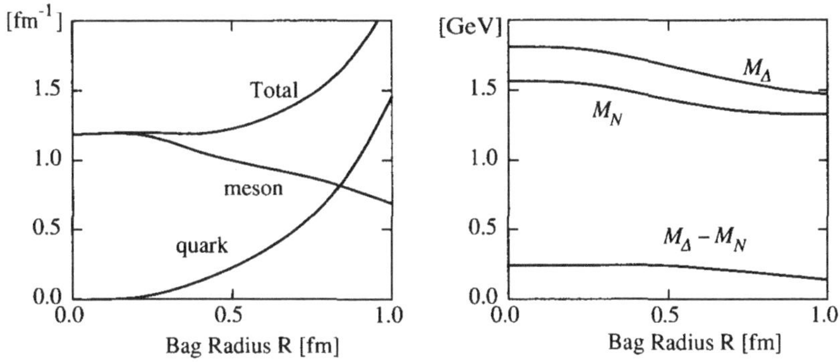

Fig. 6.9 The total moment of inertia and masses of the nucleon and delta as a function of R. For the moment of inertia the quark and meson contributions are shown separately. In the masses, the volume term $4\pi BR^3/3$ is not included.

6.3 Axial Properties

In this section, we study axial properties of the nucleon. We emphasize the role of chiral symmetry and its spontaneous breakdown. They provide useful tools not only for computing the axial-vector coupling constant g_A but also for checking consistency of solutions. To start with let us derive a relation analogous to the Gauss law, which represents the conservation of the axial flux (the flow of the axial-vector current along the radial direction).

the usual statement that the cranking method would break down in this limit [Kahana et al. 1986; Jackson et al. 1987; Hosaka et al. 1986]. We will see a similar situation for g_A

Gauss law

In the semiclassical limit where the nucleon is regarded as a slowly rotating hedgehog, the axial-vector current takes the form

$$A_i^a(\boldsymbol{x}) = (\alpha(r)\delta_{ij} + \beta(r)\hat{x}_i\hat{x}_j)\, D_{ja}\,. \tag{6.83}$$

The two terms are those that can be constructed from available tensors δ_{ij} and $\boldsymbol{x}_i$, satisfying the parity (spatial component of the axial-vector current carries positive parity). In Eq. (6.83), $\alpha(r)$ and $\beta(r)$ are functions of the radial distance r, and the rotation matrix D_{ia} is defined as in Eq. (6.44):

$$D_{ia} = \frac{1}{2}\,\mathrm{tr}\,\left[\tau_i A \tau_a A^\dagger\right]\,, \tag{6.84}$$

where A is an $SU(2)$ time dependent matrix as defined in Eq. (6.46). Let the normal flow of the axial-vector current at $r = R$ (where R is not necessarily the bag radius) be

$$\hat{x}_i\hat{x}_a A_i^a(\boldsymbol{x})|_{r=R} = (\alpha(R) + \beta(R))\, D_{ia}\hat{x}_i\hat{x}_a\,. \tag{6.85}$$

The axial flux $a(R)$ is then defined by

$$\hat{x}_i\hat{x}_a A_i^a(\boldsymbol{x})|_{r=R} \equiv a(R)\, D_{ia}\hat{x}_i\hat{x}_a\,, \tag{6.86}$$

and therefore

$$a(R) = \alpha(R) + \beta(R)\,. \tag{6.87}$$

The axial flux $a(R)$ is related to the volume integral of the axial-vector current:

$$\int_{r\leq R} d^3x\, A_i^a(\boldsymbol{x}) = \int_{r\leq R} d^3x\, \left(\alpha(r) + \frac{1}{3}\beta(r)\right) D_{ia}\,. \tag{6.88}$$

Using the current conservation for Eq. (6.83),

$$\alpha' + \beta' + \frac{2}{r}\beta = 0\,, \tag{6.89}$$

and adding $0 = \frac{r}{3}(\alpha' + \beta') + \frac{2}{3}\beta$, we find

$$\int_{r\leq R} d^3x\, A_i^a(\boldsymbol{x}) \;=\; \frac{4\pi R^3}{3}\, a(R)\, D_{ia}\,. \tag{6.90}$$

This relation is analogous to the Gauss law and is very useful in computing the axial-vector coupling constant g_A.

Definition of g_A

Let us consider the nucleon matrix element of the axial-vector current. Neglecting the energy transfer, $\omega = p'_0 - p_0 \sim 0$, in the non-relativistic limit, it takes the form

$$\langle \boldsymbol{p}'|A_i^a(\boldsymbol{x}=0)|\boldsymbol{p}\rangle = (g_A(\boldsymbol{q})\delta_{ij} + h_A(\boldsymbol{q})q_iq_j)\,\sigma_j\frac{\tau^a}{2}\,, \tag{6.91}$$

where $\boldsymbol{q} = \boldsymbol{p}' - \boldsymbol{p}$ and $q = |\boldsymbol{q}|$. In this equation, the second term is induced by the pion pole term. Current conservation (CAC) implies that

$$g_A(q) + q^2 h_A(q) = 0\,, \tag{6.92}$$

from which $h_A(q)$ is eliminated through $h_A(q) = -g_A(q)/q^2$. In the limit $q \to 0$ one can replace $\hat{q}_i\hat{q}_j \to 1/3\delta_{ij}$ (see problem 6.4), and the right hand side of Eq. (6.91) becomes $2/3 g_A\sigma_i(\tau^a/2)$. The left hand side of Eq. (6.91) is interpreted as the spatial integral over the axial-vector current. Altogether, the axial-vector coupling constant g_A is defined by [Adkins et al. 1983]

$$g_A\sigma_i\frac{\tau^a}{2} = \frac{3}{2}\int d^3x A_i^a(\boldsymbol{x})\,. \tag{6.93}$$

It is worthwhile to explain the meaning of the factor 3/2 on the right hand side of this equation, which appears when $m_\pi = 0$. For a non-zero m_π, the pion induced term in Eq. (6.91), q^2h_A, vanishes in the limit $q \to 0$, because there is no massless pion pole. Therefore, only the g_A term contributes to $\int d^3x\, A_i^a$, and the factor 3/2 is replaced by 1 as naively expected. In the chiral limit, in contrast, the h_A term has non vanishing contribution to $\int d^3x\, A_i^a$ due to the presence of the massless one pion pole.

Now the subtle point is that the asymptotic one pion contribution to the volume integral $\int d^3x\, A_i^a$ vanishes, where the precise cancellation between the g_A and h_A terms occurs. It is verified by using the asymptotic form of the pion field. The static pion field $\pi^a(\boldsymbol{x})$ satisfies the equation of motion at $r \to \infty$

$$\nabla^2\pi^a = \frac{g_{\pi NN}}{2M_N}\partial_i\left(\delta(\boldsymbol{x})\right)\sigma_i\tau^a\,, \tag{6.94}$$

where the actually extended pion source term has been effectively replaced by a point like source with the coupling strength $g_{\pi NN}$. In this equation, the Pauli matrices on the right hand side are understood to be expectation values in the nucleon state. The solution to this equation is

$$\pi^a(\boldsymbol{x}) = \frac{g_{\pi NN}}{8\pi M_N}\partial_i\left(\frac{1}{r}\right)\sigma_i\tau^a . \tag{6.95}$$

Now for sufficiently large D we have

$$\begin{aligned}\int_{r\geq D} d^3x A_i^a &= \frac{f_\pi g_{\pi NN}}{8\pi M_N}\int_{r\geq D} d^3x\,\partial_i\partial_j\left(\frac{1}{r}\right)\sigma_j\tau^a \\ &= -\frac{f_\pi g_{\pi NN}}{8\pi M_N}\int_{r\geq D} d^3x\frac{\delta_{ij}-3\hat{x}_i\hat{x}_j}{r^3}\sigma_j\tau^a . \end{aligned}\tag{6.96}$$

Taking the angle average, the integral of the last equation vanishes. However, from the tensor structure of Eq. (6.96), the δ_{ij} term is nothing but the contribution to g_A.

Turning back to Eq. (6.91), the one pion contribution to g_A is extracted by the following decomposition:

$$\begin{aligned}\langle \boldsymbol{p}'|A_i^a(0)|\boldsymbol{p}\rangle &= (g_A(q)\delta_{ij} - g_A(q)\hat{q}_i\hat{q}_j)\,\sigma_j\frac{\tau^a}{2} \\ &= \left[\frac{2}{3}g_A(q)\delta_{ij} + g_A(q)\left(\frac{1}{3}\delta_{ij} - \hat{q}_i\hat{q}_j\right)\right]\sigma_j\frac{\tau^a}{2} , \end{aligned}\tag{6.97}$$

where the second term of $1/3\delta_{ij} - \hat{q}_i\hat{q}_j$ is the one pion part and vanishes in the limit $q \to 0$.

We can summarize the argument as follows. The axial-vector coupling constant g_A has two components: one is the asymptotic contribution dominated by the one pion tail and the other is a finite range contribution from multi-pions and quarks. If $m_\pi \neq 0$, both terms contribute to the integral $\int d^3x\, A_i^a$, and therefore, it is related directly to g_A without the factor 3/2. However, if $m_\pi = 0$, the one pion part of g_A is canceled by the h_A term in the volume integral. The factor 3/2 is, therefore, needed in order to recover the missing piece. This result was originally derived by Jaffe in the linearized chiral bag model [Jaffe 1979]. It is not difficult to see that the conclusion here is valid in any chiral theories with spontaneous breakdown in the chiral limit $m_\pi = 0$.

Another intuitive way of understanding the factor 3/2 is as follows. For $m_\pi \neq 0$, the integral $\int_0^D d^3x A_i^a$ up to a finite upper limit $D \sim 1/m_\pi$ picks up all the contribution to g_A. Thus, if we perform the computation with a finite pion mass and take the limit $m_\pi \to 0$, then we can obtain the correct answer. But this does not hold if we set $m_\pi = 0$ from the beginning, since we can not reach the infinite upper limit. In that case, however, we know exactly the amount of the missing piece of g_A from chiral symmetry, which is 1/3 of the total amount. This explains the appearance of the factor 3/2 when $m_\pi = 0$.

g_A *in the leading order*

The isovector axial-vector current for quarks is defined by

$$\bar{\psi}\gamma_\mu\gamma_5\frac{\tau^a}{2}\psi = \psi^\dagger\sigma_\mu\frac{\tau^a}{2}\psi\,, \tag{6.98}$$

where $\sigma_\mu = (1, \boldsymbol{\sigma})$. Therefore, g_A is the nucleon matrix element of $\sigma_i\tau^a$, and is the quantity of order Ω^0, since it has a non-vanishing matrix element in the hedgehog state.

As anticipated in Eq. (6.90), we compute g_A without performing the volume integral $\int d^3x\, A_i^a$ explicitly. Using the identity

$$\partial_j(x_i A_j^a) = A_i^a\,, \tag{6.99}$$

the volume integral is related to the surface integral at infinity:

$$\int d^3\,x A_i^a = \int_{r\to\infty} dS_j\; x_i A_j^a\,. \tag{6.100}$$

In the limit $r \to \infty$, the axial-vector current is dominated by the one pion component as

$$A_i^a \stackrel{r\to\infty}{\longrightarrow} f_\pi \partial_i \pi^a\,. \tag{6.101}$$

Using the asymptotic solution of Eq. (6.95), the surface integral of Eq. (6.100) reduces to

$$\begin{aligned}\int dS_j \hat{x}_i A_j^a &= \int dS_j \hat{x}_i f_\pi \partial_j \pi^a \\ &= \frac{g_{\pi NN} f_\pi}{2M_N}\frac{2}{3}\sigma_i\frac{\tau^a}{2}\,.\end{aligned} \tag{6.102}$$

Comparing this equation with Eq. (6.93), we find the Goldberger-Treiman relation (3.122).

In the rigid rotor approximation, the asymptotic pion field of the rotating hedgehog is given by

$$\pi_a(\boldsymbol{x}) \stackrel{r\to\infty}{\longrightarrow} f_\pi \frac{b}{r^2} D_{aj}\hat{x}_j\,. \tag{6.103}$$

The constant b is extracted from the asymptotic behavior of the profile function $F(r)$. Quantization of the rotation allows the following replacement

$$D_{ia} \to -\frac{1}{3}\sigma_i\,\tau^a\,. \tag{6.104}$$

Thus, by comparing Eq. (6.103) and Eq. (6.95), we find

$$\frac{g_{\pi NN}}{8\pi M_N} = \frac{1}{3} f_\pi b\,, \tag{6.105}$$

which, together with the Goldberger-Treiman relation, leads to [Hosaka and Toki 1986]

$$g_A = \frac{8\pi}{3} f_\pi^2\, b\,. \tag{6.106}$$

In the chiral bag model, we are also interested in the separate contributions from quarks and pions, as well as the total value. The quark contribution g_A^q is computed by the integral of the quark axial-vector current, which is expressed by the surface integral as given in Eq. (6.90). This is now related directly to g_A^q without the factor 3/2 unlike Eq. (6.93), since the integral is restricted to the finite region $r \le R$. Therefore, we obtain

$$g_A^q = -\frac{8\pi R^3}{9}\, a_q(R)\,, \tag{6.107}$$

where $a_q(R)$ is the quark axial flux at the bag surface. The pion contribution g_A^π is then just the difference: $g_A^\pi = g_A - g_A^q$.

In Fig. 6.10 we show g_A as a function of R with the quark and pion contributions shown separately. It is an almost flat function for small $R \lesssim 0.4$, due to the smoothness condition in the limit $R \to 0$. It begins, however, to increase around $R \gtrsim 0.5$ up to $R \sim 1.2$ fm. In the region $R \lesssim 1$ fm, however, the calculated value of g_A is small as compared with the experimental value $g_A(\exp) = 1.25$. The fractioning of g_A within $R \lesssim 1$ fm is similar to that of the baryon number; at bag radii $R \sim 0.6$ fm, the quark and meson contributions are almost equal. What appears peculiar here is that the pion contribution does not disappear in the limit $R \to \infty$; it approaches 1/3 of the total value in this limit. In this respect, contents of g_A differ very much from other physical quantities, which are entirely carried by quarks in this limit. The non-vanishing one pion contribution in the large R limit is a consequence of exact chiral symmetry ($m_\pi = 0$), where the massless pion contribution is relegated to the infinitely far region.

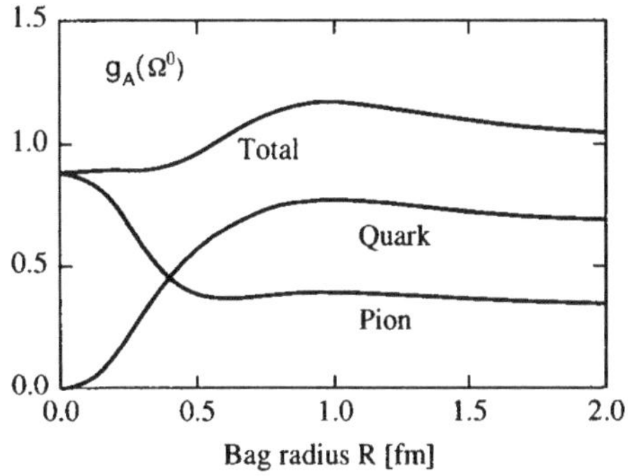

Fig. 6.10 The axial-vector coupling constant g_A in the leading order as a function of R. Quark and pion contributions are shown separately.

The fact that g_A is small is a common problem in various chiral models including the Skyrme, chiral bag and chiral quark soliton models, when only the leading order contribution in the cranking method is calculated. In fact, it is possible to find a parameter set f_π and e which reproduces the experimental value of g_A. However, in that case, the masses of the nucleon and delta become too heavy. It is not possible to reproduce both masses and g_A simultaneously.

In the chiral bag model model, the problem is more serious, since in the leading order of the semiclassical expansion the chiral bag value of g_A does not reduce to the MIT bag value in the limit $R \to \infty$. Indeed, the quark contribution in this limit is $g_A^q = 0.65$, which is only 3/5 of the MIT bag value, where 2/5 of g_A is missing¶.

The origin of the missing part may be traced back to the nucleon matrix element of the collective rotation $\langle N|D_{ia}|N\rangle = 1/3\langle N|\sigma_i\tau_a|N\rangle$, which is identified with the $(1/N_c$ times) leading term of the the quark model matrix

¶Strictly speaking, g_A in the chiral limit ($m_\pi \to 0$) does not reduce to that of the MIT bag model in the limit $R \to \infty$, where the massless pion has the non-vanishing contribution $g_A^\pi = 1/2g_A^q$ due to the conservation of the axial flux.

element

$$\begin{aligned}\langle p\uparrow|\sum_{i=1}^{N_c}\sigma_3(i)\tau_3(i)|p\uparrow\rangle &= \frac{1}{3}(N_c+2)\langle p\uparrow|\sigma_3^N\tau_3^N|p\uparrow\rangle \\ &= \frac{1}{3}(N_c+2) \qquad (6.108)\end{aligned}$$

where σ_3^N and τ_3^N are operators for the nucleon. When $N_c=3$, the missing term is 2/3. This explains the discrepancy between the chiral bag result of leading order and the MIT bag result. More details will be discussed in Chapter 7.

Quark contribution of order Ω^1

The necessity of terms of order Ω^1 was first pointed out by Wakamatsu and Watabe in the context of the chiral quark soliton model [Wakamatsu and Watabe 1993] Subsequently such higher order terms were investigated in the chiral bag model by the present authors [Hosaka and Toki 1994; Hosaka and Toki 1995]. By now the importance of those terms have been investigated in detail [Wakamatsu 1994; Christov et al. 1995].

Consider the quark matrix element of the isospin axial-vector current

$$\begin{aligned}\int d^3x\bar{\psi}\gamma_i\gamma_5\frac{\tau^a}{2}\psi &= \int d^3x\psi^\dagger\sigma_i\frac{\tau^a}{2}\psi \\ &= \langle\psi|\sigma_i\frac{\tau^a}{2}|\psi\rangle\,. \qquad (6.109)\end{aligned}$$

Substituting the rotating quark wave function of Eq. (6.62) in the current, we obtain for the $a=i=3$ component the non-vanishing matrix element of order Ω^1

$$\begin{aligned}\langle\psi|\sigma_3\frac{\tau^a}{2}|\psi\rangle &= -\frac{1}{4}\sum_{mn}\left\{\frac{\langle m|A^\dagger\sigma_3\tau_3A|n\rangle\langle n|\tau_a|m\rangle}{E_n-E_m}\Omega_a\right. \\ &\left.+\ \Omega_a\frac{\langle m|\tau_a|n\rangle\langle n|A^\dagger\sigma_3\tau_3A|m\rangle}{E_n-E_m}\right\}. \qquad (6.110)\end{aligned}$$

Using the identity $A^\dagger\tau_aA=D_{ia}\tau_i$ with the rotation matrix D_{ia} defined by

Eq. (6.44), we get

$$\langle\psi|\sigma_3\frac{\tau^a}{2}|\psi\rangle = -\frac{1}{4}\sum_{mn}\left\{\frac{\langle m|\sigma_3\tau_b|n\rangle\langle n|\tau_a|m\rangle}{E_n-E_m}D_{b3}\Omega_a + \frac{\langle m|\tau_a|n\rangle\langle n|\sigma_3\tau_b|m\rangle}{E_n-E_m}\Omega_a D_{b3}\right\}. \quad (6.111)$$

Here we have preserved the ordering of the two operators D_{b3} and Ω_a as appeared in Eq. (6.110). A priori, there is no way to fix the ordering. Here, we find that the ordering shown in Eq. (6.111) reproduces the desired higher order correction. In particular, it reproduces the correct large R limit of g_A such that the chiral bag value approaches that of the MIT bag model.

As shown explicitly in Appendix E, there are symmetry relations in the matrix elements:

$$\begin{aligned}\langle m|\tau_a|n\rangle &= -\langle n|\tau_a|m\rangle\,,\\ \langle m|\sigma_3\tau_b|n\rangle &= \langle n|\sigma_3\tau_b|m\rangle\,,\end{aligned}$$

which are used to simplify the expression of Eq. (6.111) as

$$\langle\psi|\sigma_i\frac{\tau^a}{2}|\psi\rangle = -\frac{1}{4}\sum_{mn}\frac{\langle m|\sigma_3\tau_b|n\rangle\langle n|\tau_a|m\rangle}{E_n-E_m}\left(D_{b3}\Omega_a-\Omega_a D_{b3}\right). \quad (6.112)$$

Applying the Wigner-Eckart theorem to the matrix element $\langle m|\sigma_3\tau_b|n\rangle$ as explained in Appendix E, we find the expression

$$\langle\psi|\sigma_3\frac{\tau_3}{2}|\psi\rangle = \frac{1}{9\sqrt{2}\Lambda_{\rm tot}}\sigma_3\frac{\tau_3}{2}\sum_{K,K'\geq 0}\frac{\langle K||[\sigma\,\tau]^1||\widetilde{K'}\rangle\langle\widetilde{K'}||\tau||K\rangle}{\tilde{E}_{K'}-E_K}e^{-\eta^2(\tilde{E}_{K'}-E_K)^2}\,, \quad (6.113)$$

where $[\sigma\,\tau]^1$ is the rank one tensor formed from the two vectors $\boldsymbol{\sigma}$ and $\boldsymbol{\tau}$. Here we recognize the selection rule $|K-K'|\leq 1$. In obtaining this result, we have quantized D_{ai} and $\boldsymbol{\Omega}$ with keeping the ordering in Eqs. (6.111) and (6.112), which yields the combination

$$\begin{aligned}\epsilon_{3ab}D_{a3}\Omega_b &\rightarrow -\frac{1}{6\Lambda_{\rm tot}}\epsilon_{3ab}\sigma_a\sigma_b\tau_3\\ &= -\frac{i}{3\Lambda_{\rm tot}}\sigma_3\tau_3\,. \end{aligned} \quad (6.114)$$

Eq. (6.113) is then identified with $g_A^q(\Omega^1)\sigma_3\tau^3/2$.

Now, we introduce the quantity multiplied by $\Lambda_{\rm tot}$,

$$g(F;\eta) \equiv \Lambda_{\rm tot}\, g_A^q(\Omega^1)\,. \tag{6.115}$$

As in the previous case for the moment of inertia, the double mode sum $g(F;\eta)$ is separated into the vacuum and valence parts:

$$g(F;\eta) = g_0(F;\eta) + g_{\rm val}(F;\eta)\theta(\tfrac{\pi}{2} - F)\,. \tag{6.116}$$

It turns out that the vacuum term $g_0(F;\eta)$ is linearly divergent as proportional to $1/\eta$. A detailed numerical study shows that the divergences has $\sin 2F$ dependence, which is the same as that of the zeroth order. We show in Fig. 6.11 (a) the divergent coefficient of $g_0(F;\eta)$ as a function of η.

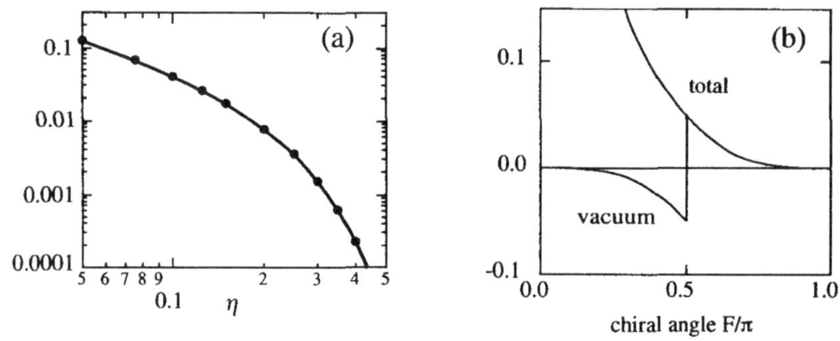

Fig. 6.11 (a) The divergent coefficient of the $\sin 2F$ term of $g_0(F;\eta)$ as a function of η, and (b) Normalized quark axial flux $g(F)$ as a function of the chiral angle F. For $F \le \pi/2$ the vacuum contribution is separately shown.

The finite term is extracted in a way analogous to the leading order. The function $g_0(F;\eta)$ is an odd function of F, and therefore, the subtraction condition is

$$g_{\rm vac}(F) \equiv \lim_{\eta\to 0}\left[g_0(F;\eta) - \frac{1}{2}\sin 2F\frac{\partial g_0}{\partial F}(0;\eta)\right]\,. \tag{6.117}$$

The subtracted finite term behaves as $\mathcal{O}(F^3)$ when $F \to 0$, or equivalently, $\mathcal{O}((\pi - F)^3)$ when $F \to \pi$. The valence contribution in Eq. (6.116) rapidly converges and so $g_{\rm val}(F) \equiv \lim_{\eta\to 0} g_{\rm val}(F;\eta)$. The resulting function

$$g(F) = g_{\rm vac}(F) + \theta(F - \tfrac{\pi}{2})g_{\rm val}(F) \tag{6.118}$$

is plotted in Fig. 6.11 (b). When $F \to 0$, $g(F)$ increase indefinitely, but the physical g_A is finite, since it is divided by Λ_{tot} which also diverges in this limit.

Pion contribution of order Ω^1

As the quark field is modified under the influence of the Coriolis force, the quark axial flux at the bag surface receives a correction of order Ω^1 as well as g_A. Accordingly the pion field is modified such that the continuity of the axial flux at the bag surface is preserved up to the same order of Ω^1.

The correction to the pion field is accommodated within the hedgehog ansatz. Explicitly, the axial flux continuity at the bag surface including the zeroth and the first order terms is written as

$$\left[\left(f_\pi^2 + \frac{2\sin^2\tilde{F}}{e^2r^2}\right)\tilde{F}'\right]_{r=R} = \hat{x}_i\hat{x}_a\left[A_i^a(\Omega^0) + A_i^a(\Omega^1)\right]_{r=R}, \qquad (6.119)$$

where the quark axial flux of higher order is computed from g_A^q by virtue of the relation (6.107). In Eq. (6.119) the tilded chiral angle includes the correction term of order Ω^1: $\tilde{F}(r) = F(r) + \delta F(r)$. The improved function $\tilde{F}(r)$ obeys the same equation of motion as that for the hedgehog skyrmion, $F(r)$.

We would like to make a remark here. As seen in Eq. (6.115), g_A and therefore the axial flux of order Ω^1 is multiplied by the total moment of inertia Λ_{tot}. Since the pion moment of inertia Λ_π contains the integral over $\tilde{F}(r)$, the equation of motion for $\tilde{F}(r)$ is an integro-differential equation of self-consistent type. In order to avoid the complexity of the self-consistent method, Λ_π may be computed by using the zeroth order pion field $F(r)$ rather than $\tilde{F}(r)$. The effect of the fully self–consistent method is of higher order Ω^2.

Total value of g_A and the limit $R \to \infty$

Now we show the total g_A including the pion and quark contributions up to order Ω^1. As in the leading order, the total value of g_A is computed from the asymptotic pion tail, in this case, of $\tilde{F}(r)$. The fractioning of g_A into the quark and the meson sectors is also calculated in the same way as before: the quark contribution is computed from the quark axial flux including the terms of order Ω^0 and Ω^1 using Eq. (6.107). We show the results in Fig. 6.12

as functions of R. Up to bag radii $R \sim 0.4$ fm, the correction term of order Ω^1 is negligible. The situation, however, changes dramatically as the bag radius R is increased. For $R \gtrsim 0.5$ fm, the correction term is enhanced and at $R \sim 0.6$ fm, the experimental value $g_A = 1.25$ is reproduced. The important role of the higher order term was originally noticed by Wakamatsu and Watabe in the context of the NJL soliton model [Wakamatsu and Watabe 1993].

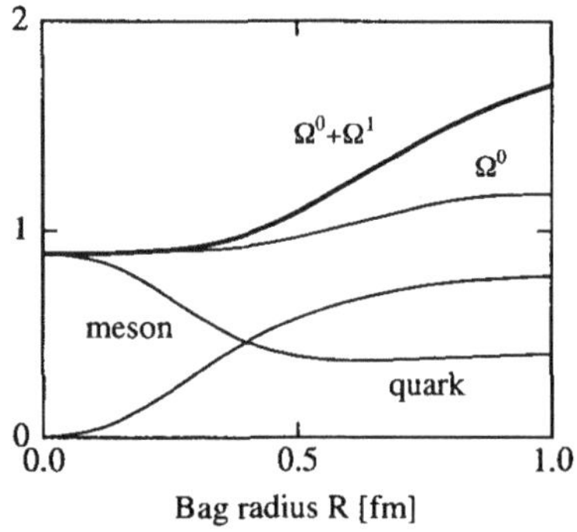

Fig. 6.12 The total axial-vector coupling constant g_A including the contribution of order Ω^1 as a function of R. For the Ω^0 term, the quark and meson contributions are separately shown.

It is interesting that the experimental value of g_A is reproduced at the chiral bag radius $R \sim 0.6$ fm, which is close to the conventionally known cutoff value adopted in various nuclear and meson processes. As we will see in the next section, static properties of the nucleon including electromagnetic properties are well reproduced at this bag radius. In the chiral bag model with vector mesons, we will see once again that the bag radius $R \sim 0.6$ provides a good description.

Finally, we look closely at the limit $R \to \infty$ and demonstrate that the chiral bag value of g_A, or the quark contribution g_A^q in that limit rigorously reduces to the MIT bag value ‖. Including the terms of order Ω^0 and Ω^1, the volume integral of the quark axial-vector current is written in a schematic

‖Once again, we note that the asymptotic value of the total g_A in the chiral bag model is 1.5 times the MIT bag value due to the asymptotic one pion contribution.

way as

$$\langle\psi|\sigma_3\frac{\tau^3}{2}|\psi\rangle = \left(\sum_m G_{mm} + \sum_{mn}\frac{G_{mn}T_{nm}}{E_n - E_m}\frac{1}{\Lambda_{\rm tot}}\right)\sigma_i\frac{\tau^3}{2}\,. \qquad (6.120)$$

Here G_{nm} and T_{nm} are the radial part of the matrix elements of $\langle n|\sigma_i\tau^a|m\rangle$ and $\langle n|\tau^a|m\rangle$, respectively. With this notation the quark moment of inertia is written as

$$\Lambda_q = \frac{1}{2}\sum_{mn}\frac{T_{mn}T_{nm}}{E_n - E_m}\,. \qquad (6.121)$$

In the limit $R \to \infty$ and $F(R) \to 0$, the 0^+ and the 1^+ states are degenerate. Consequently, only two terms survive in Eq. (6.120); one is the valence quark contribution in the first term and the other is the term of transition from the lowest 0^+ to the lowest 1^+ state in the second term:

$$\langle\psi|\sigma_3\frac{\tau^3}{2}|\psi\rangle \to G_{0^+}\left(1 + \frac{1}{E_{1^+} - E_{0^+}}\frac{1}{\Lambda_{\rm tot}}\right)\sigma_i\frac{\tau^3}{2}\,, \qquad (6.122)$$

where we have used the fact that

$$T_{0^+\to 1^+}(R \to \infty) = \langle 0^+|\tau_3|1^+\rangle = \langle 1^+|1^+\rangle = 1\,. \qquad (6.123)$$

The total moment of inertia in this limit is also dominated by the single transition term:

$$\Lambda_{\rm tot} \quad\to\quad \Lambda_q \;\to\; \frac{N_c}{2}\frac{1}{E_{1^+} - E_{0^+}}\,. \qquad (6.124)$$

Thus we have

$$g_A \to G_{0^+}\left(1 + \frac{2}{N_c}\right)\,. \qquad (6.125)$$

This is precisely the MIT bag result for general N_c, where the factor $1+2/N_c$ is the quark model value of Eq. (6.108).

Isoscalar axial-vector coupling constant $g_A^{(0)}$

This quantity is naively related to the spin fraction of the nucleon carried by the intrinsic quark spin (helicity), since $\bar{\psi}\gamma_i\gamma_5\psi \sim \psi^\dagger\sigma_i\psi$. Experimentally, the nucleon matrix element of the isoscalar axial-vector current is measured in the deep inelastic region [Ashman et al. 1988; Ashman et al. 1989; Adeva et al. 1993; Anthony et al. 1993; Abe et al. 1995]. Because of the axial $U_A(1)$ anomaly, the extrapolation of the measured quantity down to a low energy scale and its interpretation as the quark helicity has been doubted [Altarelli and Ross 1988]. A frequently quoted number is [Cheng 1996; Ji 1995]:

$$g_A^{(0)}(Q^2 \sim 10\ \mathrm{GeV}^2) = 0.31 \pm 0.07\,. \tag{6.126}$$

As usual, the isoscalar axial-vector current is the sum of the quark and meson parts. In the original chiral bag model, however, the pion current vanishes exactly, since it is not possible to create the right quantum numbers only out of pions with few derivatives **. One would attempt to introduce the isoscalar pseudo-scalar particle η_0 in order to allow for a non-zero contribution from the meson sector. Physically, this is identified with the mixed state of $\eta(547)$ and $\eta'(958)$ with the mixing angle ~ -20 degrees as being dominated by $\eta'(958)$ [PDG 2000]. In fact, $U_A(1)$ anomaly resolves the degeneracy of the η_0 and η_8 states, pushing the mass of η' up to $m_{\eta'} = 958$ MeV. In the chiral bag model, such a pseudo-scalar particle can be introduced in order to restore the $U_A(1)$ symmetry broken at the bag surface. The broken symmetry at the bag surface must be restored, since it is related to quark confinement and has nothing to do with the $U_A(1)$ anomaly.

In this spirit, $g_A^{(0)}$ was computed in the chiral bag model with η_0 †† [Rho et al. 1989; Hosaka and Weise 1989]. It turned out then that the effect of η_0 was small. In the effective theory, the main effect of the $U_A(1)$ anomaly is to generate the large mass of η'. Then, its dynamical role at low energy is in any way suppressed. Because of this relatively minor role of η_0, in the following discussions, we do not consider η_0 but just work out this problem in the original chiral bag model with only the pion [Dreiner et al. 1989].

The isoscalar axial-vector current, therefore, has only the quark component:

$$A_\mu = \frac{1}{2}\bar{\psi}\gamma_\mu\gamma_5\psi\theta_V\,. \tag{6.127}$$

The computation of the nucleon matrix element is performed in a way analogous to g_A of order Ω^1, where the leading order term of $g_A^{(0)}$ vanishes

**If we allow for higher derivative terms, it is possible to make a current with the right quantum numbers such as $\epsilon_{\mu\nu\alpha\beta}\partial_\nu B_\alpha B_\beta$. In Ref. [Cohen and Banerjee 1989] the role of such a higher order term was examined and was found to be small

††In the $SU(2)$ sector, there is no distinction between η_0 and η'.

in the hedgehog state: $\langle h|\sigma_i|h\rangle = 0$. Now $g_A^{(0)}$ is defined by

$$\begin{aligned}\frac{1}{2}\sigma_3\, g_A^{(0)} &= \int d^3x\, A_3(\boldsymbol{x}) \\ &= -\frac{N_c}{4}\sum_{mn}\left\{\frac{\langle m|A^\dagger\sigma_3 A|n\rangle\langle n|\tau^a|m\rangle}{E_n - E_m}\,\Omega_a \right. \\ &\qquad \left. +\, \Omega_a\,\frac{\langle m|\tau^a|n\rangle\langle nA^\dagger\sigma_3 A|m\rangle}{E_n - E_m}\right\} \\ &= -\frac{N_c}{2}\sum_{mn}\frac{\langle m|\sigma_3|n\rangle\langle n|\tau_3|m\rangle}{E_n - E_m}\,\Omega_3\,. \end{aligned} \tag{6.128}$$

Here on the left hand side σ_3 is the spin matrix acting on the nucleon state, and in deriving the third equation we have used the fact that the isospin rotation matrix A commutes with the (quark) spin matrix σ_i. Because $\Omega \sim N_c^{-1}$, the N_c dependence of $g_A^{(0)}$ is just N_c^0 as expected.

Now the right hand side is computed in a straightforward manner as described in Appendix E. Here we just show the numerical result in Fig. 6.13. As before, the chiral bag model interpolates the Skyrme model result of $g_A^{(0)} = 0$ and the MIT bag model result $g_A^{(0)} = 0.654$. The latter is essentially the valence quark model result but with the reduction factor through the relativistic (lower component) effect. Note that the R dependence shown here is similar to that of g_A of order Ω^1. If we compare the present results with the experimental data, we once again find that the experimental value of Eq. (6.126) is reproduced at the bag radius $R \sim 0.6$ fm.

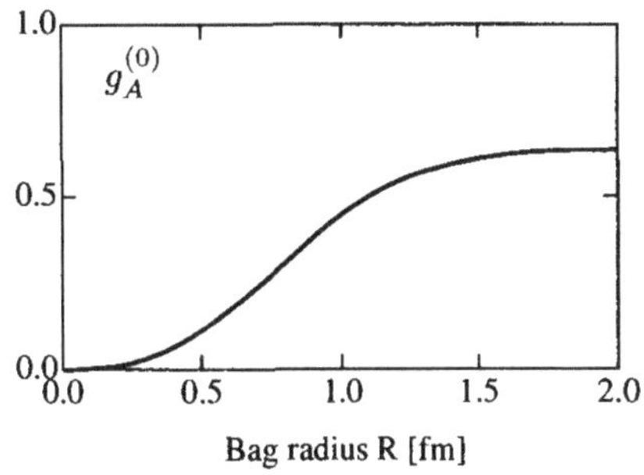

Fig. 6.13 The isoscalar axial-vector coupling constant $g_A^{(0)}$ as a function of R.

Cheshire cat violation at order Ω^1

A theoretical background of the strong R dependence of g_A and $g_A^{(0)}$ at order Ω^1 can be examined by looking at the relation between quark and meson theories. Specifically, for g_A, it turned out that one-to-one correspondence between the two theories appeared to be lost at order Ω^1, indicating an explicit violation of the Cheshire cat hypothesis at that order. A detailed investigation on this matter was performed by Wakamatsu using the chiral quark soliton model [Wakamatsu 1995]. We will discuss it briefly in this subsection.

Let us start with the lagrangian of the chiral quark soliton model [Diakonov et al. 1988; Diakonov et al. 1989; Wakamatsu and Yoshiki 1991; Watabe and Toki 1992; Meissner and Goeke 1991]:

$$\mathcal{L}_{\rm CQSM} = \bar{\psi}(i\partial_\mu \gamma^\mu - MU_5(x))\psi\,, \tag{6.129}$$

where $U_5(x) = \exp(i\boldsymbol{\tau}\cdot\boldsymbol{\phi}(x)\gamma_5/f_\pi)$ and M the mass of constituent quarks. There is a soliton solution with $B = 1$ (with three valence quarks) to leading order of Ω^0 by assuming the hedgehog ansatz $\boldsymbol{\phi}(x) = \hat{r}F(r)$. Now we compute the vacuum matrix element of a current Γ^μ,

$$\begin{aligned}\langle\Gamma^\mu\rangle &= \int d[\psi\bar{\psi}]\int d^4z\ \bar{\psi}(z)\Gamma^\mu\psi(z)\ \exp[\int d^4x\,\mathcal{L}_{\rm CQSM}] \\ &= -i\,{\rm Sp}\left[\frac{1}{i\partial_\mu\gamma^\mu - MU_5}\Gamma^\mu\right].\end{aligned} \tag{6.130}$$

Here Sp is the functional trace including traces over Dirac, isospin and color indices. The formal equivalence between the matrix element in the quark theory of Eq. (6.130) and that of the corresponding meson theory is established by showing that the current in Eq. (6.130) evaluated by a suitable derivative expansion coincides with the Noether current of the following effective lagrangian

$$e^{iS_{\rm eff}[U]} = \int d[\psi\bar{\psi}]\exp[\int d^4x\,\mathcal{L}_{\rm CQSM}]\,. \tag{6.131}$$

In the rigid quantization, the slowly rotating hedgehog ansatz, $U_5(x) = A(t)e^{\boldsymbol{\tau}\cdot\hat{r}\gamma_5 F(r)}A^\dagger(t)$, is substituted in Eq. (6.130) and we find

$$\langle\Gamma^\mu\rangle = -i\,{\rm Sp}\left[\frac{1}{i\partial_t - H + \Omega}\Gamma^\mu_A\right], \tag{6.132}$$

where

$$\begin{aligned} H &= -i\boldsymbol{\alpha}\cdot\boldsymbol{\nabla} + M\gamma_0 U_5(\boldsymbol{x})\,, \\ \Gamma^\mu_A &= A^\dagger \gamma_0 \Gamma^\mu A\,, \\ \Omega &= A^\dagger \dot{A} = \frac{i}{2}\boldsymbol{\tau}\cdot\boldsymbol{\Omega}\,. \end{aligned} \tag{6.133}$$

Usually, the trace of the temporal (time) part is replaced by the integral over the spectrum:

$$\langle \Gamma^\mu \rangle = -iT \int_{-\infty}^{+\infty} \frac{d\omega}{2\pi} \mathrm{Tr}\left[\frac{1}{\omega - H + \Omega}\Gamma^\mu_A\right]\,, \tag{6.134}$$

where T is an infinitely large time interval. The formula of Eq. (6.134) produces precisely the current evaluated in the corresponding meson theory, which would be the Skyrme model when the derivative expansion is suitably truncated.

Now we show that the formula (6.134) imposes a strong restriction on the form of currents when expanded in powers of Ω. Following Ref. [Wakamatsu 1995], we introduce the symmetry operation

$$\mathcal{T} = R_2(\pi)\, i\gamma^1\gamma^3 K\,, \tag{6.135}$$

where $R_2(\pi) = e^{i\pi\tau_2/2}$ is the 180 degree rotation along the isospin y-axis, and K is the complex conjugation operator. The transformation of Eq. (6.135) differs from the ordinary time reversal operation by the isospin rotation $R_2(\pi)$. Transformation properties under (6.135) are

$$\mathcal{T} H \mathcal{T}^{-1} = +H\,, \tag{6.136}$$

$$\mathcal{T} \Omega \mathcal{T}^{-1} = -\Omega\,. \tag{6.137}$$

Furthermore, relevant currents transform as

$$\mathcal{T}\gamma^\mu \mathcal{T}^{-1} = -\gamma^\mu\,, \tag{6.138}$$

$$\mathcal{T}\gamma^\mu\gamma_5\tau_a \mathcal{T}^{-1} = \gamma^\mu\gamma_5\tau_a\,. \tag{6.139}$$

These properties of Eqs. (6.136) – (6.139) establish the power series of the currents as

$$\langle\bar\psi\gamma^\mu\psi\rangle = \mathcal{O}(\Omega^1) + \mathcal{O}(\Omega^3) + \cdots (\Omega^{\text{odd power}})\,, \tag{6.140}$$

$$\langle\bar\psi\gamma^\mu\gamma_5\tau_a\psi\rangle = \mathcal{O}(\Omega^0) + \mathcal{O}(\Omega^2) + \cdots (\Omega^{\text{even power}})\,. \tag{6.141}$$

Therefore, one concludes, specifically, that the axial-vector coupling constant g_A is an even function of Ω, without the term of order Ω^1. This is in contradiction with what we have seen in previous subsections where the existence of the term of order Ω^1 has been emphasized. Where does this discrepancy come from?

The caveat stems from the formula (6.134). The crucial observation made in Ref. [Wakamatsu 1995] is that the usual temporal formula does not respect the non-commutative nature of the collective coordinate operators A and Ω. In a properly defined approach, the nucleon matrix element of the current Γ_μ is computed by the following time ordered functional integral:

$$\begin{aligned}\langle\Gamma_\mu\rangle &= \int d^3z_0 \int d[\psi\bar{\psi}] \exp\left[\int_{-T/2}^{T/2} d^4x\mathcal{L}_{\mathrm{CQSM}}\right] \\ &\times \quad J_N(\boldsymbol{x},T/2)\,\psi(z_0)^\dagger\Gamma_A^\mu(t_0)\psi(z_0)\,J_N(\boldsymbol{y},-T/2)\,, \qquad (6.142)\end{aligned}$$

where $z_0 = (t_0, \boldsymbol{z})$ and J_N is the interpolation operator for the nucleon written as N_c quarks. When evaluating this functional integral, time ordering of the collective coordinates in $\mathcal{L}_{\mathrm{CQSM}}$ of the exponent and those in Γ_A^μ must be carefully done as follows:

$$\begin{aligned}\langle\Gamma_\mu\rangle &= \int d^3z_0 \int d[\psi\bar{\psi}]\,J_N(\boldsymbol{x},T/2) \exp\left[\int_{t_0}^{T/2} d^4x\mathcal{L}_{\mathrm{CQSM}}\right] \qquad (6.143) \\ &\times \quad \psi(z_0)^\dagger\Gamma_A^\mu(t_0)\psi(z_0) \exp\left[\int_{-T/2}^{t_0} d^4x\mathcal{L}_{\mathrm{CQSM}}\right] J_N(\boldsymbol{y},-T/2)\end{aligned}$$

After a rather lengthy calculation, it was found that at order Ω^1 there is a term containing the factor $[D_{ai},\Omega_j]$, which precisely reduces to the result of the cranking formula in the previous sections. It is clear that the commutator $[D_{ai},\Omega_j]$ was set to zero in the naive formula (6.134) where the operator nature of these coordinates was ignored.

Incidentally, it was also verified that the term of order Ω^1 does not contradict any symmetries of strong interaction including, for example, the G-parity transformation. Such a question was raised in Refs. [Schechter and Weigel 1995a; Schechter and Weigel 1995b], where a somewhat unusual operator for particle–anti-particle conjugation was adopted, which led the authors to the erroneous conclusion.

Concluding this subsection, we emphasize that the term of order Ω^1 appears in a quark theory, while such a term disappears in the corresponding meson theory if the cranking method for the rigid mesonic soliton is applied. Thus the equivalence between the quark and meson theories appears to be explicitly violated at order Ω^1 in the rigid quantization scheme for the mesonic soliton. The correction of order Ω^1 to g_A is purely quark originated, the strength of which is then controlled by the bag radius R in the chiral bag model. Phenomenologically, the small g_A problem appears to be solved only when such a quark effect is substantial at the medium bag radius $R \sim 0.6$ fm.

6.4 Non-rigid Quantization of the Skyrmion

Coriolis coupling

So far we have sticked to the rigid quantization for the meson part. However, as it was demonstrated in section 6.1, there are quantum corrections when field fluctuations are taken into account appropriately. In particular, corrections from the Coriolis coupling are of interest (see Eq. 6.25).

Practically, however, such quantum corrections has not been considered seriously. One reason is rather technical; the leading order correction, the modification of the zero point energy due to the fluctuations, is rather difficult to compute. Another reason is that it is likely that (actually in a schematic calculation shown below) the effect is just a renormalization of of some physical quantities, which can be incorporated in effective theories by the adjustment of model parameters. In a renormalizable theory, it amounts to the renormalization of coupling constants. Here, this does not happen generally, since the effective theory is not usually renormalizable. There might be a criticism that in such a case it is not sensible to go beyond tree level calculations. Nevertheless, in this section we wish to demonstrate how computations beyond the rigid quantization are performed. Our method is similar to the one for the particle-rotor model where physical picture is best described [Bohr and Mottelson 1975]. We especially focus on the role of the Coriolis coupling, but the recoil term is not considered.

Our starting point is a general non-rigid ansatz for the skyrmion field with the rotational motion introduced separately:

$$U(t, \boldsymbol{x}) = A(t)u(t, \boldsymbol{x})A^\dagger(t)\,. \tag{6.144}$$

Substituting this expression in the Skyrme model lagrangian of Eq. (5.76), we obtain

$$\mathcal{L}_{\rm Sk} = \int d^3x \left(-\frac{f_\pi^2}{4}\,{\rm tr}\,(l_\mu l^\mu) + \frac{1}{32e^2}\,{\rm tr}\,[l_\mu,\, l_\nu]^2\right) - \frac{1}{2}\boldsymbol{\Omega}\cdot\boldsymbol{Q} + \frac{\Lambda}{2}\Omega^2\,, \qquad (6.145)$$

where $l_\mu = u^\dagger \partial_\mu u$. (In the equation below we use also $r_\mu = u\partial_\mu u^\dagger$.) In Eq. (6.145), $\boldsymbol{Q}$ is the isovector charge

$$\boldsymbol{Q} = \int d^3x\,{\rm tr}\left[i\boldsymbol{\tau}\left(-\frac{f_\pi^2}{2}l_0 + \frac{1}{8e^2}[l_\nu,\,[l_0,\,l_\nu]] + (l\to r)\right)\right]\,, \qquad (6.146)$$

and the moment of inertia Λ is given in Eq. (6.50) with the lower bound $R \to 0$. It should be noted that by allowing the time dependence for the intrinsic field u, there appears the Coriolis coupling of the form $\boldsymbol{\Omega}\cdot\boldsymbol{Q}$. This is the driving force for the higher order corrections.

Masses and g_A

In order to compute physical quantities, we specify the quantum states of the total angular momentum $\boldsymbol{K}$ and the parity. Two relevant states are the 0^+ hedgehog ground state denoted by $|0\rangle$ and the 1^+ excited states denoted by $|K=1\ K_z;n\rangle$, where n specifies quantum numbers other than $K\ K_z$. Furthermore, we assume that the expectation value of the energy in the $|0\rangle$ state gives the mass of the static hedgehog ‡‡

$$M_H = \langle 0|\int d^3x\left(-\frac{f_\pi^2}{4}\,{\rm tr}\,(l_i\, l^i) - \frac{1}{32e^2}\,{\rm tr}\,[l_i,\, l_j]^2\right)|0\rangle\,. \qquad (6.147)$$

In the rigid rotor approximation, we get the standard formula for the energy of the rotating hedgehog as

$$M_\Omega = M_H + \frac{\Lambda}{2}\Omega^2\,. \qquad (6.148)$$

Influenced by the Coriolis coupling term in (6.145), an improved solution $|\widetilde{0}\rangle$ in the laboratory (rest) frame is given by

$$|\widetilde{0}\rangle = A|0\rangle - \frac{1}{2}\sum_{a,n}\frac{A|1K_z;n\rangle\langle 1K_z;n|\boldsymbol{Q}\cdot\boldsymbol{\Omega}|0\rangle}{E_1 - E_0}\,. \qquad (6.149)$$

‡‡In the collective vibration model which will be used in what follows, the zero point energy must be subtracted in order to establish Eq. (6.147).

Note that, although it is not shown explicitly, the energy E_1 may depend on n but not on K_z. Now the mass of the rotating hedgehog is modified to be

$$M_\Omega = M_H + \frac{\Lambda_{\rm tot}}{2}\Omega^2\,, \tag{6.150}$$

where $\Lambda_{\rm tot} = \Lambda + \delta\Lambda$ with the correction to the moment of inertia given by

$$\delta\Lambda = \frac{1}{2}\sum_n \frac{|\langle 1\,3; n|Q_3|0\rangle|^2}{E_1 - E_0}\,. \tag{6.151}$$

Here we have used the fact that Q_3 excites only the $K_z = 3$ component of $|1K_z\rangle$. The deformation of the intrinsic state due to the Coriolis term gives an extra term for the moment of inertia. The masses of physical baryons are then obtained by quantizing the angular velocity by the rule $\boldsymbol{\Omega} \to \boldsymbol{\sigma}/2\Lambda_{\rm tot}$ and $\Omega^2 \to J(J+1)/4\Lambda_{\rm tot}^2$.

Now we derive an expression for the axial vector coupling constant g_A. For this, we follow the procedure in section 6.3. It is defined by

$$\frac{2}{3}\sigma_i\tau^a g_A = \langle N|\int d^3x\, A_i^a|N\rangle\,. \tag{6.152}$$

For $a = i = 3$, the matrix element for the $|\tilde{0}\rangle$ state is given as

$$\begin{aligned}\frac{2}{3}\sigma_i\tau^a g_A &= \langle\tilde{0}|\int d^3x\, A_3^3|\tilde{0}\rangle \\ &= \int d^3x\,\Big\{\langle 0|A_3^3|0\rangle D_{33} \\ &\quad - \frac{1}{2}\sum_{a,n}\frac{\langle 0|A_3^b|1a;n\rangle\langle 1a;n|Q_a|0\rangle}{E_1 - E_0} D_{3b}\Omega_a \\ &\quad - \frac{1}{2}\sum_{a,n}\frac{\langle 0|Q_a|1a;n\rangle\langle 1a;n|A_3^b|0\rangle}{E_1 - E_0}\Omega_a D_{3b}\Big\}\,. \end{aligned} \tag{6.153}$$

Here we have used the identity $A^\dagger A_i^a A = D_{ab}A_i^b$ with $D_{ab} = \frac{1}{2}\,{\rm tr}\,[\tau_a A\tau_b A^\dagger]$ and preserved the ordering of the operators Ω_a and D_{3b} as before.

Collective model

For actual computations, we describe the quantum states $|KK_Z\rangle$ as collective vibrational modes. This is not strictly an exact method but still provides a simple way to estimate the new terms. Consider three modes of the 1^+ vibration, which are introduced through the change in the coordinate as

$$x_i \to X_i = \sum_j T_{ij}^{(m)} x_j , \tag{6.154}$$

where the matrix $T_{ij}^{(m)}$ is for the m-th vibration and is given by

$$T_{ij}^{(m)} = \frac{1}{(1+q_m^2)^{1/3}} \left(\delta_{ij} + q_m \epsilon_{mij}\right) . \tag{6.155}$$

Here the quanta $q_m = q_m(t)$ are time dependent dynamical variables. The factor $1/(1+q_m^2)^{1/3}$ is introduced so that the transformation conserves the volume. We ignore possible nodal excitations of the 1^+ state, since they are in the higher energy region. The fact that Eqs. (6.154) and (6.155) give $K = 1$ excitations is verified from the following decomposition:

$$\begin{aligned} u(X_i) &= \exp i[\boldsymbol{\tau} \cdot \hat{X} F(|\boldsymbol{X}|)] \\ &\sim \exp i[\boldsymbol{\tau} \cdot \hat{x} F(r) + (\boldsymbol{\tau} \times \hat{x})_m h(r)] , \end{aligned} \tag{6.156}$$

where $F(r)$ is the chiral angle function for the static hedgehog solution and $h(r)$ is a radial function written in terms of F.

We substitute Eq. (6.154) in the lagrangian (6.145) including all the m-th excitations and expand it up to the second order in $q_m(t)$ for the harmonic approximation. We find

$$\mathcal{L}_{\rm Sk} = \frac{\Lambda}{2}\Omega^2 - \frac{1}{2}\Omega \cdot \boldsymbol{Q} + \sum_m \left(\frac{\mu}{2}\dot{q}_m^2 - M_H q_m^2\right) - M_H . \tag{6.157}$$

Here the mass parameter for the collective vibration is given by

$$\mu = \int d^3x\, r^2 \left[\frac{2f_\pi^2}{3}\frac{s^2}{r^2} + \frac{1}{e^2}\frac{s^2}{r^2}\left(F'^2 + \frac{s^2}{r^2}\right)\right] , \tag{6.158}$$

where $s = \sin F$. The vibrational motion is now quantized as a harmonic oscillator with the oscillator parameter ω given by

$$\omega = \sqrt{\frac{2M_H}{\mu}} . \tag{6.159}$$

The 0^+ and 1^+ states are then given as $n_q = 0$ and $n_q = 1$ states, respectively, where n_q is the number of the vibration quanta.

In the harmonic approximation, the isospin charge and the axial-vector current are given as

$$\begin{aligned} Q_a &= \tau_a - \mu \sum_m \dot{q}_m \ \mathrm{tr}\, A^\dagger \tau_a A \tau_m \,, \\ &\to \tau_a - 2\mu \dot{q}_a \,, \end{aligned} \tag{6.160}$$

$$\int d^3x A_i^a = G \sum_j D_{aj} (\delta_{ji} + \sum_m q_m \epsilon_{mji}) \,, \tag{6.161}$$

where G is the radial integral of the axial density given by

$$G = \int d^3x \left\{ \frac{f_\pi^2}{3} \left(F' + \frac{2cs}{r} \right) + \frac{2}{3e^2} \left(\frac{cs}{r} F'' + \frac{s^2}{r^2} F' + \frac{cs^2}{r^3} \right) \right\} \,, \tag{6.162}$$

with $s = \sin F$ and $c = \cos F$. In Eq. (6.160), the second equation follows by setting $A = 1$, which is valid in the present collective vibration model where $\dot{q}_m$ is the relevant transition operator for $|0\rangle \to |1\rangle$. From Eq. (6.160) we find

$$\langle 1a | Q_a | 0 \rangle = -2\mu \langle 1 | \dot{a} | 0 \rangle = -i\sqrt{2\mu\omega} \,, \tag{6.163}$$

where we have used the elementary formula of harmonic oscillation.

The correction to the moment of inertia is now given by

$$\delta\Lambda = \frac{1}{2} \frac{|\langle 1\, 3 | Q_3 | 0 \rangle|^2}{\omega} = \mu \,. \tag{6.164}$$

Here the sum in the cranking formula (6.151) reduces only to a single term in the collective vibration model when nodal excitations are ignored. The total moment of inertia is thus

$$\begin{aligned} \Lambda_{\mathrm{tot}} &= \Lambda + \delta\Lambda \\ &= \frac{53.3}{e^3 f_\pi} + \frac{46.2}{e^3 f_\pi} = \frac{99.5}{e^3 f_\pi} \,, \end{aligned} \tag{6.165}$$

where we have scaled the moment of inertia by introducing the dimensionless quantity $y = e f_\pi r$, and have given the resulting parameter independent numbers of the integral explicitly [Adkins et al. 1983; Zahed and Brown 1986]. We note that the correction to the moment of inertia is of the same order in $1/N_c$ as that of the leading term. Incidentally, the oscillator parameter is factored out by $e f_\pi$ and is given as

$$\omega = 1.78\, e f_\pi \,. \tag{6.166}$$

The axial-vector coupling constant g_A is computed as follows. Substituting Eq. (6.161) in Eq. (6.153), one obtains

$$\begin{aligned}\langle\tilde{0}|\int d^3x\, A_3^3|\tilde{0}\rangle &= G\{D_{33} \\ &- \frac{1}{2}\sum_{a,b}\frac{\langle 0|q_a|1a\rangle\langle 1a|Q_a|0\rangle}{E_1-E_0}\epsilon_{a3b}D_{3b}\Omega_a \\ &- \frac{1}{2}\sum_{a,b}\frac{\langle 0|Q_a|1a\rangle\langle 1a|q_a|0\rangle}{E_1-E_0}\epsilon_{a3b}\Omega_a D_{3b}\Big\}\,. \quad (6.167)\end{aligned}$$

Here we have used the trivial relations such as $\langle 0|q_m|0\rangle = \langle 0|1\rangle = 0$. Using the matrix elements $\langle 0|q|1\rangle = \sqrt{1/2\mu\omega}$ and $\langle 1a|Q_a|0\rangle = -i\sqrt{2\mu\omega}$, one finds

$$\begin{aligned}\langle\tilde{0}|\int d^3x A_3^3|\tilde{0}\rangle &= G\left(D_{33} - \frac{i}{\omega}\epsilon_{3ba}D_{3b}\Omega_a\right) \\ &\to -\frac{G}{3}\left(1+\frac{1}{\omega\Lambda_{\rm tot}}\right)\sigma_3\tau_3\,, \quad (6.168)\end{aligned}$$

and so

$$g_A = -\frac{G}{2}\left(1+\frac{1}{\omega\Lambda_{\rm tot}}\right)\,. \quad (6.169)$$

Note that $\omega \sim \mathcal{O}(N_C^0)$ from Eq. (6.166), $\Lambda_{\rm tot} \sim \mathcal{O}(N_C^1)$ and so the second term of Eq. (6.169) is of order N_C^{-1}.

It is possible to estimate the order of the correction term of g_A from experiments. The oscillator parameter ω is interpreted as the mass splitting of the $(I,J) = (1/2,3/2)$ and $(I,J) = (3/2,1/2)$ baryons and the nucleon. The low lying states of such baryons are $P_{13}(1720)$ and $P_{31}(1750)$, although the latter is a single star resonance [PDG 2000]. Thus $\omega \sim 0.8$ GeV ~ 4 fm^{-1}. The moment of inertia may be determined by the nucleon-delta mass splitting and is about 1 fm. The correction term would then be $1/\omega\Lambda_{\rm tot} \sim 0.25 = 25$ %. Thus the higher order correction to g_A would have a significant effect also in the Skyrme model as well as in chiral models with explicit quarks.

One can provide some concrete numbers by fixing the parameters f_π and e suitably. For instance, these parameters are chosen so as to reproduce the masses of the nucleon and delta after including the vibrational effect, i.e., as in Eq. (6.165). This yields $f_\pi = 75.4$ MeV and $e = 6.36$. Using them, we find $\Lambda_{\rm tot} = 99.6/e^3 f_\pi = 1.01$ fm^{-1} and $\omega = 1.78 e f_\pi = 4.33$ fm. The value of ω is close to the one determined from experiments as in the above. Thus the next to leading order term of g_A amounts to $1/\omega\Lambda_{\rm tot} = 0.23 = 23$ %. Unfortunately, however, the total value of g_A turns out to be still too small:

$$\begin{aligned} g_A(0) &= 0.448, \quad g_A(1) = 0.103 \\ g_A(\text{tot}) &= g_A(0) + g_A(1) = 0.551\,. \end{aligned} \tag{6.170}$$

It is not possible to get the experimental value $g_A = 1.25$ by varying the two parameters, f_π and e, with keeping other physical quantities in reasonable agreement with experiments.

From the examples shown in this section, the non-rigid effects are in general significant. Practically, however, the correction terms do not always improve the theoretical predictions.

6.5 Electromagnetic Properties

Currents charges and magnetic moments

In this section, we discuss electromagnetic properties of the nucleon in detail. Let us first briefly summarize the definitions of currents, charges and magnetic moments. The nucleon electromagnetic current, $J_I^\mu = (J_I^0, \boldsymbol{J}_I)$, has isoscalar ($I = S$) and isovector ($I = V$) components. The electric charges Q_I and the magnetic moments μ_I are defined by

$$\begin{aligned} Q_I &= \int d^3x J_I^0\,, \\ \mu_I \frac{\boldsymbol{\sigma}}{2M_N} &= \frac{1}{2}\int d^3x\, \boldsymbol{x} \times \boldsymbol{J}_I \;\; (= \boldsymbol{\mu})\,. \end{aligned} \tag{6.171}$$

Magnetic moments are measured in units of nuclear magneton $e/2M_N$ where $M_N = 938$ MeV. Q_I and μ_I are related to those of the proton and neutron, $Q_{p,n}$ and $\mu_{p,n}$, by

$$\begin{aligned} Q_p &= Q_S + Q_V = 1\,, \\ Q_n &= Q_S - Q_V = 0\,, \\ \mu_p &= \mu_S + \mu_V = 2.79\,, \\ \mu_n &= \mu_S - \mu_V = -1.91\,, \end{aligned} \tag{6.172}$$

where we have given empirical values. Equivalently,

$$Q_S = Q_V = \frac{1}{2}, \quad \mu_S = 0.44, \quad \mu_V = 2.35\,. \tag{6.173}$$

The currents J_I^μ are given as the symmetry currents of $U_V(1)$ for the isoscalar and $SU_V(2)$ for the isovector parts. The isovector current and the quark part of the isoscalar current are just the Noether currents of the corresponding symmetries, while the pion part of the isoscalar current emerges as a topological current. Explicitly, they are

$$J_S^\mu = \frac{1}{2N_c}\bar{\psi}\gamma^\mu\psi\theta_V + \frac{1}{2}B^\mu\theta_{\bar{V}} \tag{6.174}$$

$$\begin{aligned} \boldsymbol{J}_V^\mu &= \frac{1}{2}\bar{\psi}\gamma^\mu\boldsymbol{\tau}\psi\theta_V \qquad (6.175)\\ &+ \frac{1}{2}\,\mathrm{tr}\left[i\boldsymbol{\tau}\left(-\frac{f_\pi^2}{2}L^\mu + \frac{1}{8e^2}[L^\nu,[L^\mu,L^\nu]] + (L \to R)\right)\right]\theta_{\bar{V}}\,, \end{aligned}$$

where the baryonic current B^μ is given in Eq. (5.71), and the left and right currents are defined by

$$L_\mu = U^\dagger\partial_\mu U\,, \quad R_\mu = U\partial_\mu U^\dagger\,. \tag{6.176}$$

Electric charges

The isoscalar charge is just the baryon number (divided by two), and its properties have been discussed in Chapter 5. The total isoscalar charge is normalized rigorously to be one-half independent of the size of the bag:

$$Q_S = \frac{1}{2}(B_q + B_\pi) = \frac{1}{2}\,. \tag{6.177}$$

The total isovector charge is calculated using the rotating wave functions given in Eqs. (6.46) and (6.62). Note that the isovector charge is induced by rotation and is of order Ω^1. This is compared with the isoscalar charge, which exists at the classical level and is of order Ω^0. The isovector charge is related to the moment of inertia:

$$\boldsymbol{Q}_V = \Lambda_{\mathrm{tot}}\boldsymbol{\Omega}'\,. \tag{6.178}$$

Here Ω' is defined by

$$\Omega' = i\,\mathrm{tr}\,(\boldsymbol{\tau}\dot{A}A^{\dagger})\,, \tag{6.179}$$

which is replaced by the nucleon isospin operator upon quantization (compare with Eq. (6.67) and see also Appendix C),

$$\Omega' \to \boldsymbol{\tau}/2\Lambda_{\mathrm{tot}}\,. \tag{6.180}$$

Therefore, we find

$$\boldsymbol{Q}_V = \frac{\boldsymbol{\tau}}{2}\,, \tag{6.181}$$

which establishes the exact normalization of the isovector charge.

The fractioning of the isovector charge into the quark and pion sectors follows that of the moment of inertia. In Fig. 6.14 we show the fractioning of the isoscalar and isovector charges. It is interesting to see their difference; in particular, the isovector charge is dominated by the pion contribution up to about $R \lesssim 0.8$ fm. For instance, at $R \sim 0.5$ fm, the isoscalar charge is almost equally fractionized into the quark and pion sectors, while almost 80 % of the isovector charge is provided by the pion. This seems to be a general feature of the chiral bag model, and differs from the assumption made in Ref. [Brown et al. 1986], where the fractioning of the isovector charge was postulated to follow that of the isoscalar charge.

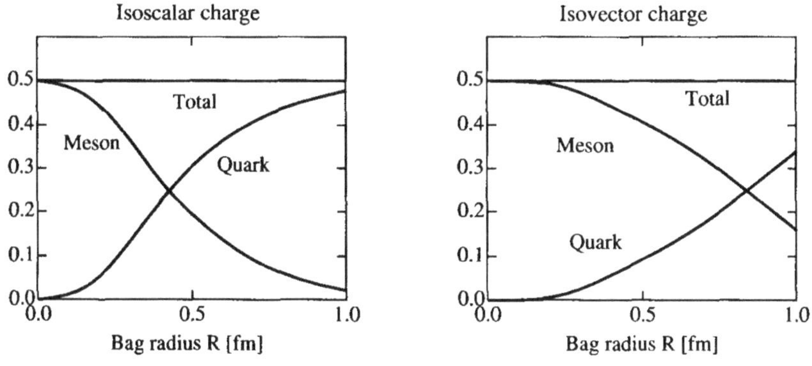

Fig. 6.14 Fractioning of isoscalar and isovector charges as functions of R.

Isoscalar magnetic moment

Spatial components of the isoscalar current J_S^i vanishes for the static hedgehog. This is because the pionic current B^i contains a time derivative and the quark current $\bar{\psi}\gamma^i\psi$ contains the vanishing matrix element of $\langle h|\sigma_i|h\rangle = 0$. Thus non-vanishing terms emerge when the hedgehog begins to rotate, and are of order Ω^1.

The pion contribution is computed by

$$\mu_S^\pi \frac{\boldsymbol{\sigma}}{2M_N} = \frac{1}{4}\int_{r\geq R} d^3x\, \boldsymbol{x} \times \boldsymbol{B}\,, \tag{6.182}$$

where the spatial part of the baryon current $\boldsymbol{B}$ is

$$\boldsymbol{B} = \frac{1}{2\pi^2}\frac{\sin^2 F}{r}F'\,\hat{x} \times \boldsymbol{\Omega}\,. \tag{6.183}$$

After quantizing the angular velocity $\boldsymbol{\Omega} \to \boldsymbol{\sigma}/2\Lambda_{\text{tot}}$, we find

$$\mu_S^\pi = \frac{M_N}{3\pi\Lambda_{\text{tot}}}\int_R^\infty r^2 dr \sin^2 F F'\,. \tag{6.184}$$

The quark part is computed by substituting the rotating quark wave function of Eq. (6.62) in Eq. (6.174):

$$\begin{aligned}
\mu_S^q \frac{\boldsymbol{\sigma}}{2M_N} &= \frac{1}{4N_c}\int_{r\leq R} d^3x\, \boldsymbol{x} \times \bar{\psi}\boldsymbol{\gamma}\psi \\
&= \frac{1}{4N_c}\sum_{mn}\frac{\langle m|\boldsymbol{Q}\cdot\boldsymbol{\Omega}|n\rangle\langle n|\boldsymbol{x}\times\boldsymbol{\alpha}|m\rangle}{E_n - E_m} \\
&= \frac{-i\sqrt{2}\,\boldsymbol{\sigma}}{24\Lambda_{\text{tot}}}\sum_{\substack{K,K'\geq 0\\ |K-K'|\leq 1}}\frac{\langle K||\hat{m}||\widetilde{K'}\rangle\langle\widetilde{K'}||\tau||K\rangle}{\tilde{E}_{K'} - E_K}e^{-\eta^2(\tilde{E}_{K'}-E_K)^2}\,.
\end{aligned} \tag{6.185}$$

Here $\hat{m} \equiv [\boldsymbol{x}\ \boldsymbol{\alpha}]^1$ is the isoscalar magnetic moment operator of rank one. Angular integrations for those matrix elements are carried out as presented in Appendix F. In the last expression of Eq. (6.185), the trivial sum over N_c quarks has been included. The radial integral in the matrix element $\langle K'||\hat{m}||K\rangle$ can not be carried out analytically. Numerical integrations have been performed using the gaussian integration for moments [Abramowitz and Stegun 1970]. Since the magnetic moment density rises as proportional to r^1 near the origin, combining with the integral weight of r^3 for magnetic moments, the formula for the fourth moment can efficiently be adopted.

As before, we define the dimensionless and normalized quantity:

$$m(F;\eta) \equiv \frac{1}{R^2}\frac{-i\sqrt{2}}{6} \sum_{K,K'\geq 0} \{\cdots\}, \tag{6.186}$$

where $\sigma/4\Lambda_{\rm tot}$ in Eq. (6.185) has been removed. The computation of the double mode sum can be performed in the same manner as for Λ_q. The quark contribution μ_S^q is then related to m by

$$\mu_S^q = \frac{R^2 M_N}{2\Lambda_{\rm tot}} m\,. \tag{6.187}$$

Again the double mode sum of Eq. (6.185) is separated into the valence and vacuum parts:

$$m(F;\eta) = m_{\rm vac}(F;\eta) + \theta(\tfrac{\pi}{2} - F)\, m_{\rm val}(F;\eta)\,. \tag{6.188}$$

It turns out that both parts are absolutely convergent [Jackson et al. 1987]. To see this, we show in Fig. 6.15 (a) the η dependence of the vacuum part $m_{\rm vac}(F;\eta)$ at $F = 0.1\pi$ and 0.2π. Calculated values (solid line) are well fitted by a regular polynomial function with a finite number of terms. Physical values are then obtained simply by taking the limit $\eta \to 0$:

$$m_{\rm vac,\ val}(F) \equiv \lim_{\eta\to 0} m_{\rm vac,\ val}(F;\eta)\,. \tag{6.189}$$

The vacuum part $m_{\rm vac}(F)$ is an odd function of F, and behaves as $\mathcal{O}(F^3)$, when $F \to 0$, or equivalently, $\mathcal{O}((F-\pi)^3)$ when $F \to \pi$. This F^3 behavior is again expected from the Skyrme model and is consistent with the smoothness condition in the limit $R \to 0$. In Fig. 6.15 (b), we show $m(F) \equiv m_{\rm vac}(F) + m_{\rm val}(F)$ as a function of F, where the vacuum and the valence contributions are shown separately.

The isoscalar magnetic moment, $\mu_S = \mu_S^q + \mu_S^\pi$, is plotted in Fig. 6.16 as a function of R together with its various components. The total value is only weakly dependent on R, in particular, at small bag radii $R \lesssim 0.4$ fm. The calculated values are about 0.3 in the entire region of $0 \leq R \lesssim 1$ fm, which is small as compared with the experimental value, $\mu_S(\exp) = 0.44$. The discrepancy here, however, is not very important for the physical magnetic moments of the proton and neutron, since they are dominated by the isovector component. This is explained by the $1/N_c$ counting for magnetic moments. From Eqs. (6.184) and (6.185), we see that the isoscalar

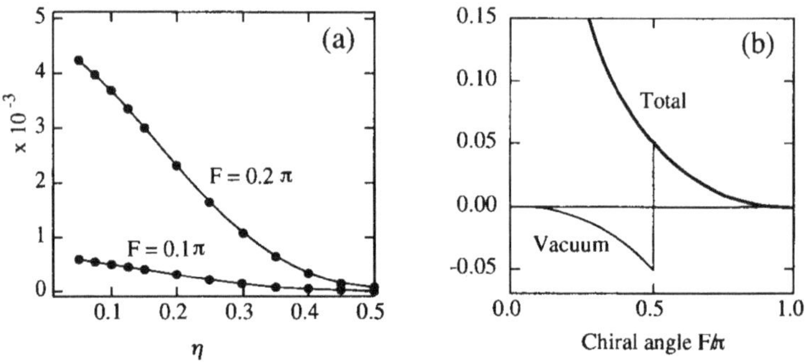

Fig. 6.15 (a) The η dependence of the double mode sum for the quark part of the isoscalar magnetic moment, and (b) the normalized quark contribution to the isoscalar magnetic moment as a function of F. For $F \leq \pi/2$ the vacuum contribution is separately shown.

magnetic moment is of order one. On the other hand, the isovector magnetic moment is of order N_c^2 (see the next subsection). As for the fractioning into the quark and meson sectors, the meson contribution dominates up to the bag radius $R \lesssim 0.6$ fm.

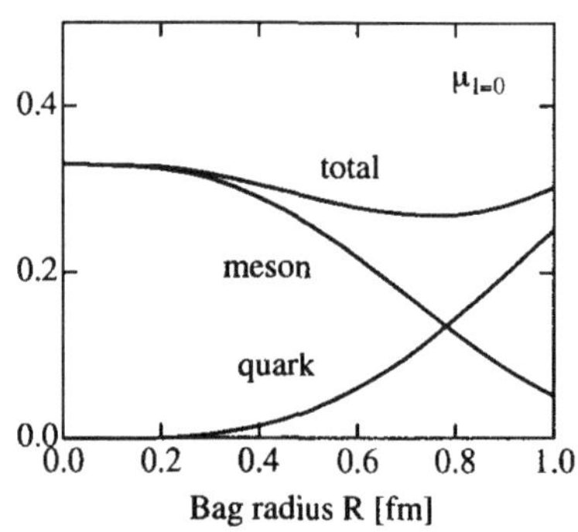

Fig. 6.16 The isoscalar magnetic moment as a function of R. The quark and meson contributions are shown separately.

Isovector magnetic moment

Zeroth order

The spatial components of the isovector current $\boldsymbol{J}_V^a$ has the spin-isospin structure $\boldsymbol{\sigma}\tau_a$, which has a non-vanishing matrix element for the hedgehog state $|h\rangle$ (at the classical level) and is a quantity of order Ω^0.

The pion contribution is computed from

$$\mu_V^\pi \frac{\boldsymbol{\sigma}\tau^a}{2M_N} = \frac{1}{4}\int_{r\geq R} d^3x\, \boldsymbol{x}\times \boldsymbol{J}_V^a\,, \tag{6.190}$$

where the spatial part of the isovector current $\boldsymbol{J}_V^a$ is given in Eq. (6.175). After performing the angular integral for the $(a,i)=(3,3)$ component, we obtain

$$\mu_V^\pi \frac{\sigma_3\tau_3}{2M_N} = \frac{4\pi f_\pi^2}{9}\int_R^\infty r^2 dr\left[1+\frac{1}{e^2 f_\pi^2}\left(F'^2+\frac{\sin^2 F}{r^2}\right)\right](3D_{33})\,, \tag{6.191}$$

where the rotation matrix D_{ab} is given by Eq. (6.44). Here we see that μ_V^π is the quantity of order N_c^2. Similarly, we will see that the quark contribution is also of order N_c^2.

The quark contribution is computed from

$$\mu_V^q \frac{\sigma_3\tau_3}{2M_N} = \frac{1}{4}\int_{r>R} d^3x\, \langle\bar{\psi}|(\boldsymbol{x}\times\boldsymbol{\gamma})_3\tau_3|\psi\rangle\,, \tag{6.192}$$

where the quark state $|\psi\rangle$ is the rotating hedgehog of Eq. (6.62). The leading order term of this matrix element is the single mode sum:

$$\begin{aligned}\mu_V^q \frac{\sigma_3\tau_3}{2M_N} &= -\frac{1}{8}\lim_{\eta\to 0}\sum_n \mathrm{sgn}(E_n)\int_0^R d^3x\, \bar{\psi}_n(\boldsymbol{x}\times\boldsymbol{\gamma})_3\tau_a\psi_n\, e^{-\eta^2|E_n|^2}\, D_{3a}\\ &+\ \theta(\pi/2-F)\frac{1}{4}\int_0^R d^3x\bar{\psi}_{0^+}(\boldsymbol{x}\times\boldsymbol{\gamma})_3\tau_a\psi_{0^+}\, D_{3a}\,,\end{aligned} \tag{6.193}$$

where the valence contribution is separately shown. The summation here includes the sum over N_c and therefore, the right hand side is of order N_c. Thus, we verify that the isovector magnetic moment is of order N_c^2, when multiplied by the mass of the nucleon $M_N \sim \mathcal{O}(N_c)$. In the vacuum term, the gaussian regulator is adopted for rapid convergence. After performing angle integration, we find the result as given in Appendix F. The remaining radial integral over r has been performed numerically by using the gaussian integration for moments.

The mode sum of Eq. (6.193) has been studied by the Stony Brook group using the Strutinsky's smearing method [Jackson et al. 1987]. It is also possible to perform the discrete mode sum directly. We define the normalized quantity

$$\begin{aligned} m(F;\eta) &= \frac{1}{R}\int_0^R d^3x\,\langle\bar{\psi}|(\boldsymbol{x}\times\boldsymbol{\gamma})_3\tau_3|\psi\rangle \\ &= -\frac{N_c}{2R}\lim_{\eta\to 0}\sum_n \mathrm{sgn}(E_n)\int_0^R d^3x\,\bar{\psi}_n(\boldsymbol{x}\times\boldsymbol{\gamma})_3\tau_3\psi_n\,e^{-\eta^2|E_n|^2} \\ &\quad + \frac{N_c}{R}\theta(\pi/2-F)\int_0^R d^3x\bar{\psi}_{0^+}(\boldsymbol{x}\times\boldsymbol{\gamma})_3\tau_3\psi_{0^+} \qquad (6.194) \\ &\equiv m_0(F;\eta)+m_{\mathrm{val}}(F;\eta)\,. \end{aligned}$$

Here m_0 and m_{val} are the vacuum and valence contributions, and the trivial sum over N_c has been performed. The quark contribution μ_V^q is then related to m by $\mu_V = (M_N R/6)m$.

It turns out that the vacuum term contains linear divergences ($\sim 1/\eta$) which have the $\sin^2 F$ dependence. Thus they can be subtracted by imposing a condition analogous to that for the moment of inertia in Eq. (6.77):

$$m_{\mathrm{vac}}(F) \equiv \lim_{\eta\to 0}\left[m_0(F;\eta) - \frac{1}{2}\sin^2 F\frac{\partial^2}{\partial F^2}m_0(0;\eta)\right]. \qquad (6.195)$$

The divergent behavior of the coefficient of the $\sin^2 F$ term, $c_{\mathrm{div}}(\eta) \equiv (\partial^2/\partial F^2)m_0(0;\eta)$ is shown in Fig. 6.17 (a), where the calculated values are fitted to the function

$$c_{\mathrm{div}}(\eta) = \frac{c_{-1}}{\eta^1} + c_{\ln}\ln\eta + c_0 + \eta c_1 + \cdots. \qquad (6.196)$$

The subtracted finite term of Eq. (6.195) behaves as $\mathcal{O}(F^4)$, when $F\to 0$ or as $\mathcal{O}((\pi-F)^4)$ when $F\to\pi$. Once again, this behavior of $m_{\mathrm{vac}}(F)$ is expected from the Skyrme model and is consistent with the smoothness condition in the limit $R\to 0$. The total value $m(F) = m_{\mathrm{vac}}(F) + \theta(F - \pi/2)\,m_{\mathrm{val}}(F)$ (per quark, $m(F)\to m(F)/N_c$) is plotted in Fig.6.17 (b) as a function of F.

The result for the leading term of the isovector magnetic moment, $\mu_V = \mu_V^q + \mu_V^\pi$, is shown in Fig. 6.18. The total value is almost independent of the bag radius R within $0 \le R \lesssim 1$ fm. The calculated values of $\sim 1.8 - 1.9$ are, however, small as compared with the experimental value of $\mu_V(\mathrm{exp})$

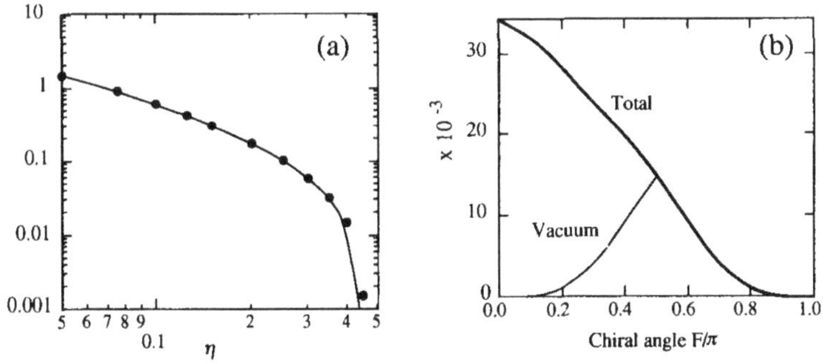

Fig. 6.17 (a) The divergent coefficient of the $\sin^2 F$ term of the normalized quark contribution to the isovector magnetic moment, $m_0(F;\eta)$, in the leading order Ω^0, and (b) The normalized quark contribution to the isovector magnetic moment of leading order Ω^0. For $F \leq \pi/2$ the vacuum contribution is separately shown.

$= 2.35$. The small isovector magnetic moment at the zeroth order is a general problem, like for g_A, in chiral models. As shown in the figure, for $R \gtrsim 1.5$ fm calculated value increases as proportional to R^1, which is just a dimensional consequence of the magnetic moment. In the large R region, the linear slope of $\mu_V(R)$ is of more interest. To leading order, it does not approach the MIT bag value but is only 3/5 of that, just as for g_A. This problem is solved by including the next to leading order term ($\sim \Omega^1$).

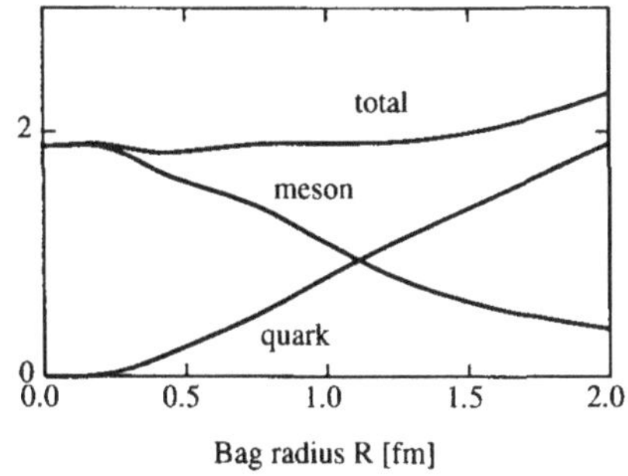

Fig. 6.18 The isovector magnetic moment in the leading order Ω^0 as a function of R. The quark and meson contributions are shown separately.

Next to leading order

Let us compute the next to leading order term of

$$\int d^3x\bar{\psi}(\boldsymbol{x}\times\boldsymbol{\gamma})_3\tau_3\psi = \langle\psi|(\boldsymbol{x}\times\boldsymbol{\alpha})_3\tau_3|\psi\rangle \equiv \langle\psi|\Gamma_3\tau_3|\psi\rangle\,, \tag{6.197}$$

$$\begin{aligned}\langle\psi|\Gamma_3\tau_3|\psi\rangle &= -\frac{1}{2}\sum_{mn}\left\{\frac{\langle m|A^\dagger\Gamma_3\tau_3A|n\rangle\langle n|\tau_a|m\rangle}{E_n-E_m}\Omega_a\right.\\ &\quad\left.+\,\Omega_a\frac{\langle m|\tau_a|n\rangle\langle n|A^\dagger\Gamma_3\tau_3A|m\rangle}{E_n-E_m}\right\}\,.\end{aligned} \tag{6.198}$$

Using the technique explored for g_A, we obtain

$$\begin{aligned}&\langle\psi|\Gamma_3\tau_3|\psi\rangle\\ &= -\frac{1}{2}\sum_{mn}\frac{\langle m|\Gamma_3\tau_b|n\rangle\langle n|\tau_a|m\rangle}{E_n-E_m}\left(D_{b3}\Omega_a-\Omega_aD_{b3}\right)\,.\end{aligned} \tag{6.199}$$

A detailed method for further computation of matrix elements is presented in Appendix F. Here we present the result similar to Eq. (6.185)

$$\begin{aligned}\langle\psi|(\boldsymbol{x}\times\boldsymbol{\alpha})_3\tau_3|\psi\rangle &= \frac{-iN_c}{9\Lambda_{\rm tot}}\sigma_3\,\tau_3 \\ &\times\sum_{\substack{K,K'\geq 0\\ |K-K'|\leq 1}}\frac{\langle K||\hat{m}||\widetilde{K'}\rangle\langle\widetilde{K'}||\tau||K\rangle}{\tilde{E}_{K'}-E_K}e^{-\eta^2(\tilde{E}_{K'}-E_K)^2}\,.\end{aligned} \tag{6.200}$$

In this expression, $\hat{m}\equiv\left[[\boldsymbol{x}\;\boldsymbol{\alpha}]^1\,\tau\right]^1$ is the rank one tensor for the isovector magnetic moment operator. Furthermore, we have quantized the products of D_{ij} and Ω_i while keeping their ordering as given in Eq. (6.199) (See also the calculation of g_A in subsection 6.3.).

Following the analysis of the isoscalar magnetic moment, we define the dimensionless quantity:

$$m(F;\eta) = \frac{-i}{9R^2}\sum_{K\geq 0}\{\cdots\} \tag{6.201}$$

and separate it into the valence and vacuum parts:

$$m(F;\eta) = m_0(F;\eta) + \theta(\pi/2-F)\,m_{\rm val}(F;\eta)\,. \tag{6.202}$$

Numerical studies indicate that the vacuum part is linearly divergent, while the valence part converges [Hosaka and Toki 1995]. The divergent terms have the same $\sin^2 F$ dependence as the leading order term. In Fig. 6.19 (a) the divergent coefficient of the $\sin^2 F$ term of m_0 is shown as a function of η, together with the fit to the function of Eq. (6.196). The finite vacuum term is then extracted by imposing the same subtraction condition as Eq. (6.195). In Fig. 6.19 (b), we show $m(F) = m_{\rm vac}(F) + \theta(\pi/2 - F)\, m_{\rm val}(F)$ as a function of F.

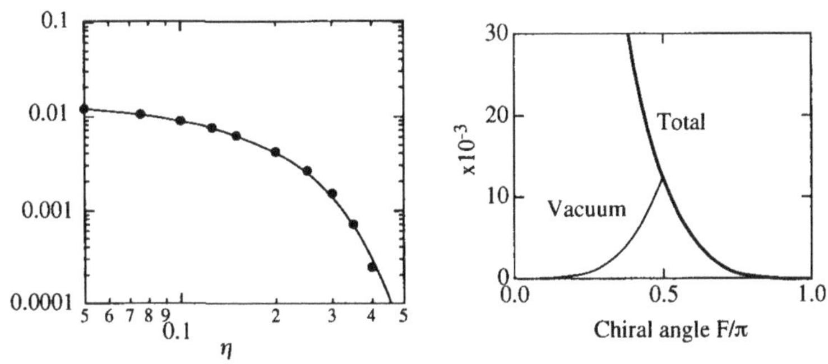

Fig. 6.19 (a) The divergent coefficient of the $\sin^2 F$ term of the normalized quark contribution to the isovector magnetic moment, $m_0(F;\eta)$, in the leading order Ω^0, and (b) The normalized quark contribution to the isovector magnetic moment of leading order Ω^0. For $F \leq \pi/2$ the vacuum contribution is separately shown.

In order to complete the calculation up to order Ω^1, the pion contribution to this order has to be included. This is calculated by using the rotationally improved pion field $\tilde{F}(r) = F(r) + \delta F(r)$ as given in Eq. (6.119). The results for the isovector magnetic moment including all the terms up to order Ω^1 is shown in Fig. 6.20. The total value is weakly dependent on R for small R, while it starts to increase for $R \gtrsim 0.5$ fm, where the higher order term becomes significant. At $R \sim 0.6$ fm, the term of order Ω^1 is only about 10 % of the total value, which is relatively small as compared to the g_A case. Thus the agreement with the experimental value is not very much improved by including the higher order terms. The role of the higher order term, however, is crucial in order to supply the missing value of 2/5 and to reproduce the correct large R behavior, just as in the case of g_A.

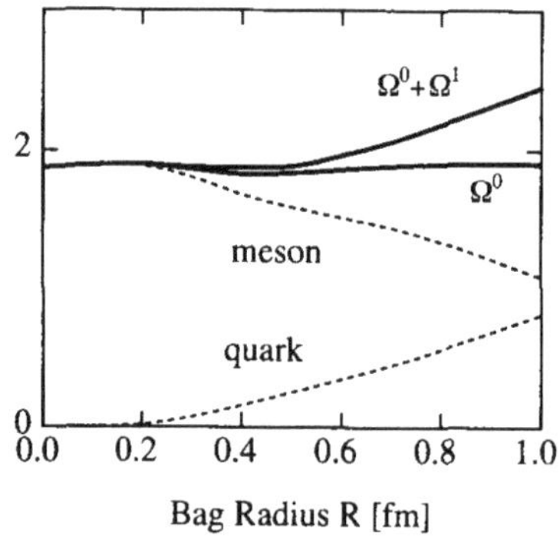

Fig. 6.20 The total isovector magnetic moment including the terms of order Ω^0 and Ω^1 as a function of R. For the leading order term, the quark and meson contributions are shown separately.

6.6 Chiral Bag with Vector Mesons

Introduction

In this section we introduce vector mesons in the chiral bag model. There are many indications of the importance of vector mesons and hence support such an extension of the model. The most prominent feature is the vector meson dominance of electromagnetic properties. This old wisdom has been proved to work pretty well down to the length scale of ~ 0.5 fm, or even less for some phenomena, such as electromagnetic form factors of hadrons including nucleons and pions [Sakurai 1969], and the photo-disintegration of the deuteron [by M. Rho and Wilkinson 1979; Rho 1984; Brown et al. 1986]. Diagrammatic description of these phenomena is shown in Fig. 6.21. Another example is the nucleon-nucleon interaction which is nicely described by the one boson exchange model including rho and omega mesons [Machleidt et al. 1987].

Chiral models with vector mesons have been extensively studied so far. In relation to the present study of the chiral bag model, one of the earliest attempts was made by Vento who introduced the omega meson in the little bag model in order to cure the instability of the little bag model [Vento 1981]. His attempt was only partially successful, since vacuum polarization was not considered. Since then, many works have been performed, among which those relevant to our study are:

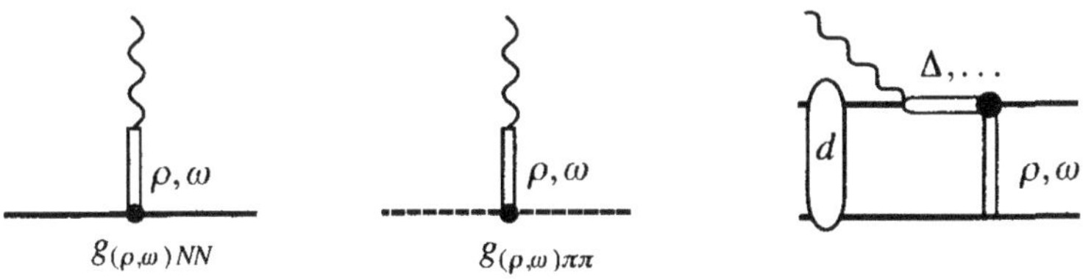

Fig. 6.21 Typical phenomena in which vector mesons such as ρ and ω play important role: (a) nucleon form factor, (b) pion form factor and (c) photodisintegration of the deuteron.

(1) The Skyrme model with the omega meson by Adkins and Nappi in which the Skyrme term was replaced by the $\omega_\mu B^\mu$ term [Adkins and Nappi 1984]
(2) The chiral bag model with omega meson by Yoro and Tatsumi [Yoro and Tatsumi 1987], Klabucar and Brown [Klabucar and Brown 1986], and Hosaka [Hosaka 1992]
(3) Hidden local symmetry (HLS) approach to the rho meson developed by the Kyoto-Nagoya group [Bando et al. 1985a; Bando et al. 1985b; Bando et al. 1988]
(4) Extended model of inclusion of ρ, a_1 and ω mesons by Yamawaki, Wakamatsu-Weise, Meissner and the present authors [Yamawaki 1987; Wakamatsu and Weise 1988; Meissner 1988; Hosaka et al. 1989; Hosaka et al. 1990].

Let us describe a brief history of the chiral bag model with vector mesons. The omega meson was rather successfully introduced in the Skyrme model [Yoro and Tatsumi 1987; Klabucar and Brown 1986]. Electromagnetic properties such as charge radii of the nucleon were improved in the isoscalar channel. However, subsequent attempts of introducing the rho meson were rather disappointing [Takashita et al. 1988]. When combined with the HLS model, R (bag radius) dependence of physical quantities became very unstable, as opposed to a naive expectation that the R dependence would become weaker when vector mesons are included, since the meson lagrangian was expected to be improved. It turned out that the inclusion of the axial-vector a_1 meson as well as the ρ meson as necessary [Hosaka et al. 1990]. Although the a_1 is heavier than the ρ, $m_{a_1} \sim 1260$ MeV, its inclusion appears to be natural from the chiral symmetry point of view.

Introduction of Vector Mesons

We consider two methods of introducing vector mesons [Bando et al. 1988; Meissner 1988]: one is the massive Yang-Mills (MYM) approach and the other is the hidden local symmetry (HLS) approach. In fact they are equivalent when ρ and a_1 mesons are introduced simultaneously [Yamawaki 1987; Wakamatsu and Weise 1988; Meissner 1988; Hosaka et al. 1989].

Consider the non-linear sigma model of $SU(2)_R \times SU(2)_L$. In the MYM approach, the symmetry group is simply gauged, introducing the corresponding (chiral) gauge bosons, ρ and a_1. The axial-vector meson a_1 is needed in this approach in order to preserve chiral symmetry. The non-zero masses of the vector mesons are then introduced by hand, breaking the gauge symmetry explicitly. Thus the gauge symmetry here is just a guiding principle to introduce the vector mesons. The resulting lagrangian in the MYM scheme is written as [Wakamatsu and Weise 1988]

$$\begin{aligned} \mathcal{L}_{\pi\rho a_1}^{MYM} &= \frac{a}{a-1}\frac{f_\pi^2}{4}\,\mathrm{tr}\,(D_\mu U D^\mu U^\dagger) \\ &+ \frac{1}{2g^2}\,\mathrm{tr}\,(V_{\mu\nu}^2 + A_{\mu\nu}^2) - a f_\pi^2\,\mathrm{tr}\,(V_\mu^2 + A_\mu^2)\,. \end{aligned} \tag{6.203}$$

Here the vector meson fields $V_\mu = -(i/2)\,\tau_a V_\mu^a$ and $A_\mu = -(i/2)\,\tau_a A_\mu^a$ describe the ρ and a_1 mesons, respectively. The corresponding field tensors are

$$\begin{aligned} V_{\mu\nu} &= \partial_\mu V_\nu - \partial_\nu V_\mu + [V_\mu, V_\nu] + [A_\mu, A_\nu]\,, \\ A_{\mu\nu} &= \partial_\mu A_\nu - \partial_\nu A_\mu + [A_\mu, V_\nu] + [V_\mu, A_\nu]\,, \end{aligned} \tag{6.204}$$

and g in the kinetic term of Eq. (6.203) is the gauge coupling constant. The covariant derivatives are defined by

$$\begin{aligned} D_\mu U &= \partial_\mu U + [V_\mu, U] - \{A_\mu, U\}\,, \\ D_\mu U^\dagger &= \partial_\mu U^\dagger + [V_\mu, U^\dagger] + \{A_\mu, U^\dagger\}\,. \end{aligned} \tag{6.205}$$

The meaning of the parameter a will be explained shortly. It is noted that the lagrangian in Eq. (6.203) has been derived from a quark theory with spontaneous breakdown of chiral symmetry, the Nambu-Jona-Lasinio model [Nambu and Jona-Lasinio 1961a; Nambu and Jona-Lasinio 1961b], through bosonization [Eguchi 1976; Wakamatsu and Weise 1988; Dhar et al. 1985; Ebert and Reinhardt 1986].

The corresponding lagrangian in the HLS scheme is given by

$$\begin{aligned}\mathcal{L}^{HLS}_{\pi\rho a_1} &= -\frac{a}{a-1} f_\pi^2 \,\mathrm{tr}\, \tilde{A}_\mu^2 + \frac{1}{2g^2}\,\mathrm{tr}\,(\tilde{V}_{\mu\nu}^2 + \tilde{A}_{\mu\nu}^2) \\ &- a f_\pi^2 \,\mathrm{tr}\,[\tilde{V}_\mu + \frac{1}{2}(\partial_\mu \xi \xi^\dagger + \partial_\mu \xi^\dagger \xi)]^2 \\ &- a f_\pi^2 \,\mathrm{tr}\,[\tilde{A}_\mu + \frac{1}{2}(\partial_\mu \xi \xi^\dagger - \partial_\mu \xi^\dagger \xi)]^2 .\end{aligned} \tag{6.206}$$

Note that the same combinations as v_μ and a_μ of Eqs. (4.74) and (4.75) appear in this equation. The $\tilde{V}$ and $\tilde{A}$ are introduced to distinguish the ρ and a_1 mesons of the HLS scheme from those of the MYM scheme. The $SU(2)$ unitary matrix ξ is related to U through $U = \xi^2$. Note that the ξ variable is the basic element of the non-linear sigma model as discussed in Chapter 4 (subsection 4.2).

In the lagrangians of Eqs. (6.203) and (6.206), the π-a_1 mixing arises through the coupling $\partial^\mu \phi_a A_\mu^a$, which must be eliminated by introducing the physical a_1 field A'_μ as

$$\begin{aligned} A'^a_\mu &= A^a_\mu + \frac{1}{a f_\pi}\partial_\mu \phi^a \qquad \text{for} \quad \text{MYM}, \\ A'^a_\mu &= A^a_\mu - \frac{a-1}{a f_\pi}\partial_\mu \phi^a \qquad \text{for} \quad \text{HLS}. \end{aligned} \tag{6.207}$$

Accordingly, the pion kinetic term is correctly normalized as

$$\frac{f_\pi^2}{4}\,\mathrm{tr}\,\partial_\mu U \partial^\mu U^\dagger$$

both for the MYM and HLS schemes. Furthermore, the parameter a relates the masses of the physical ρ and a_1 mesons by

$$a = \left(1 - \frac{m_\rho^2}{m_{a_1}^2}\right)^{-1} . \tag{6.208}$$

For $a = 2$ as implied from vector meson dominance in electromagnetic couplings, the Weinberg mass relation $m_{a_1}^2 = 2m_\rho^2$ [Weinberg 1967b] and the KSRF relation [Kawarabayashi and Suzuki 1966; Riazuddin and Fayyazuddin 1966]

$$m_\rho = \sqrt{2} f_\pi g \tag{6.209}$$

follow. Using the empirical mass of the ρ, $m_\rho = 770$ MeV, we find $g = 5.85$. This number is consistent with the decay width of $\rho \to 2\pi$.

For the models with ρ and a_1, there is an equivalence theorem between the MYM and HLS schemes [Yamawaki 1987; Hosaka et al. 1989; Wakamatsu and Weise 1988]. Explicitly, they are related to each other by the Stueckelberg transformation [Stueckelberg 1941]:

$$\begin{aligned} \tilde{V}_\mu + \tilde{A}_\mu &= \xi(V_\mu + A_\mu + \partial_\mu)\xi^\dagger , \\ \tilde{V}_\mu - \tilde{A}_\mu &= \xi^\dagger(V_\mu - A_\mu + \partial_\mu)\xi . \end{aligned} \tag{6.210}$$

Therefore, physics in the MYM and HLS schemes are completely equivalent.

It is known that the meson theory of Eq. (6.203) (or that of Eq. (6.206)) does not allow for stable soliton solutions [Ezawa and Yanagida 1986]. The Skyrme term can not be added, since it mimics the kinetic term of the ρ meson [Bando et al. 1988]. Thus we shall add another vector meson ω in order to stabilize soliton solutions. Since ω is regarded as a gauge boson of U(1), the relevant lagrangian is uniquely determined up to the second order derivative:

$$\mathcal{L}_\omega = -\frac{1}{4}\omega_{\mu\nu}^2 + \frac{a}{2}g^2 f_\pi^2 \omega_\mu^2 + \frac{N_c}{2} g\, \omega_\mu B^\mu . \tag{6.211}$$

The coupling $\omega_\mu B^\mu$ arises from the gauged Wess-Zumino term [Kaymakcalan et al. 1984]. The simultaneous gauging of $SU(2) \times SU(2) \times U(1)$ introduces only one coupling constant g which is related to ωNN coupling constant $g_{\omega NN}$ by $g_{\omega NN} = (N_c/2)\, g$, as read from Eq. (6.211). To summarize, we have the meson lagrangian for the π, ρ, a_1 and ω,

$$\mathcal{L}_{\pi\rho a_1 \omega} = \mathcal{L}_{\pi\rho a_1}^{\text{MYM or HLS}} + \mathcal{L}_\omega . \tag{6.212}$$

Nucleon properties

Hedgehog and nucleon profiles

Replacing the Skyrme lagrangian in the original chiral bag model by that with vector mesons given in Eq. (6.212), we have the lagrangian for the chiral bag with vector mesons [Hosaka et al. 1990]:

$$\mathcal{L} = \mathcal{L}_q\, \theta_V + \mathcal{L}_{\text{int}}\, \delta_V + \mathcal{L}_{\pi\rho a_1 \omega}\, \theta_{\bar{V}} . \tag{6.213}$$

We comment on the meson-quark coupling in the second term. In principle, it is possible to include surface couplings between vector mesons and quarks

(vqq coupling) as well as the πqq coupling. However, we do not consider them here for the following reasons. First, symmetry principles such as current conservation do not necessarily require surface couplings of the type vqq; the only coupling required is the πqq coupling from chiral symmetry. Secondly, some of the surface vqq couplings are not compatible with the linear boundary condition for quarks; the surface ρqq coupling of the vector type $\bar{\psi}\gamma^\mu \rho_\mu^a \tau_a \psi \delta_V$ disappears from the consistency of the linear boundary condition [Takashita et al. 1988].

In order to obtain the nucleon state, we follow the method presented in the previous sections. We first obtain the hedgehog solution, in which the meson fields take the form

$$\begin{aligned} U &= \exp[i\boldsymbol{\tau}\cdot\hat{\boldsymbol{x}}F(r)], \\ V_i^a &= \epsilon_{aij}\hat{x}_j \frac{G(r)}{r}, \\ A_i^a &= A(r)P_{ai} + B(r)\hat{x}_a\hat{x}_i, \\ \omega_0 &= \omega(r), \end{aligned} \tag{6.214}$$

and other components of vector mesons are set zero. Here, $P_{ai} \equiv \delta_{ai} - \hat{x}_a\hat{x}_i$ is orthogonal to the unit vectors $\hat{x}_a$ and $\hat{x}_i$. In Eq. (6.214) the tensor structure of, for example, the ρ meson is determined by the parity,

$$\begin{aligned} V_i^a(\boldsymbol{x}) &\overset{\text{parity}}{\longrightarrow} -V_i^a(-\boldsymbol{x}) \\ &= -\epsilon_{aij}(-\hat{x}_j)\frac{G(r)}{r} \\ &= \epsilon_{aij}\hat{x}_j\frac{G(r)}{r} \quad \text{(invariant)} \end{aligned} \tag{6.215}$$

such that the configuration of Eq. (6.214) has the positive parity.

The unknown profile functions $F(r)$, $G(r)$, $A(r)$, $B(r)$ and $\omega(r)$ are obtained by minimizing the corresponding energy, whose equations of motion are presented in Appendix G. The resulting profile functions for bag radii, $R = 0$ and $R = 0.6$ fm (where $F(R) = \pi/2$), are presented in Fig. 6.22. The two profile functions $G(r)$, $A(r)$ and $B(r)$ with and without tildes are related by the Stueckelberg transformation

$$\begin{aligned} \frac{\tilde{G}-1}{r} &= \frac{G-1}{r}\cos F + A\sin F\,, \\ \tilde{A} &= -\frac{G-1}{r}\sin F + A\cos F\,, \end{aligned} \tag{6.216}$$

$$\tilde{B} = B + F'.$$

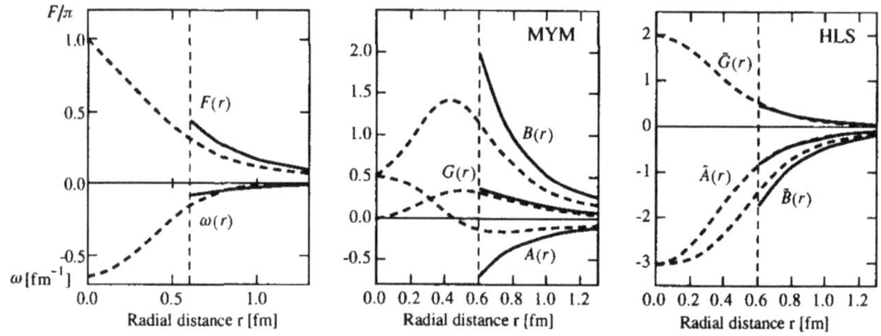

Fig. 6.22 Profile functions $F(r)$, $G(r)$, $A(r)$, $B(r)$ and $\omega(r)$ for the bag radii $R = 0$ and $R = 0.6$ fm. For the vector meson configurations G, A and B, the solutions in the HLS and MYM schemes are shown separately.

The physical nucleon is constructed, as before, by rotating the hedgehog solution in the cranking method. In this section we denote the time dependent rotation matrix by $C(t)$ instead of $A(t)$ in order to distinguish it from the A variable for the a_1 meson. The rotating hedgehog is then given by

$$\begin{aligned} \psi &\to C\psi, \\ U,\, V_i,\, A_i &\to CUC^\dagger,\, CV_iC^\dagger,\, CA_iC^\dagger . \end{aligned} \tag{6.217}$$

Accordingly, nonvanishing components are induced as

$$\begin{aligned} V_0 &= C\tau_a \frac{\Omega_b}{2}(P_{ab}u_1(r) + \hat{x}_a\hat{x}_b u_2(r))C^\dagger , \\ A_0 &= C\tau_a \epsilon_{aij}\hat{x}_i \frac{\Omega_j}{2}\varphi(r)C^\dagger , \\ \omega_i &= \epsilon_{ijk}\hat{x}_j \frac{\Omega_k}{2}\omega_A(r) , \end{aligned} \tag{6.218}$$

where $u_1(r), u_2(r), \varphi(r)$ and $\omega_A(r)$ are profile functions for the excitation fields. Note that the tensor structures of V_0 and A_0 are just opposite to those of V_i and A_i in Eq. (6.214) because the time and spatial components transform differently under the parity transformation: $V_\mu(\boldsymbol{x}) \stackrel{\text{parity}}{\longrightarrow} +V^\mu(-\boldsymbol{x})$ and $A_\mu(\boldsymbol{x}) \stackrel{\text{parity}}{\longrightarrow} -A^\mu(-\boldsymbol{x})$.

Substituting Eqs. (6.217) and (6.218) in the lagrangian of Eq. (6.213), we find the lagrangian in the rotating frame which is similar to Eq. (6.47):

$$\mathcal{L}_{\rm rot} = \mathcal{L}_h - \frac{1}{2}\boldsymbol{\Omega}\cdot\boldsymbol{Q} - \frac{\Lambda_{\rm mes}}{2}\Omega^2 . \tag{6.219}$$

The difference from Eqs. (6.47) is the hedgehog part ($\mathcal{L}_h$) and the moment of inertia for mesons ($\Lambda_{\rm mes}$). A rather lengthy expression for $\Lambda_{\rm mes}$ is given in Appendix G together with the equations of motion for the excitation fields.

Profile functions for excited fields are obtained so as to minimize the moment of inertia. The solutions at $R = 0$ and 0.6 fm are plotted in Fig. 6.23 for the MYM and HLS schemes. Again, we observe that they are related by the Stueckelberg transformation:

$$\begin{aligned} \tilde{u}_1 - 2 &= (u_1 - 2)\cos F + \phi \sin F , \\ \tilde{\phi} &= -(u_1 - 2)\sin F + \phi\cos F , \\ \tilde{u}_2 &= u_2 . \end{aligned} \tag{6.220}$$

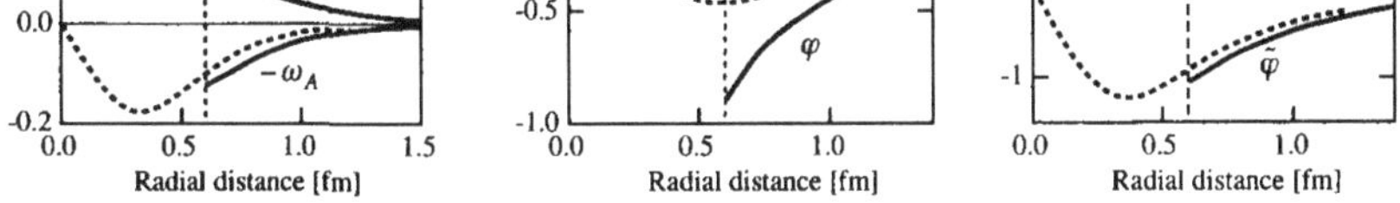

Fig. 6.23 Vector meson excitations $u_2(r)$, $\omega_A(r)$, $u_1(r)$ and $\varphi(r)$ for the bag radii $R = 0$ and $R = 0.6$ fm. For the vector meson configurations u_1 and φ, the solutions in the HLS and MYM schemes are shown separately.

Masses of the nucleon and delta

One can calculate the second order rotational energy in the same way as in section 5.6. In Fig. 6.24 the masses of the hedgehog, nucleon and delta, M_H, M_N and M_Δ, are plotted as functions of R. As in the original chiral bag model, the masses are smooth functions of R and are almost independent of R. The result here, however, differs in two respects from that of the original chiral bag (section 6.2). First, masses are smaller than in the original chiral bag model by about 300 MeV. Previously, we have argued that large classical masses would be partly reduced by the zero point energy corrections of order Ω^0. It does not seem, however, likely that the masses of 1.5 GeV for the nucleon and 1.7 GeV for the delta could be reduced solely by such corrections. Hence, the reduction of the masses in the presence of the vector mesons is appreciated. Secondly, the mass splitting of the nucleon and delta (ΔM) is improved. Although at $R = 0$ it is overestimated ($\Delta M \sim 400$ MeV), at $R \sim 0.6$ fm, $\Delta M \sim 280$ MeV, which is well compared with the experimental value of $\Delta M = 294$ MeV. At this intermediate bag radius, gluonic effects are still negligibly small. For more discussions see Refs. [Hosaka et al. 1990; Hosaka and Toki 1996].

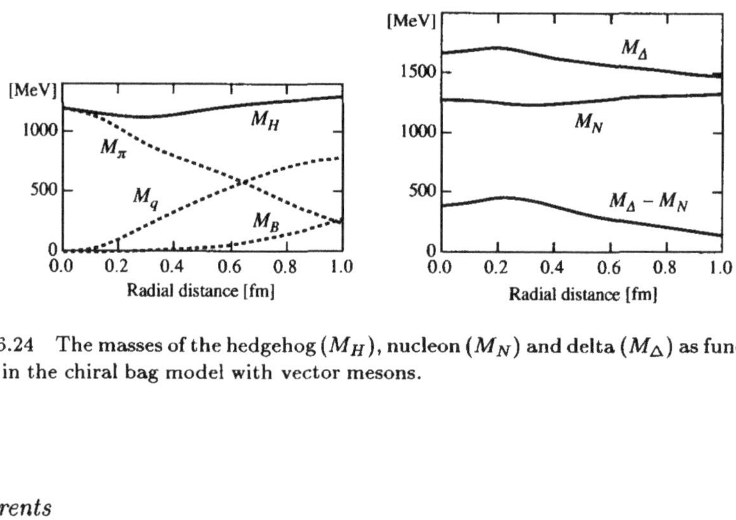

Fig. 6.24 The masses of the hedgehog (M_H), nucleon (M_N) and delta (M_Δ) as functions of R in the chiral bag model with vector mesons.

Currents

Next we investigate electromagnetic properties of the nucleon. Definitions of various currents and related observables are given in section. 6.5. Here we present only new results introduced by the vector mesons. The isoscalar current is the symmetry current associated with the $U(1)_V$ transformation:

$$\begin{aligned} \psi &\to e^{iQ\epsilon(x)}\psi\,, \\ \omega_\mu &\to \omega_\mu - \frac{1}{N_c g}\partial_\mu \epsilon(x)\,, \end{aligned} \qquad (6.221)$$

$$U,\ V_\mu,\ A_\mu \quad \to \quad U,\ V_\mu,\ A_\mu\,,$$

where $\epsilon(x)$ is the transformation parameter and $Q = (1/2)N_c$ is for the isoscalar charge. Thus the isoscalar current is given by

$$J_S^\mu = \frac{1}{2N_c}\bar{\psi}\gamma^\mu\psi\,\theta_V - \frac{m_v^2}{N_c g}\omega^\mu\,\theta_{\bar{V}}\,. \tag{6.222}$$

Note that the mesonic part is given solely by the ω meson, just as implied by the vector meson dominance.

The corresponding isovector current is given by

$$(J_V)_\mu^a = \frac{1}{2}\bar{\psi}\gamma_\mu\tau^a\psi\,\theta_V + (J_V^{\rm mes})_\mu^a\,\theta_{\bar{V}}\,. \tag{6.223}$$

The mesonic current takes the form

$$\begin{aligned}(J_V^{\rm mes})_\mu^a &= f_\pi^2 V_\mu + \frac{N_c g}{64\pi^2}\epsilon_{\mu\nu\alpha\beta}\,\omega^\nu \\ &\times\ (-U^\dagger\partial^\alpha U U^\dagger\partial^\beta U + \partial^\alpha U U^\dagger\partial^\beta U U^\dagger)\,,\end{aligned} \tag{6.224}$$

for the MYM scheme and

$$\begin{aligned}(J_V^{\rm mes})_\mu^a &= \frac{f_\pi^2}{2}\left[\xi^\dagger(\partial_\mu + \tilde{V}_\mu + \tilde{A}_\mu)\xi + \xi(\partial_\mu + \tilde{V}_\mu - \tilde{A}_\mu)\xi^\dagger\right] \\ &+\ \frac{N_c g}{64\pi^2}\epsilon_{\mu\nu\alpha\beta}\,\omega^\nu \\ &\times\ (-U^\dagger\partial^\alpha U U^\dagger\partial^\beta U + \partial^\alpha U U^\dagger\partial^\beta U U^\dagger)\,,\end{aligned} \tag{6.225}$$

for the HLS scheme. Here we have introduced the notation $J^\mu = -\frac{1}{2}i\tau_a J_\mu^a$. In the MYM scheme, the vector meson dominance of current field identity shows up explicitly as seen in the first term of $f_\pi^2 V_\mu$, while in the HLS scheme, it is so only to leading order in meson fields. This is verified by expanding $\xi = \exp(i\pi/2f_\pi) = 1 + i\pi/2f_\pi + \cdots$. Once again, however, the two expressions (6.224) and (6.225) are related by the Stueckelberg transformation, and the resulting physics are completely equivalent when higher order terms are appropriately taken into account.

Electric Charges

The isoscalar charge is given by the spatial integration

$$Q_S = \int_0^R d^3x\frac{1}{2N_c}\langle\psi^\dagger\psi\rangle - \int_R^\infty d^3x\frac{m_v^2}{N_c g}\omega(r)\,. \tag{6.226}$$

We note that the meson part in the second term is dominated by the ω meson. One can verify the normalization of the isoscalar charge by relating the volume integral of the ω field to the baryon number carried by the pion through the equation of motion:

$$(\triangle - m_v^2)\omega(r) = -\frac{N_c g}{2}\frac{\sin^2 F}{2\pi^2 r^2}F' . \tag{6.227}$$

Then we get

$$Q_S = \frac{1}{2}(B_q + B_\pi) = \frac{1}{2} \tag{6.228}$$

as expected.

The isovector charge is computed as

$$Q_V = \int_0^R d^3x \, \langle \psi^\dagger \frac{\tau_3}{2} \psi \rangle + \int_R^\infty d^3x \, \rho_V^{\mathrm{mes}}(r) , \tag{6.229}$$

where the charge density of the meson sector is given by

$$\rho_V^{\mathrm{mes}}(r) = \frac{1}{\Lambda_{\mathrm{tot}}}\left[\frac{m_v^2}{6g^2}\left(u_1 + \frac{1}{2}u_2\right) - \frac{N_c g}{24\pi^2 r}\omega_A F' \sin^2 F\right] , \tag{6.230}$$

for the MYM scheme and

$$\begin{aligned}\rho_V^{\mathrm{mes}}(r) \; &= \; \frac{1}{\Lambda_{\mathrm{tot}}}\left[\frac{m_v^2}{6g^2}\left(\tilde{u}_1 \cos F + \frac{1}{2}\tilde{u}_2 + 2 - 2\cos F - \tilde{\varphi}\sin F\right)\right. \\ &\quad \left. - \frac{N_c g}{24\pi^2 r}\omega_A F' \sin^2 F\right] , \end{aligned} \tag{6.231}$$

for the HLS scheme.

The normalization of the isovector charge is verified by relating it to the moment of inertia:

$$Q_V^{\mathrm{mes}} = \frac{1}{2}\frac{\Lambda_{\mathrm{mes}}}{\Lambda_{\mathrm{tot}}} \tag{6.232}$$

as shown in Appendix G. Thus we obtain the correctly normalized isovector charge

$$Q_V = \frac{1}{2}\left(\frac{\Lambda_q}{\Lambda_{\mathrm{tot}}} + \frac{\Lambda_{\mathrm{mes}}}{\Lambda_{\mathrm{tot}}}\right) = \frac{1}{2}\frac{\Lambda_{\mathrm{tot}}}{\Lambda_{\mathrm{tot}}} = \frac{1}{2} . \tag{6.233}$$

Although we do not show the fractionization of isoscalar and isovector charges, it is very similar to that of the original chiral bag model (see Fig. 6.14). For example, at $R \sim 0.5$ fm, the isoscalar charge is almost equally fractioned into the quark and the meson sectors, while the isovector charge is dominated by the meson sector.

Magnetic Moments

Magnetic moments for the nucleon are also calculated in the same way as in the original chiral bag model but with mesonic currents replaced by those of vector mesons. The meson contribution of the isoscalar magnetic moment has the same form for the MYM and HLS schemes and is given by

$$(2M_N)^{-1}\mu_S^{\rm mes} = \frac{\pi m_v^2}{3N_c g\Lambda_{\rm tot}}\int_R^\infty r^3 dr\,\omega_A(r)\,. \tag{6.234}$$

The corresponding isovector magnetic moment to leading order of Ω^0 is computed by

$$(2M_N)^{-1}\mu_V^{\rm mes} = \frac{4\pi}{3}\int_R^\infty r^2 dr\left(\frac{m_v^2}{3g^2}G + \frac{N_c g}{12\pi^2}\omega F'\sin^2 F\right), \tag{6.235}$$

for the MYM scheme and by

$$\begin{aligned}(2M_N)^{-1}\mu_V^{\rm mes} &= \frac{4\pi}{3}\int_R^\infty r^2 dr\left[\frac{m_v^2}{3g^2}(\tilde{G}\cos F + 1 - \cos F - r\tilde{A}\sin F)\right.\\ &\qquad\left. + \frac{N_c g}{12\pi^2}\omega F'\sin^2 F\right], \end{aligned}\tag{6.236}$$

for the HLS scheme.

In Fig. 6.25, we show the magnetic moments μ_S and μ_V as functions of R together with their fractioning into the quark and meson sectors. For the isovector magnetic moment, the contributions of order Ω^0 and of order Ω^1 are included. Calculated values are still small as compared with the experimental values just like in the original chiral bag model. The inclusion of the vector mesons does not improve the magnetic moments.

Isoscalar charge radius

Here in the isoscalar channel, we see clearly the effect of vector meson dominance. In contrast, it is less obvious to see it in the isovector channel. The reasons are: first, the isovector charge radius diverges in the chiral limit [Adkins et al. 1983], and second, the separation of the isovector charge radius from the radius of the source term of the ρ is not unique due to non-linear interactions of the ρ meson.

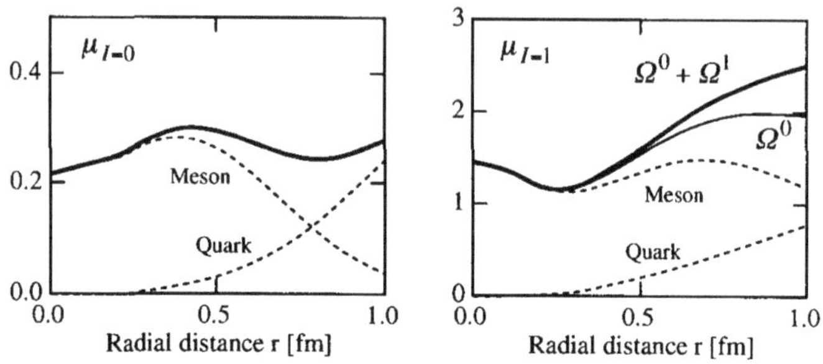

Fig. 6.25 The isoscalar and isovector magnetic moments as functions of R.

The mean square radius of the isoscalar charge distribution, $\langle r^2\rangle_S$, is computed by

$$\langle r^2\rangle_S = \int_0^R d^3x\, r^2 \frac{1}{N_c}\langle \psi^\dagger \psi\rangle - \int_R^\infty d^3x\, r^2 \frac{2m_v^2}{N_c g}\omega(r)\,. \tag{6.237}$$

In the same way, we calculate the baryon number radius $\langle r^2\rangle_B$ (the source term of ω for $r \geq R$) as

$$\langle r^2\rangle_B = \int_0^R d^3x\, r^2 \frac{1}{N_c}\langle \psi^\dagger \psi\rangle - \int_R^\infty d^3x\, r^2 \frac{\sin^2 F}{2\pi^2 r^2}F'\,. \tag{6.238}$$

The two expressions (6.237) and (6.238) are common both in the MYM and in the HLS schemes.

Using an alternative expression for the charge radius

$$\langle r^2\rangle = -6\left[\frac{d}{dq^2}\tilde{\rho}(q^2)\right]_{q^2=0} \tag{6.239}$$

with the Fourier transform of the charge density $\tilde{\rho}(q^2)$, and the equation of motion for the ω meson, one finds that

$$\langle r^2\rangle_S = \langle r^2\rangle_B + \frac{6}{m_v^2}B_{\rm mes}\,. \tag{6.240}$$

The second term expresses the vector meson dominance due to the omega meson, though it is incomplete as multiplied by the fractional baryon number in the meson sector, $B_{\rm mes}$.

Using the physical value of $m_\omega = m_v = 770$ MeV, the vector meson dominant term has a significant contribution to the isoscalar charge radius when R is not too large and B_{mes} is not too small. In the original chiral bag model, the isoscalar charge radius is identified with the baryon number radius. This corresponds to the limit $m_\omega = m_v \to \infty$, which, however, is not a good approximation in the real world of $m_v = 770$ MeV. In this respect, let us recall the following fact; in the original chiral bag model, the isoscalar charge radius and the mass of the nucleon are not simultaneously reproduced. The nucleon in the original chiral bag model is heavier and more compact than the physical nucleon. In the model with vector mesons, both are improved; using the physical masses of the vector mesons, the mass of the nucleon is reduced and the (electric) size of the nucleon becomes larger.

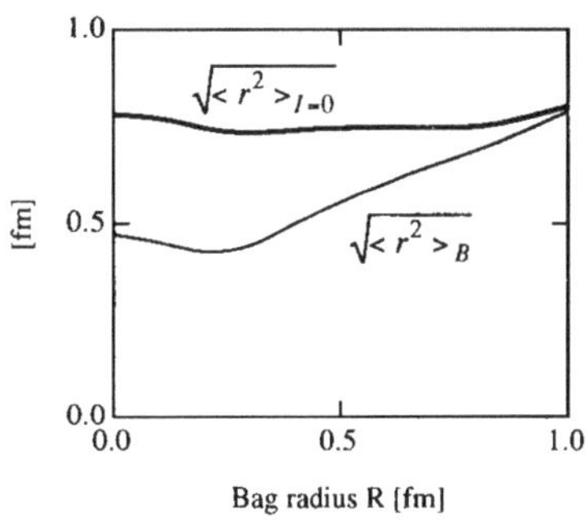

Fig. 6.26 Square roots of mean square radii of the isoscalar and baryon charges as functions of R.

In Fig. 6.26 we show the root mean square radii, $\sqrt{\langle r^2\rangle_S}$ and $\sqrt{\langle r^2\rangle_B}$, as functions of R. The calculated values of the isoscalar charge radii $\sqrt{\langle r^2\rangle_S}$ are about 0.75 fm, almost independent of R and in good agreement with the empirical value of 0.79 fm. The baryon number radius $\sqrt{\langle r^2\rangle_B}$, on the other hand, has stronger R dependence; it increases for $R \gtrsim 0.5$ fm. The difference between the two radii is due to the vector meson dominance as dictated by the second term in Eq. (6.240).

Axial Properties

The method of computing the axial-vector coupling constants g_A and $g_A^{(0)}$ has been discussed in great detail in section 6.3. What is new here is the meson contribution to g_A. The meson part of the isovector axial-vector current is given by

$$
\begin{aligned}
(J_5^{\rm mes})_\mu &= 2f_\pi^2 A_\mu + \frac{N_c\, g}{32\pi^2}\epsilon_{\mu\nu\alpha\beta}\,\omega^\nu \\
&\times\ (U^\dagger\partial^\alpha U U^\dagger \partial^\beta U + U\partial^\alpha U^\dagger U \partial^\beta U^\dagger)
\end{aligned}
\tag{6.241}
$$

for the MYM scheme, and

$$
\begin{aligned}
(J_5^{mes})_\mu &= f_\pi^2\left[\xi^\dagger(\partial_\mu + \tilde{V}_\mu + \tilde{A}_\mu)\xi - \xi(\partial_\mu + \tilde{V}_\mu - \tilde{A}_\mu)\xi^\dagger\right] \\
&+\ \frac{N_c\, g}{32\pi^2}\epsilon_{\mu\nu\alpha\beta}\,\omega^\nu \\
&\times\ (U^\dagger\partial^\alpha U U^\dagger \partial^\beta U + U\partial^\alpha U^\dagger U \partial^\beta U^\dagger)
\end{aligned}
\tag{6.242}
$$

for the HLS scheme. Using the explicit form of the axial-vector current in the hedgehog ansatz, it is straightforward to derive the following expression:

$$
g_A = -\frac{8\pi}{9}R^3 a_q(R) + 4\pi\int_R^\infty r^2 dr\,\rho_A(r)\,, \tag{6.243}
$$

where $a_q(R)$ is the quark axial flux at the bag surface as defined in Eq. (6.107), and the axial flux density of the meson sector $\rho_A(r)$ is given by

$$
\rho_A(r) = \frac{4}{3}f_\pi^2\left(\frac{2}{3}A + \frac{1}{3}B\right) - \frac{N_c g}{18\pi^2}\omega\left(\frac{\sin^2}{r^2} + \frac{\sin 2F}{r}F'\right)\,, \tag{6.244}
$$

for the MYM scheme, where $A = A(r)$ and $B = B(r)$ are the axial-vector profiles as defined in Eq. (6.214). The corresponding expression for the HLS scheme is obtained by using the Stueckelberg transformation of Eq. (6.216).

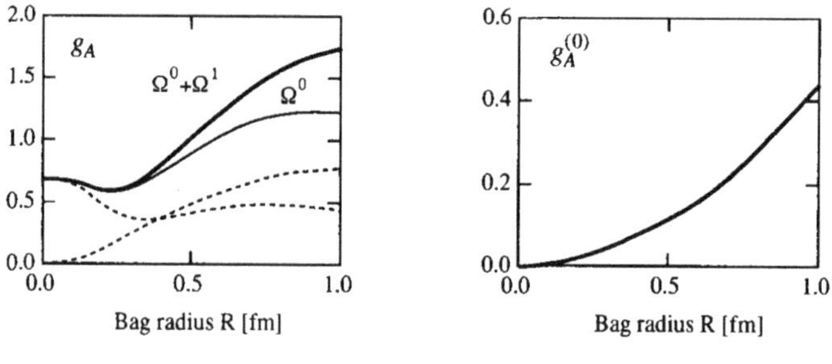

Fig. 6.27 The isovector and isoscalar axial-vector coupling constants, g_A and $g_A^{(0)}$, as functions of R.

In Fig. 6.27 (left), we show g_A as a function of R, where the first order corrections of order Ω^1 is included. As in the magnetic moments, g_A has stronger R dependence than in the original chiral bag: a narrow dip structure around $R \sim 1/4$ fm and a rapid increase beyond $R \gtrsim 0.5$ fm. For small R, calculated values are too small as compared with the experimental value. As the bag radius is increased, however, the calculated values increase rapidly and the experimental value is reproduced at $R \sim 0.6$ fm just as in the original chiral bag model. Once again, the higher order correction of order Ω^1 is important.

In Fig. 6.27, the isoscalar $g_A^{(0)}$ (right) is also plotted as a function of R. Here, only the quark contribution is calculated without the η' meson. Again the result presented here resembles that of the original chiral bag model and the experimental value of $g_A^{(0)}$ is reproduced at $R \sim 0.6$ fm. We emphasize that experimental values g_A and $g_A^{(0)}$ are reproduced almost at the same bag radius around $R \sim 0.6$ fm.

Problems

6.1

Consider the ansatz in the body-fixed frame in the scalar theory (Eq. (6.24)):

$$\phi^{\text{bf}}(t, \boldsymbol{x}) = \phi_c(\boldsymbol{x}) + g^{-1}(t)\varphi(t, \boldsymbol{x}).$$

Substituting this in the lagrangian

$$\mathcal{L} = \frac{1}{2}(\partial_t \phi)^2 - \frac{1}{2}(\boldsymbol{\nabla}\phi)^2 - V(\phi),$$

discuss the meaning (i.e., Coriolis coupling, centrifugal potential and etc.) of the terms containing time derivatives [Bohr and Mottelson 1975].

6.2

Employing the rotating hedgehog, $AU_H A^\dagger$ and $A\psi$, verify Eq. (6.47), in particular the expression of the moment of inertia given in Eq. (6.50).

6.3

Find the form of the axial densities, $\alpha(r)$ and $\beta(r)$ in Eq. (6.83), for the lagrangian

$$L = \frac{f_\pi^2}{4} \operatorname{tr} \partial_\mu U \partial^\mu U^\dagger. \tag{6.245}$$

6.4

Using the Fourier transformation, explain that in the limit $q \to 0$ the tensor $\hat{q}_i \hat{q}_j$ can be replaced by $1/3\ \delta_{ij}$.

6.5

Derive the definition of the magnetic moment in Eq. (6.171) from the hamiltonian of the electromagnetic coupling

$$H = -e \int d^3x\ \boldsymbol{J} \cdot \boldsymbol{A}, \tag{6.246}$$

Chapter 7

LARGE-N_c BARYONS

In the previous chapters we have discussed the semiclassical method for the nucleon dynamics in the chiral bag model. We attempted to treat appropriately the strong non-linear chiral interactions among pions and quarks. The effect of strong interactions that can not be reached by perturbation series can be incorporated by a (classical) mean field solution. The fluctuations around it then interact through residual interactions, the strength of which is characterized by a small parameter. In QCD, such a small parameter can be identified with the inverse of the number of colors, $1/N_c$. This identification can be a link between QCD and its effective theories.

In this chapter we discuss the $1/N_c$ expansion for baryon dynamics and see how the method works in actual cases. To start with we explain in detail the diagrammatic method of the $1/N_c$ expansion to classify interactions of mesons and baryons. The Hartree approximation for the baryon is then described in great detail. After a general introduction, we formulate the algebraic method for the baryon. This method is particularly useful in establishing relations among various models that respect spin and isospin symmetry in the large-N_c limit. This also leads naturally to the hedgehog ansatz when $N_c \to \infty$. In order to show the efficiency of this method we compute axial couplings without using particular models. Finally, we present a possible scenario for baryons which supports chiral bag like structure in the large-N_c limit.

7.1 Introduction

When a theory can not be solved directly, one would try to find a small parameter, in powers of which the theory can be expanded around a known solution (e.g., the vacuum). Often such a parameter is a coupling constant of the interaction, without which the theory reduces to a non-interacting solvable theory. The effect of the interaction is then included in perturbation series. For instance, in QED, the relevant coupling constant is $\alpha = e^2/4\pi \sim 1/137$.

In a certain class of theories, the inverse power of the number of degrees of freedom, $1/N$, can be an expansion parameter. Intuitively, one might expect that as N is increased the system approaches the classical limit. Here we see a similarity between the expansion schemes in $1/N$ and in $\hbar$. This indeed happens in QCD.

We should point out, however, a difference between the above two expansion schemes. In the perturbation theory, truncation at some order of interactions limits the number of terms to be computed. In contrast, in the $1/N$ expansion, although the interaction strength gets smaller as N is increased, the number of terms that have to be computed increases. Crudely, as N gets larger, more intermediate states should be considered, which overcome the small interaction strength.

In certain field theories, it is known that soliton solutions exist at the classical (or mean field) level. The hedgehog solutions of the Skyrme model and of the chiral bag model are such examples. They can not be reached in perturbation series of finite order. They also survive in the limit of small coupling constant. These two properties meet with the classical limit in the $1/N$ expansion.

Another interesting fact which we have seen in the preceding chapters was that essential features of the original chiral bag model persist in extended models, although details vary at quantitative level. If we look at other models such as the non-relativistic quark models, the Skyrme model and the chiral quark soliton model, such a similarity appears to be rather general, although detailed model setups look very different. This observation implies that there might be an underlying common feature in baryon dynamics. It turns out that it is indeed provided by the $1/N$ expansion, when N is identified with the number of colors, $N \to N_c$. The $1/N_c$ expansion for QCD was suggested by 't Hooft in 1974 ['t Hooft 1974]. Later, Witten analyzed in greater detail meson-baryon systems and established that baryons at low energy behaved as solitons of a non-linear meson theory [Witten 1979].

By the end of the 90's, properties of large-N_c baryons have been extensively studied by algebraic methods for spin and isospin symmetry [Dashen and Manohar 1993b; Dashen and Manohar 1993a; Jenkins 1993b; Jenkins 1993a; Dashen et al. 1994; Dashen et al. 1995; Luty and March-Russell 1994]. By choosing a particular representation of the algebra, it has become possible to make a unified view on the various effective theories such as the Skyrme model, the non-relativistic quark model and the chiral bag model. Such an algebraic structure must be a common basis of all chiral models if they respect QCD, in particular spin and isospin symmetry of light flavor sector, where model independent analysis is possible. The large-N_c algebra provides a systematic expansion in powers of $1/N_c$ for baryon physics, which in turn is related to the semiclassical method for solitons.

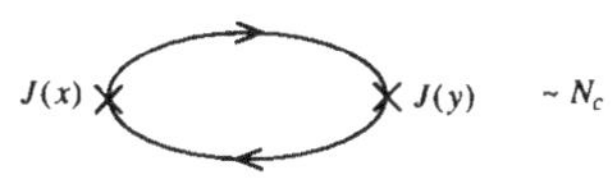

Fig. 7.1 A meson-Meson correlation function.

7.2 General Counting Rules

Meson vertices

Let us start with a two point function of a meson

$$\langle 0|J(x)J(y)|0\rangle \,. \tag{7.1}$$

Here the meson interpolating field $J(x) \sim \bar{\psi}^a \Gamma \psi^a$ is color singlet and therefore contains a sum over color indices. The matrix Γ is chosen such that the bilinear operator J reflects the meson quantum numbers. The lowest order diagram without gluon interactions is shown in Fig. 7.1. Because of the sum over the color degrees of freedom, the amplitude is proportional to N_c:

$$\langle 0|J(x)J(y)|0\rangle \sim N_c \,. \tag{7.2}$$

On the other hand the spectral representation

$$\langle 0|J(x)J(y)|0\rangle = \sum_n |\langle 0|J(0)|n\rangle|^2 \exp(-ip_n(x-y)) \tag{7.3}$$

includes the sum over physical intermediate states n without color degrees of freedom. Therefore, the color factor in Eq. (7.2) should be absorbed into the square of the matrix element $|\langle 0|J(0)|n\rangle|^2$. This implies that the meson decay constant is proportional to $\sqrt{N_c}$:

$$f_M \sim \langle 0|J(0)|M\rangle \sim \sqrt{N_c}\,. \tag{7.4}$$

When gluon interactions are inserted, the N_c dependence of the diagrams appears to grow up in powers of N_c. For instance, consider ladder series of diagrams as depicted in Fig. 7.2. Since a gluon can change quark colors in N_c different ways, the power of N_c increases as more gluons are inserted. This would imply that the theory becomes meaningless in the limit $N_c \to \infty$. As shown by 't Hooft, however, it is possible to make the theory sensible by assuming that the quark-gluon coupling constant behaves as

$$g_{gqq} \sim \frac{1}{\sqrt{N_c}}\,. \tag{7.5}$$

Having considered this all Feynman diagrams are bounded by some power of $1/N_c$. For the case shown in Fig. 7.2, all ladder diagrams scale as $(1/N_c)^{-1} = N_c^1$.

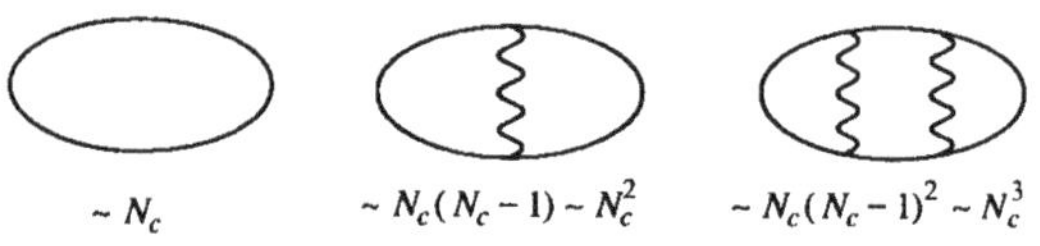

Fig. 7.2 Gluon ladder contributions to the meson-meson correlation functions.

Next we consider what happens if non-linear gluon interactions are included. Take higher order diagrams of the meson-meson correlations as depicted in Fig 7.3. The diagram on the left hand side involving a three-gluon vertex is obtained by inserting one gluon line into the diagram in the middle of Fig. 7.2. In so doing, one gqq vertex, one ggg vertex and one quark internal line (and one gluon line also) are added. This, however, does not change the power counting of $1/N_c$. At each new vertex, a coupling

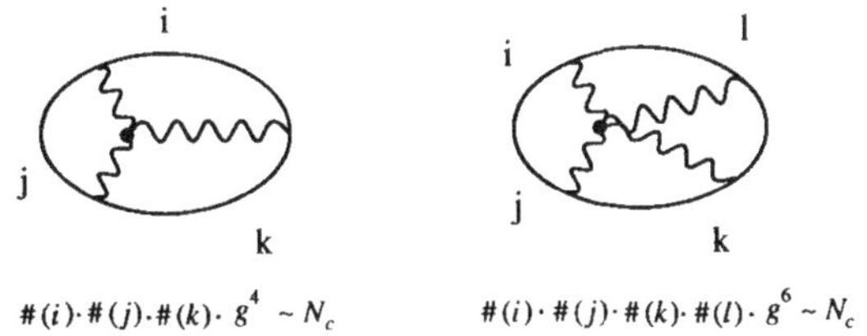

Fig. 7.3 Meson-meson correlation functions with gluon higher order interactions.

constant g is attached, contributing the factor $(1/\sqrt{N_c})^2 = 1/N_c$, while the internal quark line brings the factor N_c. All in all, there is no change in the power counting. Similarly, one can see that the diagram on the right hand side has the same $1/N_c$ dependence as the left one.

It is possible to consider more complicated diagrams in a similar manner. It is useful to introduce the so called double line notation for gluons. Since color octets are constructed as

$$3 \times 3^* = 1 + 8\,, \tag{7.6}$$

gluons behave, group theoretically, in the same way as the octet part of quark and antiquark pairs. In diagrams, this can be shown as in Fig. 7.4 (a). Using the double line notation, the power counting of $1/N_c$ is much simplified; for a given diagram, (the power of $1/N_c$) = (the number of quark loops) minus (half of the number of active gqq vertices in the double line notation). The meaning of the active gqq vertex is explained for the case of the diagram shown in Fig. 7.4 (b). There is only one active gqq vertex in there (bold lines); one can choose the lines a and b to form one such a vertex, then there is no way to form other active vertices. It is apparent that for the four-gluon ($gggg$) vertex, there are two active vertices in the double line notation.

We can now verify the following general rules:

- The leading diagrams are what are called planar diagrams in which all internal lines together with vertices are embedded in a single plane. Examples of planar diagrams are shown in Fig. 7.5 (a).
- Non-planar diagrams which need more than one plane to embed quark and gluon lines are suppressed in higher orders of $1/N_c$. Examples of non-planar diagrams are shown in Fig. 7.5 (b).

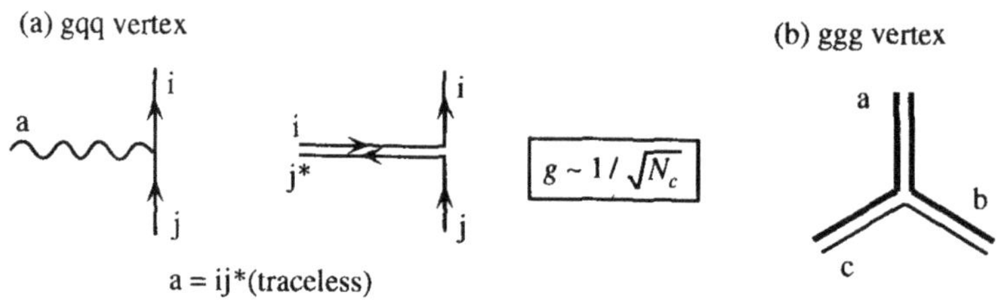

Fig. 7.4 Double line notation for gluons.

- Diagrams which involve quark internal loops (Fig. 7.5 (c)) are also suppressed in powers of $1/N_c^L$, where L is the number of quark loops.

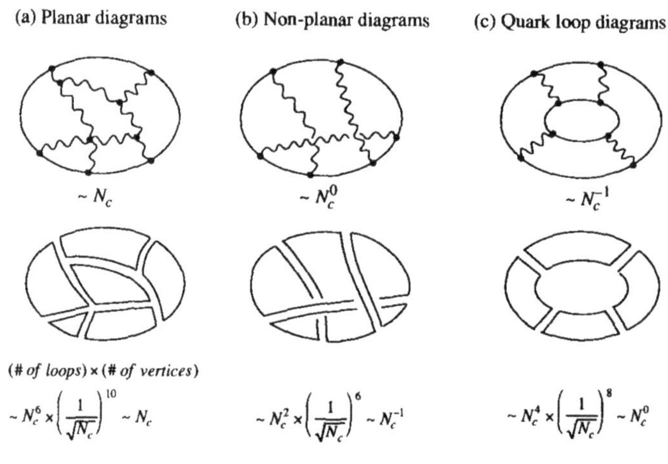

Fig. 7.5 Various diagrams and their double line representations; (a) planar (b) non-planar and (c) quark-loop diagrams. The powers of $1/N_c$ is indicated for each diagram.

From these, we can establish the $1/N_c$ behavior of a vertex that involves only mesons (see Fig 7.6). Note that an M-meson amplitude is topologically equivalent to a two meson amplitude. Hence the net $1/N_c$ behavior of these amplitudes is the same; they are all of order N_c. However, at each meson annihilation or creation point, the factor $\sqrt{N_c}$ is attached. Therefore, $1/N_c$

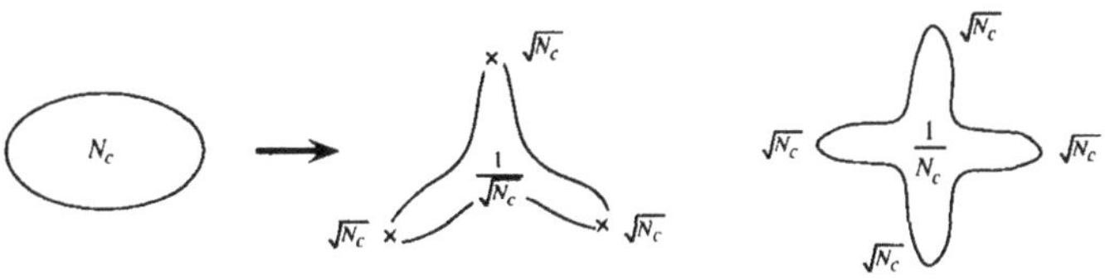

Fig. 7.6 Various meson vertices.

dependence of the M-meson vertex is given by

$$N_c \times \sqrt{N_c}^{(-M)} = (1/\sqrt{N_c})^{M-2}\,. \tag{7.7}$$

For instance, a meson decay constant f_m is regarded as a vertex having only one meson, and so $f_m \sim (1/\sqrt{N_c})^{1-2} \sim \sqrt{N_c}$. The mass of a meson is regarded as a vertex having two mesons, and so $m \sim (1/\sqrt{N_c})^{2-2} \sim 1$. Similarly, as the number of mesons entering a vertex increases, the power of $1/\sqrt{N_c}$ of the vertex increases.

Baryons

Baryons are treated differently from mesons, for they are bound states of N_c quarks [Witten 1979]. As in the non-relativistic quark model we consider baryon wave functions as products of orbital, spin and flavor parts. Since the color part is totally antisymmetric, the remaining part of the wave function behaves as a bosonic wave function. Therefore, for the ground state, all quarks can occupy in the same lowest orbit. Hence, as the number of quarks N_c increases, the size of the baryon remains unchanged, but the quark density inside the baryon increases. It often happens that high density limit is the place where the semiclassical method of the Hartree approximation can apply. The Hartree approximation together with the weak coupling $g \sim 1/\sqrt{N_c}$ can establish scaling laws for various meson-baryon observables.

Consider a non-relativistic N_c quarks interacting through a weak two body potential. Strictly, the validity of such a setup has not been justified. Consequences, however, will be tested as compared with our knowledge of baryons. We write the hamiltonian as

$$H = \sum_{i=1} \frac{p_i^2}{2m} + \frac{1}{2N_c} \sum_{ij} V(\boldsymbol{x}_i, \boldsymbol{x}_j)\,. \tag{7.8}$$

The factor $1/N_c$ in the interaction term reflects the $1/N_c$ dependence of that term correctly. For one gluon exchanges, the gqq vertex scales as $1/\sqrt{N_c}$, and so the two body potential scales as $1/N_c$. The same $1/N_c$ scaling follows in a meson exchange potential, since mqq vertices are expected to scale as $1/\sqrt{N_c}$ (see Fig. 7.7).

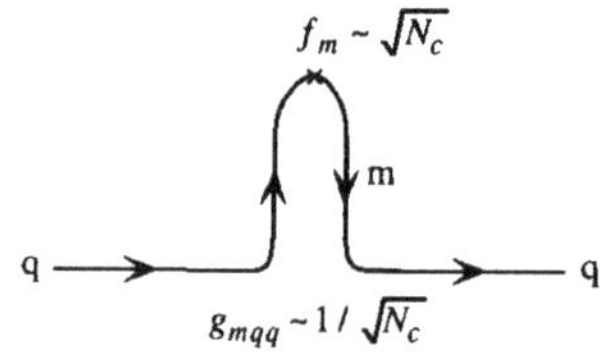

Fig. 7.7 $1/N_c$ counting for a meson-quark (mqq) vertex. The factor of the meson decay constant $\sqrt{N_c}$ is canceled by the factor $1/\sqrt{N_c}$ of the mqq vertex.

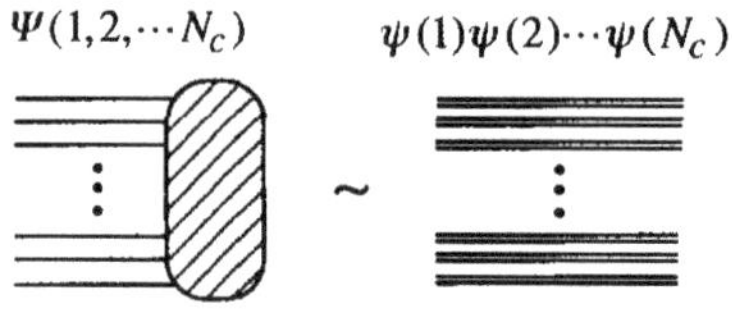

Fig. 7.8 Baryon wave function in the Hartree approximation. Single lines are for non-interacting quarks, while double lines for interacting (Hartree dressed) quarks.

In the Hartree approximation the ground state baryon wave function is assumed to be a product of the same N_c single particle states (see Fig. 7.8):

$$\Psi(1,\ldots,N_c) = \psi(1)\cdots\psi(N_c)\,. \tag{7.9}$$

Then the variational principle

$$\delta\langle\Psi|H|\Psi\rangle = 0 \tag{7.10}$$

subject to the constraint $\langle\Psi|\Psi\rangle = 1$ leads to the Hartree equation for $\psi(\boldsymbol{x})$:

$$\left(\frac{\nabla^2_{\boldsymbol{x}}}{2m} + U(\boldsymbol{x})\right)\psi(\boldsymbol{x}) = \varepsilon\psi(\boldsymbol{x})\,. \tag{7.11}$$

Here the single particle potential $U(\boldsymbol{x})$ is given by

$$
\begin{aligned}
U(\boldsymbol{x}) &= \frac{1}{N_c}\sum_i{}' \int d\boldsymbol{x}_i V(\boldsymbol{x},\boldsymbol{x}_i)|\psi(\boldsymbol{x}_i)|^2 \\
&\sim \int d\boldsymbol{x}_i V(\boldsymbol{x},\boldsymbol{x}_i)|\psi(\boldsymbol{x}_i)|^2 . \qquad (7.12)
\end{aligned}
$$

The sum in the first equation excludes the contribution from $\psi(\boldsymbol{x})$ itself, and in the second equation we approximate $N_c - 1 \sim N_c$. The first crucial observation is that the factor $1/N_c$ in the interaction is canceled by the combinatorial factor $N_c - 1$. Hence the Hartree potential for ψ is of order one. The total energy of the baryon in the Hartree approximation is then given by

$$
\begin{aligned}
E &= \langle\Psi|H|\Psi\rangle \\
&= N_c\left(\frac{1}{2m}\int d\boldsymbol{x}(\psi^\dagger\nabla^2\psi + U(\boldsymbol{x})|\psi(\boldsymbol{x})|^2)\right) \\
&= N_c\left(\frac{1}{2m}\int d\boldsymbol{x}\psi^\dagger\nabla^2\psi\right. \\
&+ \left.\int d\boldsymbol{x}d\boldsymbol{x}'|\psi(\boldsymbol{x})|^2 V(\boldsymbol{x},\boldsymbol{x}')|\psi(\boldsymbol{x}')|^2\right) . \qquad (7.13)
\end{aligned}
$$

From this we observe that the total baryon mass scales as N_c.

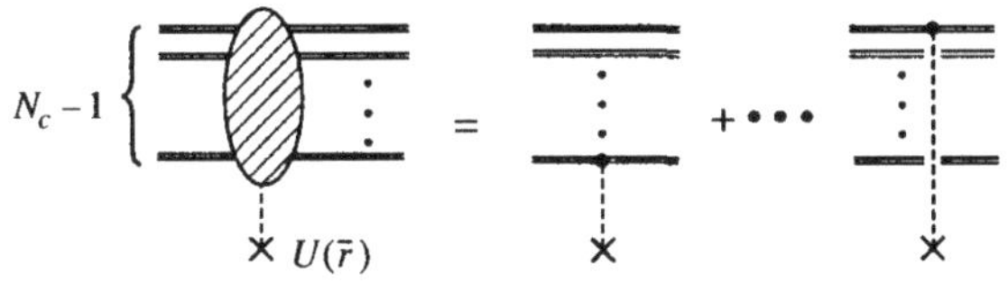

Fig. 7.9 Diagrammatic representation of the Hartree potential.

It would be instructive to present a diagrammatic interpretation of what we have done so far. The key equation is the Lippman-Schwinger equation for ψ, which is the inverse of Eq. (7.11):

$$
\psi = \phi + GU\psi , \qquad G = \frac{1}{\varepsilon - \frac{\nabla^2}{2m}} , \qquad (7.14)
$$

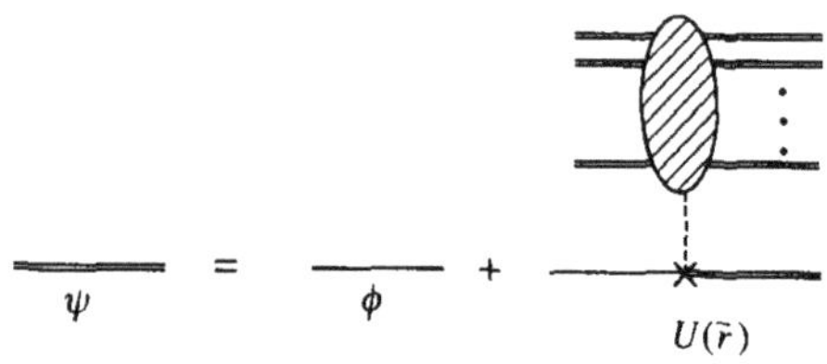

Fig. 7.10 Lippman-Schwinger equation for the Hartree approximation.

where ϕ is the unperturbed wave function satisfying

$$\left(e - \frac{\nabla^2}{2m}\right)\phi = 0\,. \tag{7.15}$$

We have written an unperturbed quark ϕ by a single line, and a Hartree-dressed quark ψ by a double line as in Fig. 7.8. Now the interaction $U(\boldsymbol{x})$ is presented diagrammatically in Fig. 7.9. Then the Hartree equation is presented as in Fig. 7.10.

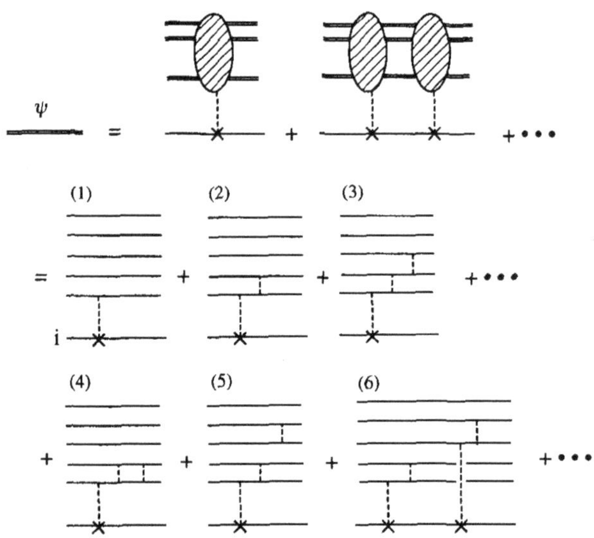

Fig. 7.11 Lippman-Schwinger equation for the baryon in the Hartree approximation.

These two ingredients enable one to write all diagrams in perturbation series. In Fig. 7.11 we show several diagrams for the baryon wave function. It is the sum of the geometric series of the Hartree potential $U(\boldsymbol{x})$. Let us count powers in $1/N_c$ in some of these diagrams.

(1): This diagram is the lowest order with respect to the two body interaction, V/N_c. After taking the sum over spectators $(j \neq i)$ it is of order $1 \sim V/N_c \times (N_c - 1)$. This provides the basic building block of the Hartree potential (shaded blobs).
(2-3): These look higher order, for the small interaction V/N_c is repeated more than twice. However, combinatorial factors cancel the small factors; for (2) $(V/N_c)^2 \times N_c(N_c - 1) \sim 1$, and for (3) $(V/N_c)^3 \times N_c(N_c-1)(N_c-2) \sim 1$. In fact, all the diagrams of this kind in which an interaction occurs only once between the same pair of quarks are of order one.
(4): This is not included in the Hartree approximation. The ladder series are genuine two body process which can not be accounted for by the Hartree approximation.
(5): This disconnected type appears in the baryon-baryon transition amplitude. The power counting, however, seems to be problematic, since the power of N_c seems to increases for more disconnected graphs. The resolution is in the fact that the connected amplitude of order N_c is exponentiated to yield all amplitudes including disconnected diagrams [Witten 1979]. This is a general property of the vacuum to vacuum amplitude *.
(6): This type is regarded as one of higher order in the geometric series.

A handy notation is available by adopting the Goldstone diagram of many body physics, where the baryon of N_c-quarks is regarded as the ground state of the many-body system [Goldstone 1957]. Therefore, nothing is drawn for the ground state baryon itself. The potential and the scattering equation are, however, depicted in Fig. 7.12. The closed loop of tadpoles means that the sum over N_c quarks is taken. The perturbative expansion of a Hartree quark then produces diagrams of the type called daisy or super daisy diagrams. In this scheme the energy of the baryon is depicted in Fig. 7.13, which is the sum of the one-body kinetic term and the two-body potential terms.

*See textbooks on field theories or many-body problems.

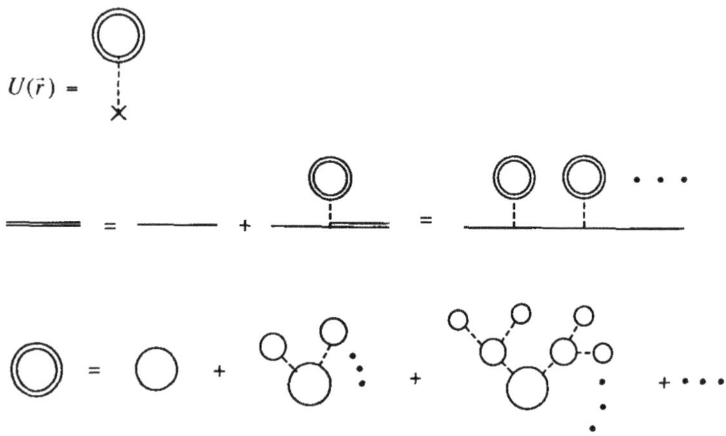

Fig. 7.12 Goldstone diagrams for the Hartree potential (top), LS equation (middle) and daisy diagrams for a Hartree-dressed quark like (bottom).

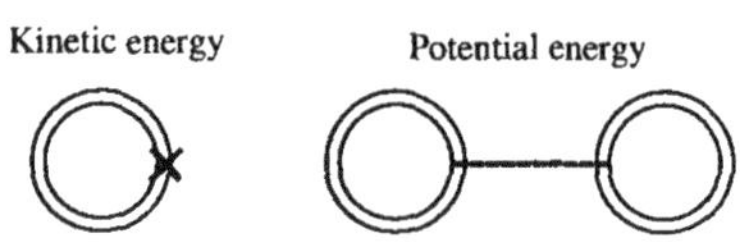

Fig. 7.13 The Goldstone diagram for the Hartree energy.

Meson and baryon couplings

Having ground state baryons as N_c quark bound states in the Hartree approximation, we can study more on meson-baryon systems.

For meson-baryon Yukawa couplings, we consider a baryon matrix element of a meson interpolating field:

$$\begin{aligned} V_{MBB}(q) &\sim \int d^4x e^{-iqx} \langle B | J_M(x) | B \rangle \,, \\ J_M &= \bar{\psi} \hat{M} \psi \,, \end{aligned} \tag{7.16}$$

where $\hat{M}$ is a suitable matrix (including Dirac, flavor matrices, derivatives and etc) carrying the mesonic quantum numbers. A diagram for the matrix element is depicted in Fig. 7.14. Note that from the conservation of energy and momentum, it is not possible for all external lines (either meson or baryons) to be on mass shell. It is not difficult to verify that the leading contribution stems from connected diagrams as in the first term of Fig. 7.14. The meson quark vertex scales as $1/\sqrt{N_c}$. This is then multiplied by the number of quarks N_c in the baryon. Hence the meson-baryon coupling scales as $\sqrt{N_c}$.

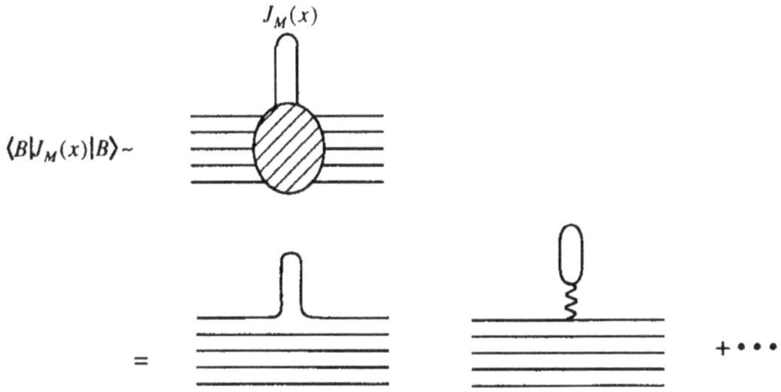

Fig. 7.14 Baryon matrix element of one meson interpolating field for a meson-baryon vertex.

Next we consider a meson-baryon scattering amplitude. It is a baryon matrix elements of two meson interpolating fields:

$$T_{MB\to M'B} \sim \int d^4x d^4y\, e^{-iqx+ipy} \langle B|J_{M'}(y)J_M(x)|B\rangle \,. \tag{7.17}$$

The corresponding diagram is shown in Fig. 7.15. Naively, one might expect that the leading contribution comes from the disconnected diagram (b) which scales as $\mathcal{O}(N_c)$, since the combinatorial factor is larger than that of (a). However, for on shell mesons and baryons the amplitude (b) vanishes for the same reason as that for the MBB coupling. In contrast, the diagram (a) is allowed, since the intermediate quark can be off mass shell (in excited states). Therefore the leading term for meson-baryon scattering is described by the connected type (a), the order of which is N_c^0.

Similarly, one can consider other meson-baryon processes and determine their $1/N_c$ dependence. One can verify

$$T_{aMB\to bMB} \sim \left(\frac{1}{\sqrt{N_c}}\right)^{a+b-2} \,. \tag{7.18}$$

Here a and b are the numbers of mesons in the initial and final states,

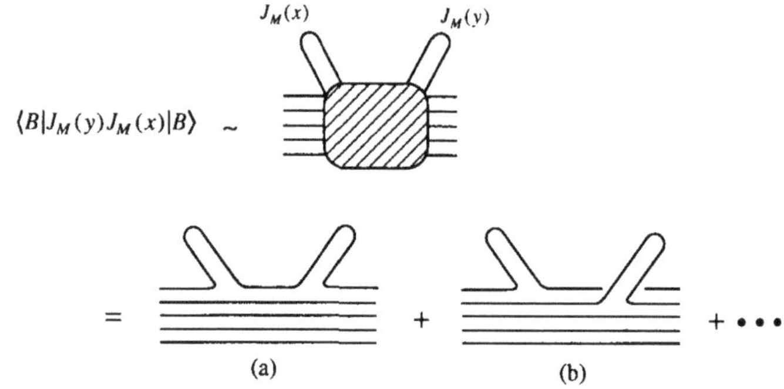

Fig. 7.15 Baryon matrix element of two meson interpolating field for meson-baryon scatterings.

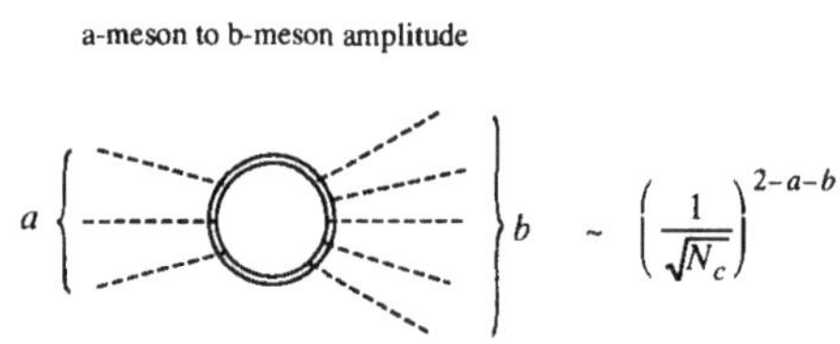

Fig. 7.16 Goldstone diagrams for a-meson baryon to b-meson baryon processes.

respectively. Goldstone diagrams for those processes are shown in Fig. 7.16, where mesons are represented by dashed lines.

7.3 Counting Rules for Solitons

In this section we investigate how the general $1/N_c$ counting rule applies to soliton solutions. We discuss two examples: the ϕ^4-theory in 1+1 dimension and the chiral bag model.

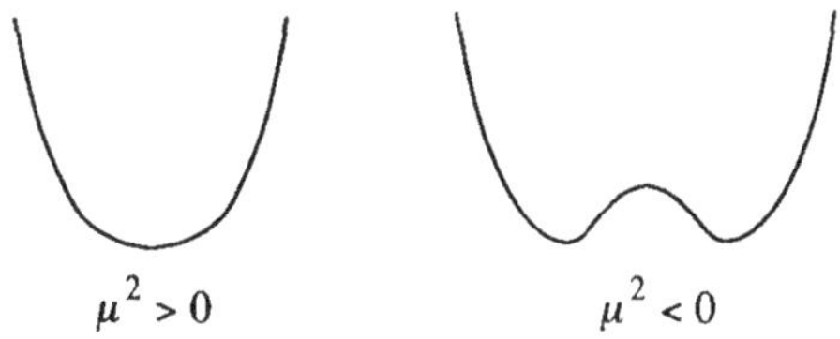

Fig. 7.17 Two distinguished forms of the potential of the ϕ^4-theory depending on the sign of μ^2.

ϕ^4-soliton in 1 + 1 dimension

Consider a scalar field theory in 1 + 1 dimensional space-time, whose lagrangian is given by

$$\mathcal{L} = \frac{1}{2}(\partial_\mu \phi)^2 - \frac{1}{2}\mu^2\phi^2 - \frac{\lambda}{4}\phi^4 . \tag{7.19}$$

The properties of this system are well known and therefore, what we discuss here is only related to the $1/N_c$ expansion. We consider the case of negative mass $\mu^2 < 0$, when the potential

$$V(\phi) = \frac{\lambda}{4}\phi^4 + \frac{1}{2}\mu^2\phi^2 \tag{7.20}$$

becomes a double well form as sketched in Fig. 7.17. A constant value $\mu^4/4\lambda$ may be added to the potential so that the potential minima are shifted to zero.

Consider solutions to the static field equation:

$$\phi'' = \mu^2 + \lambda\phi^3 . \tag{7.21}$$

Apparently there are two trivial solutions

$$\phi(x) = \pm a , \tag{7.22}$$

where

$$a = \sqrt{\frac{-\mu^2}{\lambda}} . \tag{7.23}$$

They are just the trivial vacua of the theory. We are, however, interested in non-trivial (kink) solutions which connect two vacua between $x \to \pm\infty$. Such a solution can be also found easily. Imposing the boundary condition either $\phi(\pm\infty) \to \pm a$ or $\phi(\pm\infty) \to \mp a$, one can integrate the field equation and find

$$\phi_\pm = \pm a \tanh m(x - X)\,, \tag{7.24}$$

where

$$m = \sqrt{\frac{-\mu^2}{2}}\,, \tag{7.25}$$

and X determines the position of the solutions. The solution $\phi_+(x)$ is shown in Fig. 7.18.

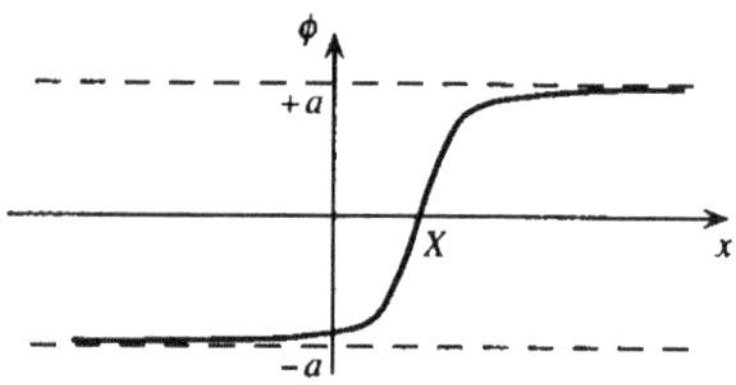

Fig. 7.18 A soliton solution of the ϕ^4-theory.

The mass of the solution takes a finite value as

$$\begin{aligned} M &= \int_{-\infty}^{+\infty} dx \left(\frac{1}{2}\phi'^2 + \frac{\lambda}{4}\left(\phi^2 + \frac{\mu^2}{\lambda}\right)\right) \\ &= \int_{-\infty}^{+\infty} dx\ \phi'^2 \\ &= a^2\sqrt{\frac{-\mu^2}{2}} \int_{-\infty}^{+\infty} dz \frac{1}{\cosh^4 z} \\ &= \frac{4}{3\sqrt{2}} \frac{(-\mu^2)^{3/2}}{\lambda}\,. \end{aligned} \tag{7.26}$$

The link to the $1/N_c$ expansion is made by assuming that

$$\mu \sim 1\,, \tag{7.27}$$

$$\lambda \sim 1/N_c\,; \tag{7.28}$$

the mass of the meson μ is of order one while the coupling constant λ is of order $1/N_c$ and is a small quantity in the limit $N_c \to \infty$. Naively, in the limit of small coupling, the theory becomes free theory. However, the soliton solution can survive no matter how small the coupling constant is. In this limit, we observe from Eqs. (7.24)-(7.26) that:

- The vacuum expectation value of ϕ at $x \to \pm\infty$ increases as $\sqrt{N_c}$. This reflects the fact that meson decay constants are of order $\sqrt{N_c}$.
- The size of the soliton is proportional to $1/\sqrt{-\mu^2}$ and is independent of N_c. This reflects the fact that baryon size is of order one.
- The mass of the soliton is of order $1/\lambda \to N_C$. This is also what we expect for baryons.

One would have reached these conclusions without explicit computations if we wrote the lagrangian in terms of the scaled field

$$\varphi = \sqrt{\lambda}\phi \,. \tag{7.29}$$

We have

$$\mathcal{L} = \frac{1}{\lambda}\left(\frac{1}{2}(\partial_\mu\varphi)^2 - \frac{1}{2}\mu^2\varphi^2 - \frac{1}{4}\varphi^4\right) . \tag{7.30}$$

We observe that the N_c dependence appears as an overall factor of the lagrangian.

Cranking method – translational motion

The collective motion of the ϕ^4 soliton is a translational motion dictated by the integral constant X in Eq. (7.24). Elevating it to a time dependent dynamical variable,

$$X \to X(t) \tag{7.31}$$

and using the ansatz

$$\phi = a \tanh m(x - X(t)) \,, \tag{7.32}$$

we find the collective lagrangian

$$\begin{aligned} L &= \int dx \left(\frac{1}{2}\dot{\phi}^2 - \frac{1}{2}(\phi')^2 - \frac{1}{2}\mu^2\phi^2 - \frac{\lambda}{4}\phi^4\right) \\ &= \int dx \left(\frac{1}{2}\dot{X}^2\phi'^2 - \cdots\right) \\ &= \frac{1}{2}M\dot{X}^2 - M \,. \end{aligned} \tag{7.33}$$

It is not difficult to verify that the moment of inertia, the coefficient of $(1/2)\dot{X}^2$, equals the classical mass M. Defining the canonical momentum conjugate to X,

$$P = \frac{\partial L}{\partial \dot{X}} = M\dot{X}\,, \tag{7.34}$$

we obtain the collective hamiltonian

$$H = \frac{1}{2M}P^2 + M\,, \tag{7.35}$$

which is the expected result in the non-relativistic limit of the theory.

In quantization, the momentum P is regarded as a quantity of order one. Hence the kinetic energy

$$K = \frac{1}{2M}P^2 \tag{7.36}$$

and the velocity

$$V = \frac{P}{M} \tag{7.37}$$

are small quantities of order $1/N_c$.

The Skyrme Model

We can establish similar scaling behavior in the Skyrme model;

$$\mathcal{L} = -\frac{f_\pi^2}{4}\,\mathrm{tr}\,[\partial_\mu U U^\dagger \partial^\mu U U^\dagger] + \frac{1}{32e^2}\,\mathrm{tr}\,[\partial_\mu U U^\dagger, \partial_\nu U U^\dagger]^2\,, \tag{7.38}$$

if we recognize

$$f_\pi \sim \sqrt{N_c}\,, \tag{7.39}$$

$$e \sim \frac{1}{\sqrt{N_c}}\,. \tag{7.40}$$

The first equation holds since f_π is the decay constant of the pion, while the second equation is understood if e is identified with the coupling constant of $\rho \to 2\pi$ (three-meson vertex). The chiral field

$$U = \exp(i\boldsymbol{\tau} \cdot \boldsymbol{\pi}/f_\pi) \tag{7.41}$$

is the quantity of order one. Therefore, both terms of the Skyrme lagrangian are of order N_c, and so is the energy of the hedgehog solution.

For the rotating hedgehog

$$A(t)U_h A^\dagger(t)\,, \tag{7.42}$$

the collective motion dictated by A is regarded as slow. In the energy of the rotating skyrmion (see Eq. (6.65)),

$$\begin{aligned}
E_{\rm rot} &= E_h + \frac{1}{2}\Lambda\Omega^2 \\
&\to E_h + \frac{1}{2\Lambda}I(I+1)\,, \\
E_h &= 4\pi\int r^2\,dr\left[\frac{f_\pi^2}{2}\left(F'^2+\frac{2}{r^2}\sin^2 F\right)\right. \\
&\qquad \left. + \frac{\sin^2 F}{2e^2r^2}\left(2F'^2+\frac{\sin^2 F}{r^2}\right)\right]\,, \\
\Lambda &= \frac{8\pi}{3}f_\pi^2\int_0^R r^2dr\,\sin^2 F\left(1+\frac{1}{e^2f_\pi^2}\left(F'^2+\frac{\sin^2 F}{r^2}\right)\right)\,,
\end{aligned}$$

it is now apparent that the classical hedgehog energy E_h and the moment of inertia Λ are of order N_c, and that the rotational velocity Ω and the rotational energy are of order $1/N_c$.

7.4 Large-N_c Algebra for Baryons

Commutation relations

Let us consider a meson-baryon process where m mesons interact with a baryon as shown in Fig. 7.19. From the $1/N_c$-counting rule, the T-matrix for this process should scale as $(1/\sqrt{N_c})^{m-2}$. However, if we write the process as a product of a T-matrix of $m-1$ mesons and mBB vertex (and a baryon propagator), it appears to behave as $(1/\sqrt{N_c})^{m-4}$, which would be of higher order. In order to be consistent with the Witten's counting rule, some of these terms cancel by imposing some suitable conditions. These are the consistency conditions among meson-baryon vertices.

To start with let us consider the pion-nucleon scattering amplitude (Fig. 7.20). The general $1/N_c$ counting rule determines that it is of order $\mathcal{O}(1)$. If we count the Born diagrams of Fig. 7.20, however, they are of order N_c. Since the πNN Yukawa coupling scales as $\sqrt{N_c}$, we may factor out the N_c dependence explicitly by writing it as $\sqrt{N_c}\mathcal{V}^\alpha_{XY}$. Here the

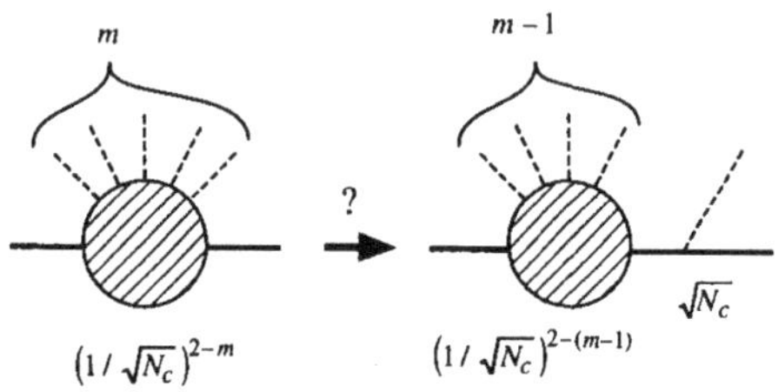

Fig. 7.19 m-meson to b-meson processes.

index α represents the quantum numbers brought into the vertex through the derivative coupling of the pion; they are the adjoint representations of angular momentum ($L_\pi = 1$) and of isospin ($I_\pi = 1$), and so $\alpha \sim (i, a)$. The indices X and Y label the quantum numbers of the nucleon, the fundamental representations of spin and isospin: $X \equiv (J(\text{spin}), I(\text{isospin}))$. Using these notations, the scattering amplitude is given by

$$\mathcal{M} \sim N_c \sum_Y \left\{ \frac{\mathcal{V}^\beta_{XY} \mathcal{V}^\alpha_{YZ}}{\omega + M_Z - M_Y} - \frac{\mathcal{V}^\alpha_{XY} \mathcal{V}^\beta_{YZ}}{\omega - M_X + M_Y} \right\}, \tag{7.43}$$

where $M_{X,Y,Z}$ are the masses of the baryons.

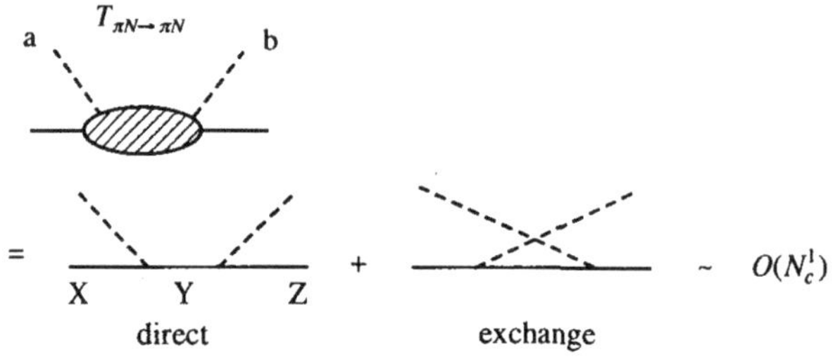

Fig. 7.20 Born diagrams for πN scattering.

These two Born terms must cancel to be consistent with the general $1/N_c$ counting. For this to happen it is sufficient that the mass differences between intermediate baryons are of order $1/N_c$, and that the following relation holds,

$$\sum_Y \left(\mathcal{V}^\beta_{XY}\mathcal{V}^\alpha_{YZ} - \mathcal{V}^\alpha_{XY}\mathcal{V}^\beta_{YZ}\right) \stackrel{N_c\to\infty}{\longrightarrow} 0\,. \tag{7.44}$$

Here the left hand side must be suppressed as $1/N_c$. Eq. (7.44) is now regarded as a matrix equation for $\mathcal{V}^\alpha$ whose components are $\mathcal{V}^\alpha_{XY}$. This is the primary consistency condition for large-N_c QCD [Dashen et al. 1994; Gervais and Sakita 1984a; Gervais and Sakita 1984b].

In the limit $N_c \to \infty$, combining the spin (J) and isospin (I) algebra for light flavors, $K = SU(2)_J \times SU(2)_I$, with that generated by Eq. (7.44), we obtain the following commutation relations:

$$\begin{aligned}[]
[\mathcal{J}^a, \mathcal{J}^b] &= i f^{abc}\mathcal{J}^c\,,\\
[\mathcal{J}^a, \mathcal{V}^\alpha] &= i(t^a)^\alpha_\beta \mathcal{V}^\beta\,,\\
[\mathcal{V}^\alpha, \mathcal{V}^\beta] &= 0\,.
\end{aligned} \tag{7.45}$$

Here $\mathcal{J}^a \in K$ and f^{abc} are the generators and the structure constants for the group K. $(t^a)^\alpha_\beta$ are the matrices of the generators $\mathcal{J}^a$ in the basis of α: $(t^a)^\alpha_\beta \equiv \langle\alpha|J^a|\beta\rangle$. Eqs. (7.45) show that $\mathcal{J}^a$ and $\mathcal{V}^\alpha$ form the non-compact group algebra $G = K \times T$, the semi-direct product of K by the Abelian group T generated by $\mathcal{V}^\alpha$ †. Because of the non-compact nature of the group, the baryon states form an infinite tower, for example those with equal spin and isospin $(I, J) = (\frac{1}{2}, \frac{1}{2}), (\frac{3}{2}, \frac{3}{2})$, etc. This is discussed in the next section.

The method of induced representations

Consider a group G and its subgroup $H \subset G$. The question is for a given representation of the subgroup H, how one can construct representations of G. In this case, we say that representations of G are induced by representations of H. In our problem, $G = K \times T$ and $H = T$, whose Lie algebra is given by Eq. (7.45). The representation of the abelian group T is easy to find, since the generators commute each other (see Eq. (7.45)) and can be diagonalized in the basis $|v\rangle$,

$$\mathcal{V}^\alpha|v\rangle = v^\alpha|v\rangle\,. \tag{7.46}$$

†The non-commutativity of the second equation of Eq. (7.45) indicates that the product of K and T is not a direct product but is a semi-direct product.

Thus $\mathcal{V}^\alpha$ are regarded as coordinates in a nine dimensional manifold $V \ni v$. The dimension nine is 3 (adjoint representation of spin) $\times$ 3 (adjoint representation of isospin).

A point v in the manifold is characterized by the orbit. It is defined as a set of points generated by K from one point $\bar{v}$. In other words, for a given *initial* point $\bar{v}^\alpha$ on an orbit, all points on the orbit are generated by

$$(k\bar{v})^\alpha = U(k)^\alpha_\beta \bar{v}^\beta \,, \tag{7.47}$$

where k is an element of K and $U(k)^\alpha_\beta$ is the unitary representation of k in the basis of α,

$$U(k)^\alpha_\beta = \langle\alpha|k|\beta\rangle \,. \tag{7.48}$$

Therefore, for a given $\bar{v}$, the dimension of the orbit is at most dim(K). The orbits are specified by the Casimir operators of the group K, the invariants under all transformations $k \in K$.

Sometimes it happens that $\bar{v}$ is invariant under the transformations ℓ which are elements of a subgroup $L \subset K$. Then the dimension of the orbit reduces to dim(K) $-$ dim(L). The subgroup which leaves $\bar{v}$ invariant is called the little group.

As an example, consider a set of points $\mathcal{V} \sim (x, y, z)$ in the three dimensional space and the rotational group. As shown in Fig. 7.21, orbits are the sphere of radius r, since all points $\bar{v}$ on the sphere are reached from a point $\bar{v}$ by rotations. The radius $r = \sqrt{x^2 + y^2 + z^2}$ is invariant under rotations. Therefore it is the Casimir operator. However, the point $\bar{v}$ is left invariant under the $U(1)$ rotation around the axis which passes through the origin and the point $\bar{v}$. Hence this $U(1)$ is the little group.

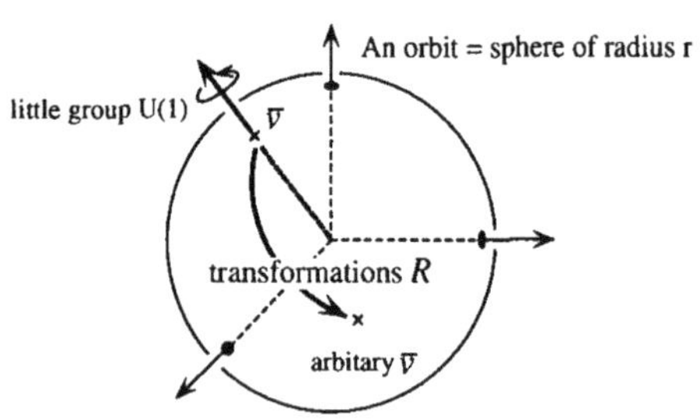

Fig. 7.21 A schematic example of an orbit.

Hedgehog representation

We demonstrate that, as the first non-trivial example of the large-N_c algebra of Eq. (7.45), the hedgehog state emerges in a natural way.

The relevant group is spin and isospin: $K = SU(2)_J \times SU(2)_I$ and T generated by $\mathcal{V}^\alpha$. The index $\alpha = (i, a)$, where $i = 1, 2, 3$ and $a = 1, 2, 3$, is formed from spin and isospin indices. Thus $\mathcal{V}^\alpha = \mathcal{V}_{(i,a)} \sim v_{(i,a)}$ parameterize a nine dimensional manifold. Under the transformations $k \in K$, however, a point $\bar{v}_{(i,a)}$ can cover at most a six dimensional manifold since the number of the group parameters $K = SU(2)_J \times SU(2)_I$ is six. The symmetry transformations k are therefore not sufficient to generate the whole nine dimensional manifold $\mathcal{V}$. The missing three dimensions are constrained by the three Casimir invariants:

$$\begin{aligned} \mathcal{C}_1 &= v_{ia} v_{ia} \,, \\ \mathcal{C}_2 &= \epsilon_{ijk} \epsilon_{abc} v_{ia} v_{jb} v_{kc} \,, \\ \mathcal{C}_3 &= v_{ia} v_{ib} v_{jb} v_{ja} \,. \end{aligned} \tag{7.49}$$

There are several points $\bar{v}$ (and therefore, orbits) that are not transformed under the transformations of non-trivial little groups. One example is $\bar{v}_{ia} = \delta_{1i}\delta_{1a}$, where the little group is $U(1)_J \times U(1)_I$. In this case, the Casimir invariants are

$$\mathcal{C}_1 = \mathcal{C}_3 = 1, \quad \mathcal{C}_2 = 0 \,. \tag{7.50}$$

Another example which is more interesting is provided by $\bar{v}_{ia} = \delta_{ia}$ where the little group is the diagonal $SU(2)_{J+I}$, and

$$\mathcal{C}_1 = \mathcal{C}_3 = 3, \quad \mathcal{C}_2 = 6 \,. \tag{7.51}$$

Observing that v_{ia} transform as

$$v_{ia} \to (t\, v\, u^\dagger)_{ia} \,, \tag{7.52}$$

where t^{ij} and u^{ab} be the 3×3 matrices for spin and isospin transformations, it is not difficult to verify that the point $\bar{v}_{ia} = \delta_{ia}$ is invariant under the diagonal subgroup $SU(2)$ of $SU(2)_J \times SU(2)_I$.

The states $|v\rangle$ are now specified by the three labels:

(1) the three Casimir invariants
(2) the coordinates on the coset K/L
(3) representations of the little group

Let the coordinates in K/L denoted by A, then

$$|v\rangle = |A, \bar{v}; \lambda\rangle\,, \tag{7.53}$$

where λ are the labels for the representation of the little group L. Here the constraints of C_1, C_2 and C_3 are determined by the point $\bar{v}$. Finally, we comment on that the representation of Eq. (7.53) is known to be irreducible [Gervais and Sakita 1984b; Gervais and Sakita 1984a].

In the case we are interested in, $K = SU(2)_J \times SU(2)_I$, $L = SU(2)_{J+I}$, and $\lambda = (K, K_z)$ for the combined angular momentum $K = J + I$. Specifically, $(K, K_z) = (0,0)$ corresponds to the spherical hedgehog. The fact that the matrices v^α are diagonal in the basis of A (see Eq. 7.46) corresponds to the semiclassical limit, where the overlap kernel reduces to the delta function:

$$\langle A', \bar{v}; (K, K_z)|A, \bar{v}; (K, K_z)\rangle \sim \delta(A' - A)\,. \tag{7.54}$$

The decomposition of the basis of the induced representation of Eq. (7.53) into components of physical baryons, irreducible representations of the symmetry group K, can be performed by the projection method. The general case for such a decomposition is found in Ref. [Gervais and Sakita 1984b; Gervais and Sakita 1984a]. Here we show it for $K = SU(2) \times SU(2)$ and $\bar{v}_{ia} = \delta_{ia}$. Let us start with a point $A = A_0 \in SU(2)$ in Eq. (7.53) and denote the state as $|A_0; KK_z\rangle$. The projection operator is defined by (see problem 7.2)

$$\int d[\omega\omega'] D^{I*}_{tt'}(\omega)\, D^{J*}_{mm'}(\omega')\, R^\tau(\omega)\, R^\sigma(\omega')\,, \tag{7.55}$$

where R^τ and R^σ are rotation operators in the isospin and the spin spaces, respectively. Now setting $\omega = AA'$ and $\omega' = A'$, the projection calculation goes as follows;

$$\begin{aligned}
&\int d[AA'] D^{I*}_{tt'}(AA')\, D^{J*}_{mm'}(A')\, R^\tau(AA')\, R^\sigma(A')\, |A_0; KK_z\rangle \\
&= \sum_{t''} \int d[AA'] D^{I*}_{tt''}(A) D^{I*}_{t''t'}(A')\, D^{J*}_{mm'}(A') \\
&\quad \times R^\tau(A)\, \underbrace{R^\tau(A') R^\sigma(A')}_{K-\text{rotation}}\, |A_0; KK_z\rangle\,,
\end{aligned} \tag{7.56}$$

where we have used the multiplication law $R(g_1g_2) = R(g_1)R(g_2)$. The combination $R^\tau(A')R^\sigma(A')$ is just the rotation operator of the little group generated by $\boldsymbol{K} = \boldsymbol{I} + \boldsymbol{J}$. It is now clear that the initial point A_0 is redundant and is absorbed into the integral variable A', and so we rewrite the state $|A_0; KK_z\rangle \to |KK_z\rangle$. Making use of

$$\begin{aligned} R^\tau(A')R^\sigma(A')|KK_z\rangle &= R^{\tau+\sigma}(A')|KK_z\rangle \\ &= \sum_{K'_z} |KK'_z\rangle\langle KK'_z|R^{\tau+\sigma}(A')|KK_z\rangle \\ &= \sum_{K'_z} |KK'_z\rangle D^K_{K'_z K_z}(A'), \end{aligned} \tag{7.57}$$

we have

$$\begin{aligned} (7.56) &= \sum_{t''} \int d[AA'] D^{I*}_{tt''}(A) D^{I*}_{t''t'}(A')\, D^{J*}_{mm'}(A') \\ &\quad \times R^\tau(A) \sum_{K'_z} D^K_{K'_z K_z}(A')|KK'_z\rangle \\ &= \sum_{t''K'_z} \int d[A']\, D^{I*}_{t''t'}(A')\, D^{J*}_{mm'}(A') D^K_{K'_z K_z}(A') \\ &\quad \times \int d[A] D^{I*}_{tt''}(A) R^\tau(A)|KK'_z\rangle \\ &= \frac{8\pi^2}{2K+1} \sum_{t''K'_z} (It''\, Jm|KK'_z)(It'\, Jm'|KK_z) \\ &\quad \times \int d[A] D^{I*}_{tt''}(A) R^\tau(A)|KK'_z\rangle . \end{aligned} \tag{7.58}$$

When $K = 0$ (and hence $K_z = K'_z = 0$), the Clebsch-Gordan coefficients lead to the constraint that the hedgehog state $K = 0$ is composed by the physical baryons with equal spin and isospin:

$$I = J\,, \quad m = -t''\,, \quad m' = -t'\,. \tag{7.59}$$

Incidentally, when $K = 1$, the state of Eq. (7.53) is decomposed into the baryons with $(I, J) = (1/2, 3/2), (3/2, 1/2), \cdots$.

7.5 Finite N_c

Quark representation

So far the discussion has been made in the limit $N_c \to \infty$. The reduction of the above procedure to the case of finite N_c is performed by the technique of *extraction* for group representations. In the literatures [Dashen et al. 1994; Gervais and Sakita 1984a], conversely, starting from finite representations of a compact group $SU(4)$, infinite representations of the non-compact group G have been obtained. This is the method of *contraction*.

The first step in the argument is based on the fact that the number of generators of $K \times T$, with $K = SU(p) \times SU(q)$ and T being a direct product of the adjoint representations of $SU(p)$ and $SU(q)$,

$$N = (p^2-1) + (q^2-1) + (p^2-1)(q^2-1) = (pq)^2 - 1\,. \tag{7.60}$$

This turns out to coincide the number of generators of the $SU(pq)$ group. Let us look at the subgroup $SU(p) \times SU(q) \subset SU(pq)$. Denoting the generators of the subgroup by $\mathcal{J}$ and the remaining generators by $\mathcal{W}$, the algebra now takes the form

$$\begin{aligned}
[\mathcal{J}^a, \mathcal{J}^b] &= i f^{abc} \mathcal{J}^c\,, \\
[\mathcal{J}^a, \mathcal{W}^\alpha] &= i D(a)^\alpha_\beta \mathcal{W}^\beta\,, \\
[\mathcal{W}^\alpha, \mathcal{W}^\beta] &= i\left(g_1^{\alpha\beta\gamma} \mathcal{W}^\gamma + g_2^{\alpha\beta c} \mathcal{J}^c\right).
\end{aligned} \tag{7.61}$$

Here f, g_1, g_2 are the structure constants. For the case of $p = q = 2$, the constants g_1 vanish and the last commutators of Eq. (7.61) can be written as

$$[\mathcal{W}^{ia}, \mathcal{W}^{jb}] = \frac{i}{4}\epsilon_{ijk}\delta_{ab} J^k + \frac{i}{4}\epsilon_{abc}\delta_{ij} I^c\,, \tag{7.62}$$

where J^i and I^a are the generators of the spin and isospin group.

Consider now the N-dimensional symmetric representation of $SU(4)$, where the matrix elements of $\mathcal{W}$ are of order N. The group contraction $SU(4) \to G = K \times T$ can be performed first by replacing $\mathcal{W}^\alpha$ by $\mathcal{V}^\alpha \equiv \mathcal{W}^\alpha/N$. In the limit $N \to \infty$ the algebra of Eq. (7.61) written in terms of $\mathcal{V}$ then reduces to the algebra of Eq. (7.45). The index N is then identified with the number of colors in what is called the "quark representation" (which refers to valence quarks) as shown in the following.

Consider the wave function of a baryon. As discussed in Chapter 2, it can be written as a direct product of orbital, spin, isospin (flavor) and color parts. Since the color part is totally antisymmetric the rest of the wave function must be totally symmetric. For ground state baryons, all quarks occupy the lowest S ($l = 0$) state, which gives a symmetric orbital wave function, and thus the spin and isospin part must be symmetric as well. The symmetric representation of $SU(4) \supset SU(2) \times SU(2)$ is specified by an index N, which can now be identified with the number of colors N_c. An example which realizes this is given by the operators,

$$\begin{aligned} J^i &= \frac{1}{2}\sum_{n=1}^{N} \sigma^i(n)\,, \quad I^i = \frac{1}{2}\sum_{n=1}^{N} \tau^i(n)\,, \\ W^{ia} &= \sum_{n=1}^{N} \sigma^i(n)\tau^a(n)\,. \end{aligned} \tag{7.63}$$

For an explicit representation of the group, we can make use of the technique of the interacting boson model in nuclear and molecular physics [Arima and Iachello 1984]. This method was applied to the quantization of the skyrmion rotation by Amado and collaborators [Oka et al. 1987; Amado et al. 1987]. Let us introduce operators which annihilate the fundamental representations of the spin and isospin group. For instance, $\alpha_{u\downarrow}$ annihilates an up quark with spin down. These operators satisfy the standard commutation relations

$$[\alpha_{tm}, \alpha^\dagger_{t'm'}] = \delta_{tt'}\delta_{mm'}\,. \tag{7.64}$$

The symmetric N_c representation of the $SU(4)$ group is then generated by creating the same N_c quarks from the vacuum. What interests us here is the hedgehog state:

$$|N_c; h_q\rangle = \frac{1}{\sqrt{N_c!}}\left(z_q^\dagger\right)^{N_c}|0\rangle\,, \tag{7.65}$$

$$z_q = \frac{1}{\sqrt{2}}(\alpha^\dagger_{u\downarrow} - \alpha^\dagger_{d\uparrow})\,, \tag{7.66}$$

where the subscript q stands for quarks. The operator z_q is exactly what is obtained by coupling the spin and isospin to the zero grand spin of the hedgehog state.

The state in Eq. (7.65) breaks rotational and iso-rotational symmetries, and therefore all states obtained by applying an isospin (or equivalently spin) rotation operator $R(A)$ on the hedgehog state of Eq. (7.65) are degenerate,

$$\begin{aligned}|N_c; h_q[A]\rangle &\equiv R(A)|N_c; h\rangle \\ &= \sqrt{\frac{1}{N_c!}}\left[\frac{1}{\sqrt{2}}\left(D^{1/2}_{uu}(A)\alpha^\dagger_{u\downarrow} + D^{1/2}_{du}(A)\alpha^\dagger_{d\downarrow}\right.\right. \\ &- \left.\left. D^{1/2}_{dd}(A)\alpha^\dagger_{d\uparrow} - D^{1/2}_{ud}(A)\alpha^\dagger_{u\uparrow}\right)\right]^{N_c}|0\rangle\,. \qquad (7.67)\end{aligned}$$

Here $D^{1/2}_{\mu\nu}(A)$ are the D-functions of rank 1/2. The semiclassical nature of the large N_c limit is reflected in the overlap function,

$$\langle N_c; h[A']|N_c; h[A]\rangle = [\cos\theta_{AA'^\dagger}]^{N_c}\,, \qquad (7.68)$$

where $\theta_{AA'^\dagger}$ represents symbolically the relative angle between the orientations A and A'. For instance, if we take $A' = 1$ and $A = (\alpha, \beta, \gamma)$ (α, β and γ are the three Euler angles for A), then

$$\begin{aligned}\langle N_c; h_q[0]|N_c; h_q[A]\rangle &= \left(\frac{D^{1/2}_{uu}(A) + D^{1/2}_{dd}(A)}{2}\right)^{N_c} \\ &= \left(\cos\frac{\alpha+\gamma}{2}\cos\frac{\beta}{2}\right)^{N_c}. \qquad (7.69)\end{aligned}$$

It is apparent that after an appropriate rescaling of the states this overlap approaches a delta function

$$\langle N_c; h[A']|N_c; h[A]\rangle \stackrel{N_c\to\infty}{\longrightarrow} \delta(A - A'). \qquad (7.70)$$

This sharp peak is characteristic of a semiclassical limit.

In order to obtain some feeling for how the quark representation works in actual computation, let us compute the axial-vector coupling constants. For this, let us briefly outline the projection theory for computing nucleon matrix elements. The nucleon state $|N\rangle$ is projected out from the hedgehog state $|h\rangle$ as in Eq. (7.58)

$$|N\rangle \propto \int d[A] D^{1/2*}_{t,-m}(A)\, R(A)\, |h\rangle\,. \qquad (7.71)$$

The nucleon matrix elements of an observable $\mathcal{O}_N$ are then computed by

$$\mathcal{O}_N = \frac{\langle N|\mathcal{O}|N\rangle}{\langle N|N\rangle}\,, \qquad (7.72)$$

where the denominator is needed for normalization.

Let us calculate the isoscalar and isovector axial-vector coupling constants for the nucleon, $g_A^{(0)}$ and g_A. In the algebraic method, the relevant operators are expressed as (see problem 7.3)

$$\begin{aligned} S_3 &\equiv \left[\alpha^\dagger\alpha\right]^{(10)}_{00} \\ &= \left(\alpha^\dagger_{u\uparrow}\alpha_{u\uparrow} - \alpha^\dagger_{d\downarrow}\alpha_{d\downarrow} + \alpha^\dagger_{u\downarrow}\alpha_{u\downarrow} - \alpha^\dagger_{d\uparrow}\alpha_{d\uparrow}\right), \qquad (7.73) \\ T_{33} &\equiv \left[\alpha^\dagger\alpha\right]^{(11)}_{00} \\ &= \left(\alpha^\dagger_{u\uparrow}\alpha_{u\uparrow} + \alpha^\dagger_{d\downarrow}\alpha_{d\downarrow} - \alpha^\dagger_{u\downarrow}\alpha_{u\downarrow} - \alpha^\dagger_{d\uparrow}\alpha_{d\uparrow}\right). \qquad (7.74) \end{aligned}$$

Here the super- and sub-scripts in $[\cdots]^{(\cdots)}_{\cdots}$ are for (S, I) and their third components S_z, I_z, respectively. It is not difficult to check that these operators have the same transformation properties as σ_3 and $\sigma_3\tau_3$. The expectation values in Eq. (7.72) for the $p\downarrow$ state are computed in a straightforward manner, and the results are [Hosaka and Walet 1995]

$$\begin{aligned} \langle p\downarrow|S_3|p\downarrow\rangle &= -1, \qquad (7.75) \\ \langle p\downarrow|T_{33}|p\downarrow\rangle &= -\frac{N_c}{3}\left(1+\frac{2}{N_c}\right). \qquad (7.76) \end{aligned}$$

The minus signs reflects that they are the matrix elements of the $p\downarrow$ state.

The result of Eq. (7.75), $\langle p\downarrow|S_3|p\downarrow\rangle = -1$, as independent of N_c is trivial, since it just reflects the normalization of the wave function. In the non-relativistic quark model this is equivalent to the statement that nucleon spin is entirely carried by the quarks. On the other hand, the result of Eq. (7.76) depends on N_c, since it counts the axial charge of all N_c quarks. Specifically, it becomes (neglecting the minus sign) $5/3$ when $N_c = 3$ as in the non-relativistic quark model, while it grows as N_c in the limit $N_c \to \infty$, the scaling as expected in the Skyrme model. Thus the quark representation is believed to interpolate the quark model and the Skyrme model results by varying N_c from 3 to ∞ [Manohar 1984].

Second Consistency Condition

The second consistency condition for $1/N_c$ expansion arises when considering the $\pi N \to \pi\pi N$ amplitude [Dashen et al. 1994]. The large N_c counting rule implies that the amplitude is of order $N_c^{-1/2}$, while the leading order tree graphs as in Fig. 7.22 appears to be of order $N_c^{3/2}$. Adding all the possible combinations of the tree diagrams, the scattering amplitude $T_{\pi N\to\pi\pi N}$ contains a factor proportional to

$$N_c^{3/2}\left[\mathcal{V}^\alpha,\, [\mathcal{V}^\beta,\, \mathcal{V}^\gamma]\right], \tag{7.77}$$

where we have again factored out the N_c dependence of the amplitude and the indices of the nucleon quantum numbers have been suppressed. The fact that the double commutator of Eq. (7.77) must vanish leads to a constraint on $1/N_c$ corrections to matrix elements.

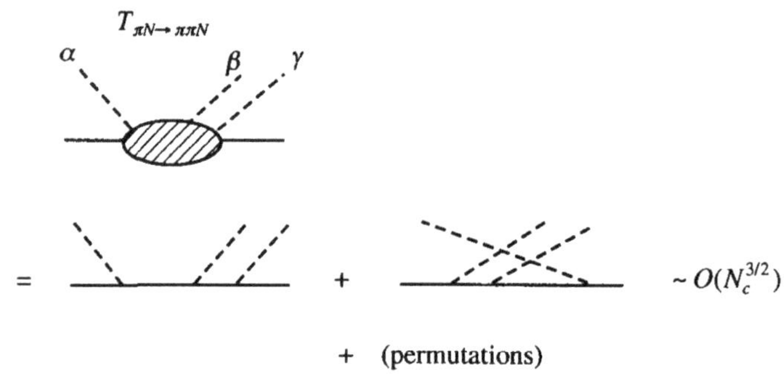

Fig. 7.22 Leading order tree diagrams for $\pi N \to \pi\pi N$.

Let us consider the axial-vector coupling constant g_A. Since it has the same spin and isospin structure as that of the πNN Yukawa coupling, it is defined by

$$\langle N|\bar\psi\gamma_i\gamma_5\tau^a\psi|N\rangle = g_A\langle N|\mathcal{V}^{ia}|N\rangle\,. \tag{7.78}$$

For $1/N_c$ corrections, the vertex function $\mathcal{V}^{ia}$ is expanded in powers of $1/N_c$,

$$\mathcal{V}^{ia} = \mathcal{V}_0^{ia} + \frac{1}{N_c}\mathcal{V}_1^{ia} + \frac{1}{N_c^2}\mathcal{V}_2^{ia} + \cdots\,. \tag{7.79}$$

Substituting this equation in the constraint (7.77) (without $N_c^{3/2}$) $= 0$ and expanding it up to order $1/N_c$, one finds

$$\left[\mathcal{V}_0^\alpha,\, \left[\mathcal{V}_1^\beta,\, \mathcal{V}_0^\gamma\right]\right] + \left[\mathcal{V}_0^\alpha,\, \left[\mathcal{V}_0^\beta,\, \mathcal{V}_1^\gamma\right]\right] = 0\,, \tag{7.80}$$

where we have used the primary consistency condition in the leading order, $\left[\mathcal{V}_0^\alpha, \mathcal{V}_0^\beta\right] = 0$. It has been shown that the solution of Eq. (7.80) is [Dashen and Manohar 1993a]

$$\mathcal{V}_1^\alpha = \text{const} \cdot \mathcal{V}_0^\alpha \,. \tag{7.81}$$

This implies that the axial-vector current operator $\mathcal{V}^\alpha$ is not renormalized up to the first order of $1/N_c$. This is similar to the Ademollo-Gatto theorem [Ademollo and Gatto 1964] which says that conserved charges are not renormalized up to the first order in symmetry breaking parameter. Thus matrix elements of the axial-vector current are summarized as

$$\langle N|\bar{\psi}\gamma_i\gamma_5\tau^a\psi|N\rangle = \tilde{g}_A\langle N|\mathcal{V}_0^{ia}|N\rangle + \mathcal{O}(1/N_c^2)\,, \tag{7.82}$$

Equation (7.82) indicates that ratios of axial-vector coupling constants, and therefore those of the πBB couplings, are not subject to corrections of order $1/N_c$. The reason that the ratio is relevant here is that in the coefficient, $\tilde{g}_A$, higher order terms in $1/N_c$ could be included depending on the dynamics considered. Therefore, modulo absolute values, the relation (7.82) explains the general success of the $SU(4)$ quark model and the Skyrme model for ratios of pion-baryon coupling constants. For example, in the convention of Ref. [Brown and Weise 1975], the ratio of $g_{\pi N\Delta}/g_{\pi NN}$ of the quark model is $\sqrt{72/25} \sim 1.70$ while that of the Skyrme model is $3/\sqrt{2} \sim 2.12$ and their difference is $0.42/2.12 \sim 20$ % level, which is roughly compatible with $1/N_c^2$. Phenomenologically, the Skyrme model result (or the large N_c limit of the quark model) appears to be in better agreement with experiment.

7.6 Other Representations and g_A

In the algebraic method, we can go slightly further in determining the absolute value of g_A. In this section we compute it up to the next to leading order in $1/N_c$. This requires some dynamical information on nucleon structure. The assumptions we use are, however, minimal and are acceptable by many baryon models.

We attempt to find representations which is considered to involve more dynamics than in the simplest quark representation of Eq. (7.65). It, however, must have the same contracted large-N_c algebra of Eq. (7.45).

Relativistic effects

First we consider the relativistic effects which come from the lower component of the quark wave function. For the ground state of $L = 0$, the lower component carries $L = 1$. Hence we need to include the effect of higher orbital angular momentum. Formally this can be achieved first by extending the rotational group $SU(2)$ to $SU(2)_S \times O(3)_L$ for spin (S) and orbital angular momentum (L) group, and then take its diagonal subgroup: $SU(2)_S \times O(3)_L \supset SU(2)_{J=S+L}$, where the generators of the diagonal group are the sum $\boldsymbol{J} = \boldsymbol{S}+\boldsymbol{L}$. The $SU(2)_J$ group is now combined with the isospin $SU(2)_I$ group to form the desired chain $SU(4)_q \supset SU(2)_J \times SU(2)_I$.

The relativistic hedgehog quarks are now written as (see the bag model wave functions in Eqs. (5.21)-(5.23))

$$\begin{aligned} |N_c; h\rangle &= \begin{pmatrix} c_1\left[[\frac{1}{2}, 0]^{J=1/2}, \frac{1}{2}\right]^{K=0} \\ c_2\left[[\frac{1}{2}, 1]^{J=1/2}, \frac{1}{2}\right]^{K=0} \end{pmatrix}^{N_c} |0\rangle \\ &\equiv \left[\begin{pmatrix} c_1 \\ c_2 \boldsymbol{\sigma}\cdot\hat{\boldsymbol{x}} \end{pmatrix} \chi^\dagger\right]^{N_c} |0\rangle \,. \end{aligned} \tag{7.83}$$

Here the coupling scheme is $[[S, L]^J, I]^K$ and $\chi^\dagger \equiv \left[[\frac{1}{2}, 0]^{J=1/2}, \frac{1}{2}\right]^{K=0} = \frac{1}{\sqrt{2}}(\alpha^\dagger_{u\downarrow} - \alpha^\dagger_{d\uparrow})$. The coefficients c_1 and c_2 satisfying $|c_1|^2 + |c_2|^2 = 1$ dictate the ratio of the upper ($L = 0$) to the lower ($L = 1$) components. Their values are determined dynamically. The last expression of Eq. (7.83) is useful in actual computation, where χ is interpreted as a column vector in spin space and σ is understood to be a matrix in that space.

The matrix elements for S_3 and T_{33} are computed in a straightforward manner. The results are

$$\langle p\downarrow|S_3|p\downarrow\rangle = -\left(|c_1|^2 - \frac{1}{3}|c_2|^2\right) \equiv -g_A^{(0)} \tag{7.84}$$

and

$$\langle p\downarrow|T_{33}|p\downarrow\rangle = -\left(|c_1|^2 - \frac{1}{3}|c_2|^2\right)\frac{N_c}{3}\left(1 + \frac{2}{N_c}\right) \equiv -g_A \,. \tag{7.85}$$

We find the same suppression factor $\left(|c_1|^2 - \frac{1}{3}|c_2|^2\right)$ for both matrix elements. It is incorporated in relativistic quark models such as the bag model and the chiral quark soliton model. As an example of the maximally relativistic case, the MIT bag model yields the reduction factor 0.654. Accordingly, $g_A^{(0)} = 0.654$ and $g_A = 5/3 \cdot 0.654 = 1.09$. If we wish to reproduce the experimental value of $g_A^{(0)}$ we need a suppression factor of about 1/4. This implies an unrealistically small value for $g_A \sim 5/3 \cdot 1/4 \sim 0.4$. The inclusion of the asymptotic one pion contribution does not help very much. It increases g_A about by 50 % [Jaffe 1979], but still it is only about $0.4 \times 1.5 \sim 0.6$.

Pion cloud

The second extension is to include the effect of the pion cloud around the nucleon. In the hedgehog ansatz, the pion carries the grand spin $K = 0$ through the coupling of its unit isospin and unit orbital angular momentum (P-wave). The relevant group is then $O(3)_I \times O(3)_L$, which can be put in another $SU(4)$ group. Therefore, the extended group for the hedgehog quarks and pions is

$$SU(4)_q \times [O(3)_I \times O(3)_L]_\pi \subset SU(4)_q \times SU(4)_\pi . \tag{7.86}$$

Once again, we pick up the diagonal subgroup, $SU(4)_q \times SU(4)_\pi \supset SU(4)_{q+\pi}$. The effect of the pion that we discuss here should be distinguished from that of the one pion tail. Rather it is related to finite range effects which are responsible for the formation of non-linear hedgehog structure.

We consider an ansatz of the direct product of N_c quarks and N_π pions. The $SU(4)_\pi$ algebra can be generated by the operator β which is similar to the quark operator α. Since the pion carries $I = L = 1$, the pion hedgehog may be a square of the representation $(I, J) = (1/2, 1/2)$:

$$\left(z_\pi^\dagger\right)^2 = \left[\frac{1}{\sqrt{2}}(\beta_{u\downarrow}^\dagger - \beta_{d\uparrow}^\dagger)\right]^2 . \tag{7.87}$$

This reproduces correctly the transformation properties of the pion which is the quark and anti-quark pair under spin and isospin rotations. If we want to distinguish the particle and anti-particle nature explicitly, one of the β bosons in Eq. (7.87) should be replaced by $\beta_{u\downarrow}^\dagger \to -\beta_{\bar{d}\downarrow}$ and $\beta_{d\uparrow}^\dagger \to -\beta_{\bar{u}\uparrow}$. The ansatz of Eq. (7.87) contains not only the pure P wave pions of $[L, I]_{K_z}^K = [1, 1]_0^0$ but also σ like component as $[0, 0]_0^0$. The σ component here, however, together with the pion component, form a representation of the linear sigma model.

With these remarks, we write the desired hedgehog state for the hybrid model as

$$|N_c; h_{q+\pi}\rangle = \frac{1}{\sqrt{N_c!}} \left(z_q^\dagger\right)^{N_c} \frac{1}{\sqrt{(2N_\pi)!}} \left(z_\pi^\dagger\right)^{2N_\pi} |0\rangle \,. \tag{7.88}$$

The number N_π can not be determined in the framework of the algebraic method. Here we only note that N_π is proportional to N_c. This is verified from a diagram as shown in Fig. 7.23. The two pion lines extending from the vertex of the number operator $\hat{n}_\pi$ are attached to two out of the N_c quark lines in a baryon. The combinatorial factor $N_c(N_c - 1) \sim N_c^2$ is multiplied by two πqq Yukawa couplings of order $1/\sqrt{N_c}$. Thus the N_c dependence of the number of the pion in the nucleon is $N_c^2 N_c^{-1} \sim N_c^1$.

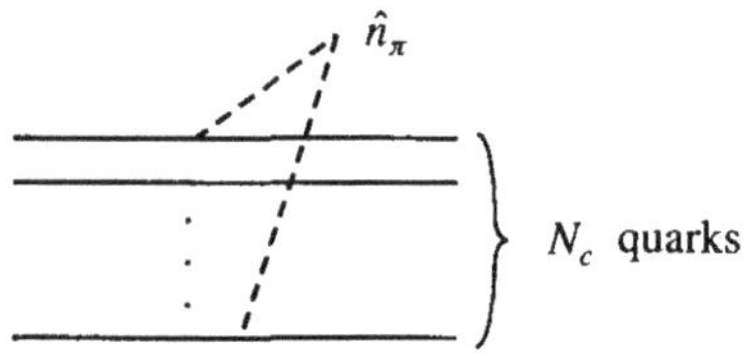

Fig. 7.23 Counting of N_c dependence of the pion number.

Now the nucleon matrix elements for S_i and T_{ia} are computed in the same manner as before. Note that these operators act on the N_c quarks as they are written by the quark operators α. The results are

$$\langle p\downarrow|S_3|p\downarrow\rangle = -\frac{N_c}{N_c + 2N_\pi} \equiv -g_A^{(0)}\,, \tag{7.89}$$

$$\langle p\downarrow|T_{33}|p\downarrow\rangle = -\frac{N_c}{3}\left(1 + \frac{2}{N_c + 2N_\pi}\right) \equiv -g_A\,. \tag{7.90}$$

These results have simple but interesting implications. The nucleon spin $g_A^{(0)}$ becomes less than unity for a finite number of pions $N_\pi \neq 0$, where a part of the nucleon spin is carried by the angular momentum of pions. When $N_\pi = 0$, Eq. (7.89) reduces to the quark model result, where the entire nucleon spin is carried by the quark spin, while when $N_\pi \to \infty$ the nucleon (not the quark) spin is entirely carried by the pion cloud [Wakamatsu 1989], the result of the Skyrme model. The same argument applies also to the second term of g_A in Eq. (7.90). In particular, it is interesting to see that the Skyrme model result is obtained in the limit $N_\pi \to \infty$, rather than

$N_c \to \infty$. This may be understood by interpreting the Skyrme soliton as a coherent superposition of infinitely many pions.

Now, let us estimate g_A and $g_A^{(0)}$, including both the relativistic and pionic effects [Johnson et al. 1990]. A simple computation can show that this amounts to multiplying the reduction factor $|c_1|^2 - \frac{1}{3}|c_2|^2$ to both Eqs. (7.89) and (7.90). A reasonable estimate for the reduction factor is ~ 0.7. For $N_c = 3$, g_A of Eq. (7.90) thus becomes

$$g_A \sim 0.7\left(1 + \frac{2}{3 + 2N_\pi}\right). \tag{7.91}$$

The 50 % enhancement due to the asymptotic one pion contribution must be added to this result. Therefore, we get for the total g_A

$$g_A \quad \sim \quad 1.5 \times 0.7\left(1 + \frac{2}{3 + 2N_\pi}\right) \sim \left(1 + \frac{2}{3 + 2N_\pi}\right). \tag{7.92}$$

The experimental value of $g_A \sim 1.25$ is then reproduced when $N_\pi \sim 2.5$. Note that the $1/N_c$ correction term in Eq. (7.92) is about 20 %, which is in good agreement with the results obtained in the previous chapter. Using the same parameters, the nucleon spin $g_A^{(0)}$ turns out to be

$$g_A^{(0)} \sim 0.7 \frac{3}{3 + 2N_\pi} \sim 0.23\,.$$

It is indeed remarkable that the algebraic method can reproduce both g_A and $g_A^{(0)}$ simultaneously.

Finally, in order to give a rough idea of what the pion number N_π means, we would like to relate it to the bag radius R of the chiral bag model. It appears natural to do so if we notice their similar role in interpolating the bag model and the Skyrme model values of g_A and $g_A^{(0)}$. Crudely, the number of the pion N_π is inversely related to the bag radius R. More concretely, we take the ratio of the leading (Ω^0) to next leading order (Ω^1) terms of g_A in the chiral bag model and compare it with the corresponding one of the algebraic method in Eq. (7.90). By taking the ratio, the unknown relativistic factor is canceled. This is not possible for $g_A^{(0)}$ because the leading order term (Ω^0) vanishes. Thus we obtain the relation

$$\frac{2}{3 + 2N_\pi} = \frac{g_A(\Omega^1)}{g_A(\Omega^0)}, \tag{7.93}$$

from which we obtain

$$N_\pi = \frac{g_A(\Omega^0)}{g_A(\Omega^1)} - 1.5\,. \tag{7.94}$$

In Fig. 7.24, we show this effective pion number N_π as a function of the bag radius R. As anticipated, N_π decreases as the bag radius is increased, and vise versa. Remarkably, the number $N_\pi \sim 2.5$ estimated in Eq. (7.92) corresponds to $R \sim 0.6$ fm.

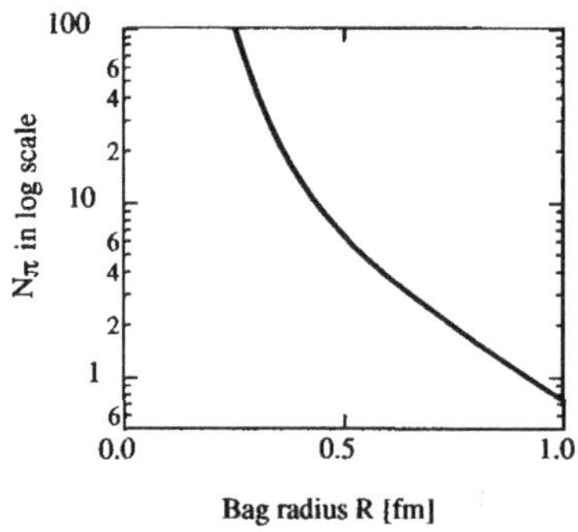

Fig. 7.24 N_π as a function of R.

7.7 Meson-Baryon System

In this section we discuss a possible scenario where the chiral bag picture emerges from a meson-baryon system in the large-N_c limit. In the limit $N_c \to \infty$, a baryon becomes infinitely heavy, and the hedgehog state appears naturally as a static classical solution. In the static solution, the rotational motion in spin or isospin space is suppressed and the infinite tower of baryon excitations with $I = J$ is degenerate. Altogether, the baryon in the large-N_c limit is regarded as a static source of meson cloud, located at a point with a fixed orientation of spin and isospin.

Properties of the bare state of the baryon are renormalized by mesonic cloud. Let us investigate meson-baryon effective theory which is valid in the large-N_c limit and study what classes of diagrams are dominant in the limit $N_c \to \infty$. For this purpose, we consider the self-energy diagram for the baryon. The simplest one is the one-loop diagram as depicted in

Fig. 7.25 (a). This energy is of order N_c^1, for each Yukawa coupling carries the factor $\sqrt{N_c}$. Therefore, this one-loop self-energy dominates in the large-N_c limit. Among higher order diagrams with arbitrary number of meson lines, we see that the following diagrams survive in the large N_c limit, *i.e.*; those containing only meson tree diagrams when the baryon lines are removed. An example of such a typical diagram is depicted in 7.25 (b). In contrast, those including purely mesonic loops as depicted in Fig. 7.25 (c) are suppressed in the large-N_c limit.

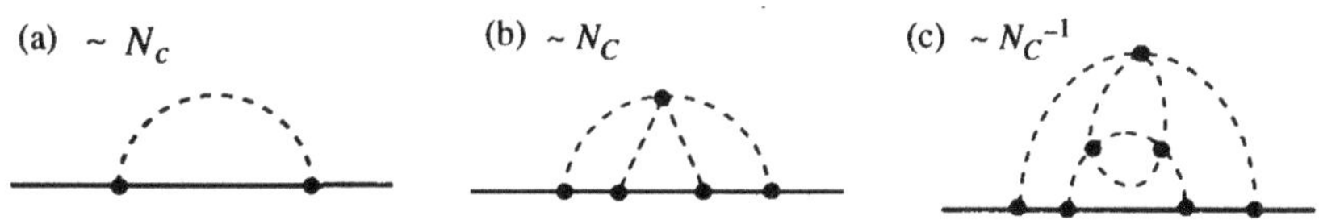

Fig. 7.25 Self energies of the nucleon due to pion loops; (a) one loop diagram, (b) diagram without meson loops and (c) diagram with meson loops.

Now diagrams without meson loops are summed all at once by solving the classical equation of motion. To be specific, let us consider the pion-nucleon system. The coupling of the pion with the nucleon assumed to take the form

$$A_i^a(\pi)\, A_i^a(B)\,, \tag{7.95}$$

the product of the axial-vector currents for the pion and the baryon. This coupling provides a source term of the pion of the delta function type, if the baryon would be point like. The nucleon axial-vector current is written in the large-N_c limit as

$$A_i^a(B) \to 3\bar{g}_A \operatorname{tr}\left[\tau_i A \tau_a A^\dagger\right] \equiv 6\bar{g}_A\, D_{ia} \tag{7.96}$$

where $\bar{g}_A$ is the bare coupling constant and $D_{ia} = \frac{1}{2}\operatorname{tr}\left[\tau_i A \tau_a A^\dagger\right]$. The pion axial-vector current is

$$A_i^a(\pi) = f_\pi \partial_i \pi^a(x) + \cdots\,, \tag{7.97}$$

where the dots represent non-linear terms. They depend on the pion lagrangian used, but the leading order term of Eq. (7.97) is uniquely determined. Smearing the baryon source of the axial-vector current, the equation of motion for the pion is written as

$$(-\Delta + m_\pi^2)\pi^a + \frac{\partial V}{\partial \pi^a} = \frac{3\bar{g}_A}{f_\pi^2} D_{ia}\partial_i J(\boldsymbol{x}), \tag{7.98}$$

where $\partial V/\partial \pi^a$ on the left hand side involves self interactions among the pions. $J(\boldsymbol{x})$ is the smeared source, the simplest choice of which would be

$$J(\boldsymbol{x}) = \begin{cases} \frac{3}{4\pi R^3} & |\boldsymbol{x}| \leq R, \\ 0 & |\boldsymbol{x}| > 0. \end{cases} \tag{7.99}$$

The cutoff R here has a role, for example, in making the energy correction finite, as shown in below. In the asymptotic region, the bare baryon looks like a point like source. For $A = 1$, $D_{ia} \to \delta_{ia}$ and solutions to Eq. (7.98) take the hedgehog form, $\pi^a = \hat{x}^a F(r)$.

Such a picture of the meson-baryon system with the smeared baryon source is reminiscent of the chiral bag model, where the cutoff R may be identified with the bag radius. Let us look at this correspondence in a bit more detail. First, we note that the pion profile function $F(r)$ looks qualitatively like what is shown in Fig. 7.26, where $F(r)$ starts zero at the origin and then ends again zero at infinity. The kink at $r = R$ is produced by the surface term $\partial J(r)/\partial r$ on the right hand side of Eq. (7.98). Then the quark bag may be interpreted as the bare baryon ($J(r)$) plus pion cloud in the interior region of $r < R$. The bare baryon corresponds to the bag with N_c valence quarks, and the pion cloud to sea quarks (vacuum polarization). This picture is the one considered by Gervais and Sakita [Gervais and Sakita 1984a; Hosaka and Toki 1988b].

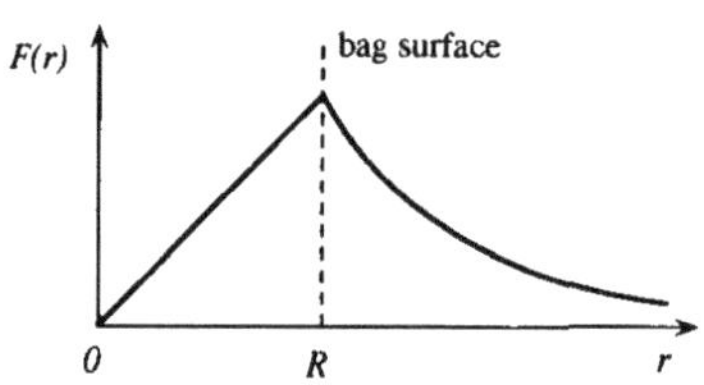

Fig. 7.26 A typical behavior of the pion profile function.

In order to see the essential point, let us consider the simplest case for the free pion, where all the non-linear interactions, $\partial V/\partial \pi^a$ of Eq. (7.98) can be neglected. The solution is easily found:

$$F_{\rm in}(r) = \frac{3\bar{g}_A}{8\pi f_\pi^2 R^3}\, r\,, \qquad r < R. \tag{7.100}$$

$$F_{\rm out}(r) = \frac{3\bar{g}_A}{8\pi f_\pi^2}\, \frac{1}{r^2}\,, \qquad r > R, \tag{7.101}$$

The energy of the pion is computed as

$$\begin{aligned} E_\pi &= 2\pi f_\pi^2 \int_0^\infty dr r^2 \left[F'^2 + 2\frac{F^2}{r^2}\right] \\ &\quad - \frac{3g_A}{2} \int_0^R d^3x J(r) \left[F' + \frac{2F}{r}\right] \\ &= \frac{27 g_A^2}{32\pi f_\pi^2 R^3} - \frac{27 g_A^2}{16\pi f_\pi^2 R^3} \\ &= -\frac{27 g_A^2}{32\pi f_\pi^2 R^3}\,, \end{aligned} \tag{7.102}$$

where the pion-baryon interaction energy is just twice the pion kinetic energy, as the virial theorem implies. As $R \to 0$, the divergent energy of Eq. (7.102) may be reabsorbed into the bare baryon mass so that the renormalized baryon mass becomes finite. In contrast, Eq. (7.101) implies that the pion-baryon coupling, or the axial-vector coupling constant g_A is not renormalized for all R.

In general, if non-linear pion interactions are present, both the mass and g_A are renormalized. In an effective meson-baryon theory, physical baryon properties are reproduced after including mesonic corrections to bare quantities. This leads to a set of renormalization group equations. Since only the mesonic corrections that are dominant in the limit $N_c \to \infty$ are included, it is called the large-N_c renormalization group [Dorey et al. 1994; Dorey and Mattis 1995]. In the above example of the free pion, the field strength of the pion becomes indefinitely large as R is decreased. This is made finite by suitably choosing the pion lagrangian, *e.g.*, that of the Skyrme model. In this case, the point $R \to 0$ is a smooth limit of the renormalization group equations, where baryon dynamics is entirely governed by the meson cloud.

In the large-N_c limit, therefore, we expect that a suitably chosen meson-baryon effective lagrangian should be able to reproduce physical quantities by adjusting bare coupling constants according to the renormalization group equation. At each R, mesonic corrections are calculated, and the bare quantities are adjusted so that they reproduce physical values. In this sense, the idea of the large-N_c renormalization group resembles the Cheshire cat hypothesis.

In this respect, the actual chiral bag model is considered to provide an approximate realization of the large-N_c renormalization program for finite N_c and with physical parameters such as pion decay constant. In such a description, R can no longer be a free parameter but may be fixed at around 0.6 fm.

Problems

7.1

Find a kink solution of the sine-Gordon theory in 1 + 1 dimension (The potential of the sine-Gordon theory is shown in Fig. 7.27.):

$$\mathcal{L} = \frac{1}{2}(\partial_\mu \phi)^2 - \frac{\alpha}{\beta^2}(1 - \cos \beta\phi)\,, \tag{7.103}$$

satisfying the boundary conditions $\phi(-\infty) = 0, \phi(+\infty) = 2\pi/\beta$. Establish the $1/N_c$ counting rules of this theory, following the analysis of the ϕ^4-theory in section 7.3.

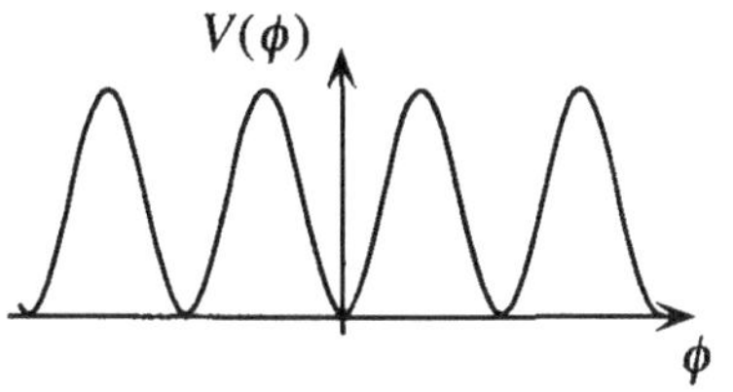

Fig. 7.27 The potential of the sine-Gordon theory.

7.2

We consider wave functions in the parameter space x. Free motion which is characterized by the eigenstate of the momentum $\hat{p} = -i\partial_x$ satisfies the translational invariance. Denote the wave function by $\psi_p(x)$. Now the eigenstate of the position X, $|X\rangle$ breaks the translational invariance. In order to recover the broken symmetry, we can perform the superposition

$$|p\rangle \sim \int dX \;\psi_p(X)|X\rangle\,. \tag{7.104}$$

(1) Show that $|p\rangle$ satisfies the eigenvalue equation $\hat{p}|p\rangle = p|p\rangle$.
(2) Define the projection operator

$$\Pi_p = \int dX \;\psi_p(X)R(X)\,, \tag{7.105}$$

where $R(X)$ is the operator for displacement by X. Show that

$$\Pi_p\Pi_{p'} \sim \delta(p - p')\Pi_p\,. \tag{7.106}$$

7.3

For a particle of spin 1/2 and isospin 1/2, one can expand the field operator in terms of the four vectors,

$$\psi(x) = \alpha_{u\uparrow}\psi_{u\uparrow}(x) + \alpha_{u\downarrow}\psi_{u\downarrow}(x) + \alpha_{d\uparrow}\psi_{d\uparrow}(x) + \alpha_{d\downarrow}\psi_{d\downarrow}(x)\,, \qquad (7.107)$$

where α_{mn} are annihilation operators appeared in Eqs. (7.73) and (7.74), and $\psi_{mn}(x)$ are normalized wave functions such that

$$\int d^3x\psi^\dagger_{mn}\psi_{m'n'} = \delta_{mm'}\delta_{nn'}\,. \qquad (7.108)$$

Compute

$$\int d^3x\psi^\dagger_{mn}\sigma_3\psi_{m'n'}\,, \quad \int d^3x\psi^\dagger_{mn}\sigma_3\tau_3\psi_{m'n'}\,, \qquad (7.109)$$

and verify Eqs. (7.73) and (7.74).

7.4

Let us consider the matrix element of Eq. (7.72). For instance the numerator can be written as

$$\langle N|\mathcal{O}|N\rangle \sim \int d[AA']D^*_N(A)D_N(A')\langle h|R^\dagger(A')\mathcal{O}R(A)|h\rangle\,. \qquad (7.110)$$

If $\mathcal{O}$ is a scalar under isospin rotation, it follows that

$$R^\dagger(A')\mathcal{O}R(A) = R^\dagger(A')R(A)R^\dagger(A)\mathcal{O}R(A) = R^\dagger(A')R(A)\mathcal{O} \qquad (7.111)$$

and find

$$\begin{aligned}\langle N|\mathcal{O}|N\rangle &\sim \int d[AA']D^*_N(A)D_N(A')\langle h|R^\dagger(A')R(A)\mathcal{O}|h\rangle \\ &\sim \int d[A]D^*_N(A)\langle h|R(A)\mathcal{O}|h\rangle\,. \qquad (7.112)\end{aligned}$$

Using this relation and Eqs. (7.73), (7.74), compute the matrix elements of Eq. (7.75) and (7.76).

Chapter 8

EXCITED BARYONS

So far we have discussed mostly properties of the ground state baryons. There we have emphasized the role of chiral symmetry. In particular, it was shown that strong interactions among the pions and quarks play important roles. In order to achieve further understanding of QCD, it is useful to study excited states also. An immediate question is then whether the similar role of chiral symmetry persists to the physics of excited baryons or not. The answer to this question is not yet fully understood.

Up to now number of approaches have been developed for the description of excited baryons. The string model inspired by the Regge theory [Regge 1959; Regge 1960; Nambu 1969; Susskind 1970; Goto 1971], covariant harmonic oscillator model [Feynman et al. 1971; Fujimura et al. 1970; Ishida 1971a; Ishida 1971b], non-relativistic quark model [Isgur and Karl 1978a; Isgur and Karl 1979a; Isgur and Karl 1979b; Hey and Kelly 1983; Glozman and Riska 1996], bag model [Chodos et al. 1974a; Chodos et al. 1974b; DeGrand et al. 1975], Skyrme model [Hayashi et al. 1984; Hayashi and Holzwarth 1984; Mattis and Karliner 1985; Mattis and Peskin 1985; Zahed and Brown 1986] and so on. They incorporate partly the dynamics of QCD, but to achieve quantitative description, number of parameters are required to reproduce properties of excited states.

In this chapter, we propose yet another point of view where excited states are regarded as a rotational states. The virtue of this lies in its simplicity and systematics. To discuss the point, we start with phenomenological analysis for masses. The systematics is found and explained by a simple non-relativistic quark model with a deformed harmonic oscillator potential. In the latter half of this chapter, we discuss electromagnetic transitions in the non-relativistic quark model for further study of baryon structure. Detailed method of calculation is explained.

8.1 Systematics in Baryon Masses

Let us look at experimental data from the particle data book [PDG 2000]. Before analyzing all the flavor $SU(3)$ sector, we investigate the nucleon sector of isospin u and d. Masses of well established nucleon excitations (with four and three stars) are shown in Fig. 8.1 (a). From the figure, we observe the following features:

(1) In the positive parity sector, the level spacing of the higher energy states is larger than that of the lower energy states.
(2) The level spacing of the negative parity sector is larger than that of the positive parity sector.

In fact, these features resemble the structure of rotational band of deformed states. To convince ourselves, in Fig. 8.1 (b), rotational spectrum of the oxygen nucleus is shown, which is well compared with the nucleon spectrum of (a).

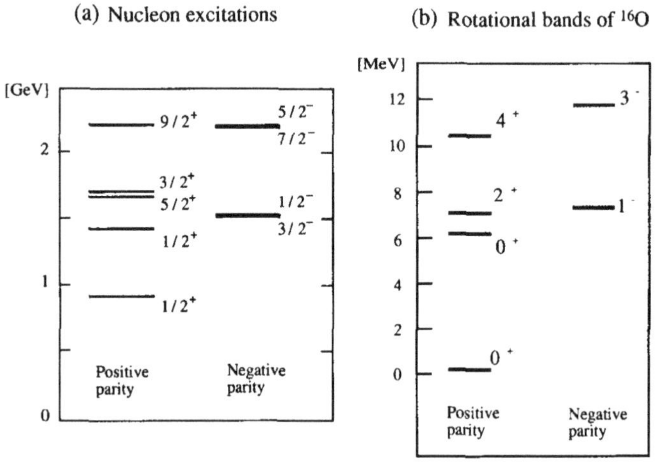

Fig. 8.1 (a) Masses of excited nucleons, and (b) rotational bands of ^{16}O.

The similarity between the baryon masses and rotational spectra is rather universal for the flavor $SU(3)$ sector, with strangeness $S = 0, -1$ (Λ and Σ) and -2 (Ξ). To see this, we plot the masses of 50 well observed states with three and four stars in Fig. 8.2. At first glance, various levels appear randomly, and it does not seem that there is systematics there.

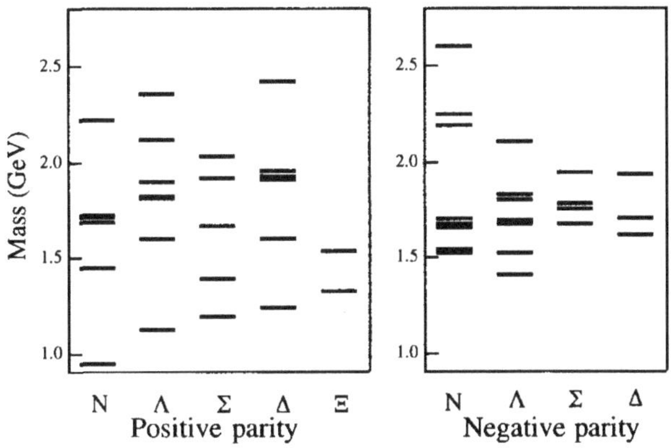

Fig. 8.2 Baryon masses of light flavors u, d and s.

To proceed further, first we classify these states into appropriate members of a multiplet of the $SU(2)_{spin} \times SU(3)_{flavor} \subset SU(6)$ group. As discussed in Chapter 2, section 2.7, we denote the multiplet as ${}^{2s+1}D_\sigma$, where the superscript $2s+1$ is for the spin degeneracy, D for the dimension of flavor $SU(3)$ representations and σ for the symmetry of the spatial wave function which is either symmetric (S) or mixed-symmetric (MS). Wave functions of the 56-plet of $SU(6)$ are products of a symmetric spatial part and a symmetric spin-flavor part. They include ${}^2 8_S$ and ${}^4 10_S$. Similarly, wave functions of the 70-plet are products of a mixed symmetric spatial part and a mixed symmetric spin-flavor part. They include ${}^2 1_{MS}$, ${}^2 8_{MS}$, ${}^4 8_{MS}$ and ${}^2 10_{MS}$. Note that for negative parity states of $l = 1$, the 56-plet with a symmetric spatial part is not allowed since it corresponds to spurious translational motion of the center of mass. Further details on quark model states has been explained in Chapter 2.

In Table 8.1, we summarize baryon states which are classified according to the above scheme [Takayama et al. 1999]. Our identification is rather primitive. First we determine an orbital angular momentum l and spin s

from the observed baryon spin and parity, j^P. It turns out that, in many cases, for a given l there appear two groups of states which are separated by about 200 MeV. We have identified the lower group with spin doublet ($s = 1/2$) and the higher group with the spin quartet ($s = 3/2$). In some cases, such an assignment becomes less obvious. For instance, we can not rule out the possibility that $\Sigma\, S_{11}(1750)$ could be a member of ${}^4 8_{MS}$, as it seems to be degenerate with $\Sigma\, D_{15}(1775)$. We have to mention also that we do not consider possible configuration mixings, which would become important for such states as $N\, S_{11}(1535) - N\, S_{11}(1650)$ where ${}^2 8_{MS}$ and ${}^4 8_{MS}$ mix [Isgur and Karl 1977; Isgur and Karl 1978b; Isgur and Karl 1978a; Isgur and Karl 1979a; Arima et al. 1992], and $\Sigma\, S_{11}(1750) - \Sigma\, D_{13}(1940) - \Sigma\, S_{11}(2000)$ where ${}^2 8_{MS}$, ${}^4 8_{MS}$ and ${}^2 10_{MS}$ mix [Fujiwara 1993].

Altogether, 62 states are listed in Table 8.1, which amount to about 80 % of all baryon states listed in the particle data table. We have not, however, included the Ω states, since their spins and parities are not determined.

In Fig. 8.3, we have rearranged the levels of Fig. 8.2 according to the following rules [Takayama et al. 1999].

(1) All excited masses are measured from the ground states of the corresponding spin-flavor channel. This removes the flavor dependence expected from the celebrated Gell-Mann-Okubo mass formula [Gell-Mann 1962; Okubo 1962; Cheng and Li 1984]:

$$\begin{aligned} \frac{m_\Sigma + 3m_\Lambda}{2} &= m_N + n_\Xi \\ m_\Omega - m_{\Xi^*} &= m_{\Xi^*} - m_{\Sigma^*} = m_{\Sigma^*} - m_{N^*}\,, \end{aligned} \tag{8.1}$$

where $N^* = \Delta$. In doing so, we expect that the major part of the flavor dependence of the excited masses is accounted for by the Gell-Mann-Okubo mass formula.

(2) We have lowered the masses of spin quartet states by 200 MeV as expected from spin correlations between the spin doublet and quartet states. We have done this both for positive and negative parity states. The value of 200 MeV can be extracted from the energy difference of $\Sigma(1189) - \Sigma(1385)$, and of $\Xi(1317) - \Xi(1530)$.

(3) The three ${}^2 8_{MS}$ states of $l = 0$, $N\, P_{11}(1710)$, $\Lambda\, P_{01}(1810)$ and $\Sigma\, P_{11}(1880)$, are also lowered by 200 MeV. This would be caused by, for example, a δ-function type interaction in order to split the spatially symmetric and mixed symmetric states [Bhaduri et al.

Table 8.1 Baryon states from the latest Particle Data Group [PDG 2000]. Those without stars are either four or three star states. We have also included some two star states as indicated in the table.

Positive Parity		N			Λ		
		$^2 8_S$	$^2 8_{MS}$	$^4 8_{MS}$	$^2 8_S$	$^2 8_{MS}$	$^4 8_{MS}$
$N=0$	$l=0$	$P_{11}(939)$	-	-	$P_{01}(1116)$	-	-
$N=2$	$l=0$	$P_{11}(1440)$	$P_{11}(1710)$		$P_{01}(1600)$	$P_{01}(1810)$	
	$l=2$	$F_{15}(1680)$		$P_{13}^{**}(1900)$	$F_{05}(1820)$		$F_{05}(2110)$
		$P_{13}(1720)$		$F_{17}^{**}(1990)$	$P_{03}(1890)$		$F_{07}^{*}(2020)$
				$F_{15}^{**}(2000)$			
	$l=4$	$H_{19}(2220)$			$H_{09}(2350)$		

Positive Parity		Σ			Δ	Ξ	
		$^2 8_S$	$^2 8_{MS}$	$^4 10_S$	$^4 10_S$	$^2 8_S$	$^4 10_S$
$N=0$	$l=0$	$P_{11}(1189)$	-	$P_{13}(1385)$	$P_{33}(1232)$	$P_{11}(1317)$	$P_{13}(1530)$
$N=2$	$l=0$	$P_{11}(1660)$	$P_{11}^{**}(1880)$	$P_{13}^{*}(1840)$	$P_{33}(1600)$		
	$l=2$	$F_{15}(1915)$		$F_{17}(2030)$	$F_{37}(1950)$		
				$F_{15}^{*}(2070)$	$F_{35}(1905)$		
				$P_{13}^{**}(2080)$	$P_{33}(1920)$		
					$P_{31}(1910)$		
	$l=4$				$H_{3\,11}(2420)$		
					$F_{37}^{**}(2390)$		

Negative Parity		N		Λ		
		$^2 8_{MS}$	$^4 8_{MS}$	$^2 8_{MS}$	$^4 8_{MS}$	$^2 1_{MS}$
$N=1$	$l=1$	$D_{13}(1520)$	$D_{15}(1675)$	$D_{03}(1690)$	$D_{05}(1830)$	$D_{03}(1520)$
		$S_{11}(1535)$	$D_{13}(1700)$	$S_{01}(1670)$	$S_{01}(1800)$	$S_{01}(1405)$
			$S_{11}(1650)$			
	$l=3$	$G_{17}(2190)$	$G_{19}(2250)$			$G_{07}(2100)$
		$D_{15}^{**}(2200)$				
	$l=5$	$I_{1\,11}(2600)$				

Negative Parity		Σ		Δ		Ξ
		$^2 8_{MS}$	$^4 8_{MS}$	$^2 10_{MS}$	* *	$^2 8_{MS}$
$N=1$	$l=1$	$D_{13}(1670)$	$D_{15}(1775)$	$D_{33}(1700)$	$D_{35}(1930)$	$D_{13}(1820)$
		$S_{11}(1750)$	$D_{13}(1940)$	$S_{31}(1620)$		
			$S_{11}^{*}(2000)$			
	$l=3$			$G_{37}^{*}(2200)$		
				$D_{35}^{*}(2350)$		
	$l=5$					

1984].

Then the resulting mass spectra show a remarkable similarity in all spin-flavor channels as shown in Fig. 8.3. The horizontal axis are labeled by the spin-flavor representations which are grouped into N, Λ, Σ and Δ states.

The negative parity baryons have a mixed symmetric (MS) spatial wave function and are all identified with members of the 70-plet. In contrast, positive parity baryons can have either a symmetric (S for 56-plet) or a mixed symmetric (MS for 70-plet) spatial wave function. There are eight states which are classified into the 70-plet in the positive parity sector. Among them $N\,P_{11}(1710)$, $\Lambda\,P_{01}(1810)$ and $\Sigma\,P_{11}(1880)$ are identified with $^{2}8_{MS}$, while the others with $^{4}8_{MS}$. Since there are not many 70-plet states, we put both spin doublet and quartet states in the same column in Fig. 8.3

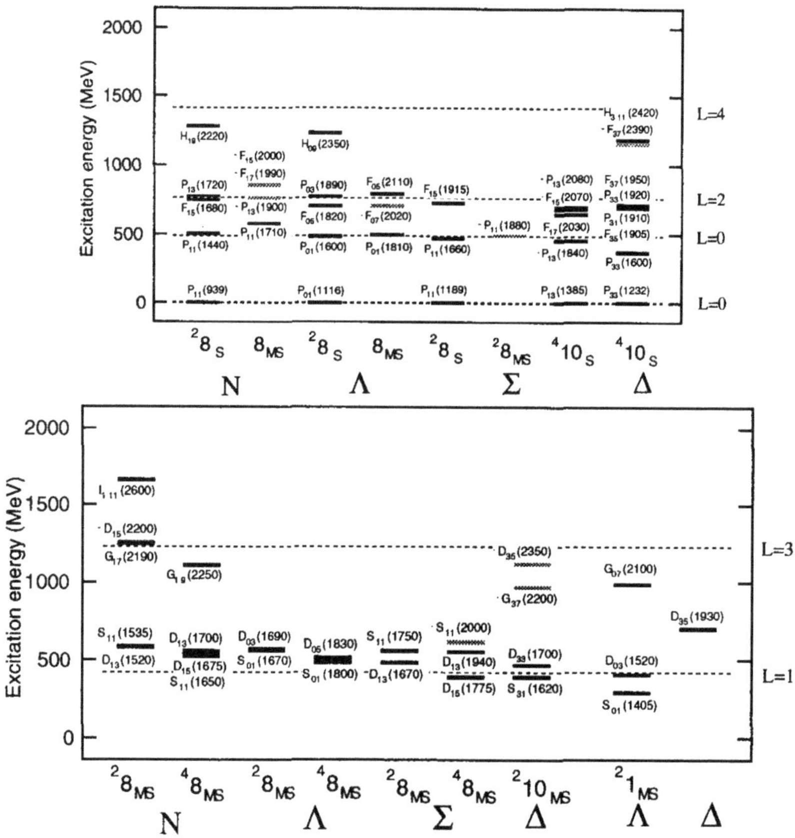

Fig. 8.3 Rearranged baryon masses.

for the N and Λ sectors. This is why we omit the superscript 2 or 4 in the labels.

There are several states which need comments:

(1) $\Lambda(1405)$, $\Lambda(1520)$ and possibly $\Lambda(2100)$
These states lie substantially lower than the other corresponding states. They are possibly dominated by the meson-baryon resonance (or molecular like) states as demonstrated many times in chiral models for mesons and baryons [Kumar and Nogami 1980; Oller and Oset 1998; Kimura et al. 2000; PDG 2000].

(2) $N(2600)$
The excitation energy of this state, $2600 - 938 \sim 1700$ MeV, is substantially less than what is expected from the simple rotational band for $l = 5$. It could be identified with an intruder state of ($N = 3, l = 5$), which appears below the ($N = 1, l = 5$) state [Takayama et al. 2000] (see problem 8.3).

(3) $\Delta(1930)$
This state can not fit to the systematics of the rotational band. There are works on this state by Cutkosky and Hendrick [Cutkosky et al. 1976] and Bowler et al. [Bowler et al. 1980], where the possibility of a quark-gluon hybrid state is discussed.

8.2 Quarks in a Deformed Oscillator Potential

Variation

It is known in nuclear physics that the typical energy levels of rotational band of Fig. 8.1 can be reproduced by a simple model of deformed harmonic oscillator potential. Let us apply this model to the baryons [Toki et al. 1983; Bhaduri et al. 1984; Hosaka et al. 1998].

Consider non-relativistic fermions moving in a harmonic oscillator potential. This looks very much different from the chiral bag model that we have investigated in Chapters 5 and 6. However, there could be a link between them through chiral symmetry. When chiral symmetry is spontaneously broken, the (current) quarks acquire the constituent mass due to the interaction with the non-zero chiral condensate in the spontaneously broken vacuum. The resulting massive constituent quarks are the building blocks of baryons.

Now we consider the hamiltonian for three massive valence quarks as the constituents of baryons:

$$H = \sum_{i=1}^{3} \left(\frac{\boldsymbol{p}_i^2}{2m} + \frac{m}{2}(\omega_x^2 x_i^2 + \omega_y^2 y_i^2 + \omega_z^2 z_i^2) \right) . \tag{8.2}$$

The harmonic oscillator potential for quark confinement is taken to be anisotropic as described by $\omega_x \neq \omega_y \neq \omega_z$. The masses of constituent quarks are assumed to be about 300 MeV for u, d and s quarks. In practice, physically relevant parameters are the oscillator parameters ω_i and the value of m is not very important as far as the energy spectrum is concerned.

Separating internal and center of mass coordinates through the following coordinate transformation

$$\begin{aligned} \boldsymbol{R} &= \frac{1}{\sqrt{3}}(\boldsymbol{x}_1 + \boldsymbol{x}_2 + \boldsymbol{x}_3) , \\ \boldsymbol{\rho} &= \frac{1}{\sqrt{2}}(\boldsymbol{x}_1 - \boldsymbol{x}_2) , \\ \boldsymbol{\lambda} &= \frac{1}{\sqrt{6}}(\boldsymbol{x}_1 + \boldsymbol{x}_2 - 2\boldsymbol{x}_3) , \end{aligned} \tag{8.3}$$

we obtain the hamiltonian for the internal ρ and λ degrees of freedom as

$$H^{\text{int}} = \frac{\boldsymbol{p}_\rho^2}{2m} + \frac{m}{2}\left(\omega_x^2 \rho_x^2 + \omega_y^2 \rho_y^2 + \omega_z^2 \rho_z^2\right) + (\rho \to \lambda) . \tag{8.4}$$

Eigenstates of the internal hamiltonian of Eq. (8.4) are labeled by a set of quantum numbers $(n_x^\rho, n_y^\rho, n_z^\rho;\ n_x^\lambda, n_y^\lambda, n_z^\lambda)$, whose eigenenergies are given by

$$\begin{aligned} E^{\text{int}} &= (n_x^\rho + \tfrac{1}{2})\omega_x + (n_y^\rho + \tfrac{1}{2})\omega_y + (n_z^\rho + \tfrac{1}{2})\omega_z + (\rho \to \lambda) \\ &= (N_x + 1)\omega_x + (N_y + 1)\omega_y + (N_z + 1)\omega_z , \end{aligned} \tag{8.5}$$

where $N_x = n_x^\rho + n_x^\lambda$, $N_y = n_y^\rho + n_y^\lambda$, $N_z = n_z^\rho + n_z^\lambda$. Thus the intrinsic energy of Eq. (8.5) is regarded as a function of N_x, N_y, N_z and ω's.

So far all steps have been straightforward. We are now parting from the standard method. The key idea is to regard the oscillator parameters ω_i as dynamical degrees of freedom. This could be justified if the potential is interpreted as a mean field of an underlying microscopic theory, where it is determined by a self-consistent condition of a coupled system.

Practically we perform energy minimization under the variation of ω_i's for a given oscillator state of (N_x, N_y, N_z). However, unless we have some additional conditions, the variation leads to the trivial result; $E^{\rm int} \to 0$, when $\omega_x = \omega_y = \omega_z \to 0$. In order to avoid this collapse of the system, we impose the volume conservation condition

$$\omega_x \omega_y \omega_z = \omega^3 = \text{const} . \tag{8.6}$$

In nuclear physics this condition is very well satisfied due to the saturation property of nuclei. It is not clear whether the same condition holds for baryons or not. For instance, in the MIT bag model, the size of the baryon were determined by pressure balance between quarks and the cavity in the vacuum. As discussed in problem 8.1, however, a simple dynamical consideration of such a pressure effect leads to the same deformation as obtained with the volume conservation condition.

Now the minimum energy of Eq. (8.5) is given by

$$E^{\rm int} = 3(N_x+1)^{1/3}(N_y+1)^{1/3}(N_z+1)^{1/3}\omega , \tag{8.7}$$

when

$$\omega_x : \omega_y : \omega_z = \frac{1}{N_x+1} : \frac{1}{N_y+1} : \frac{1}{N_z+1} . \tag{8.8}$$

The inverse relation of Eq. (8.8) is a bit more convenient, since it gives the ratio of deformation lengths along the three principal axes:

$$a_x : a_y : a_z = N_x+1 : N_y+1 : N_z+1 , \tag{8.9}$$

where $a_{x,y,z}$ are the root mean square lengths along the x, y, z directions.

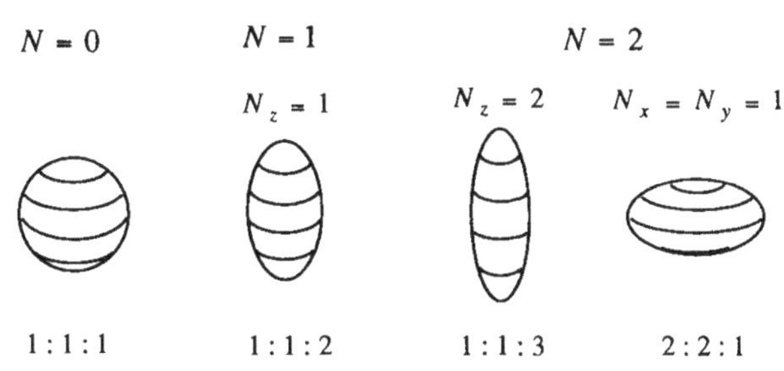

Fig. 8.4 Deformation of excited baryons.

Table 8.2 Various quantities of prolately deformed states. Dimensional quantities are scaled by an appropriate power of ω.

Positive parity					
N	d	E^{int}	$E^{\text{spherical}}$	$1/2\mathcal{I}$	$\langle l^2 \rangle$
0	1	–	3	–	0
2	3	4.33	5	0.0721	8
4	5	5.13	7	0.0329	24
6	7	5.74	9	0.0191	48

Negative parity					
N	d	E^{int}	$E^{\text{spherical}}$	$1/2\mathcal{I}$	$\langle l^2 \rangle$
1	2	3.78	4	0.126	3
3	4	4.76	6	0.0467	15
5	6	5.45	8	0.0246	35

Several examples are shown in Fig. 8.4. The ground state with $N = 0$ ($N_x = N_y = N_z = 0$) yields the spherically symmetric state just as in the conventional quark model. For excited states of $N = 1$ ($N_x = N_y = 0$, $N_z = 1$ when the z-axis is chosen as the symmetry axis), the minimum energy state deforms prolately with the ratio of short to long axes $1 : 2$. The energy of this state is 3.780ω as compared to 4ω of the spherical state of $N = 1$. For excited states of $N = 2$, there are two cases. One is $N_x = N_y = 0$, $N_z = 2$ which leads to prolate deformation with the ratio of short to long axes $1 : 3$. The energy is 4.327ω as compared to 5ω of the spherical state with $(n, l) = (0, 2)$ or $(1, 0)$. Another is $N_x = N_y = 1, N_z = 0$ which yields an oblate shape with the ratio of short to long axes $1 : 2$, whose energy is 4.762ω. In general, for a given $N = N_x + N_y + N_z$, the prolate shape has the minimum energy. Therefore, in the following discussions we consider only the prolate deformation of ($N_x = N_y = 0, N_z = N$) with the symmetry axis chosen along the z direction. We summarize in Table 8.2 various physical quantities of prolately deformed states.

Table 8.3 Excitation energies of the $N = 2$ and $N = 1$ rotational bands. Energy values in units of MeV are computed by using $\omega = 644$ MeV.

l	$N=2$ $E(N=2,l)-E(N=0)$ in ω	in MeV	l	$N=1$ $E(N=1,l)-E(N=0)$ in ω	in MeV
0	0.750	483	1	0.654	421
2	1.183	752	3	1.914	1230
4	2.192	1410	5	4.182	2690

Cranking formulae

Mass formula

Since the deformed states break rotational symmetry, it should be recovered for physical baryons which are eigenstates of angular momentum l. This can be performed by the cranking method [Bohr and Mottelson 1975; Murthy et al. 1984; Bhaduri 1988]. The key quantities are the moment of inertia $\mathcal{I}$ and the angular momentum fluctuation $\langle l^2 \rangle$ of the deformed state. In the harmonic oscillator model, both of them can be computed analytically. For a prolately deformed state with z-axis symmetry ($\omega_x = \omega_y \neq \omega_z$), they are given by (see problem 8.2)

$$\begin{aligned} \mathcal{I}_N &= \frac{N_z+1}{\omega_z} + \frac{N_y+1}{\omega_y} \\ &= \frac{1}{(N+1)^{1/3}}\left(N^2+2N+2\right)\frac{1}{\omega}, \qquad (8.10) \\ \langle l^2 \rangle_N &= \left(\left(\frac{\omega_x}{\omega_z}\right)^2 - 1\right) = N(N+2), \qquad (8.11) \end{aligned}$$

The rotational band is then constructed for each N (compare with Eq. (6.70):

$$E_{N\,l} = E_N^{\mathrm{int}} - \frac{1}{2\mathcal{I}_N}\langle l^2 \rangle_N + \frac{1}{2\mathcal{I}_N} l(l+1) . \qquad (8.12)$$

Numerical values for $\mathcal{I}$ and $\langle l^2 \rangle$ are shown in Table 8.2, and the rotational energies $E_{N\,l}$ for the $N = 2$ and $N = 1$ bands are shown in Table 8.3.

We note that in the formula (8.12), the energy subtraction due to angular momentum fluctuation, the second term of Eq. (8.12), is very important.

In the present model it can be computed exactly, to be contrasted with field theoretical problems involving solitons as discussed in chapter 6, where the computation of such terms is generally very difficult. In the present case the quantization of the three quark system can be performed exactly and therefore, such quantum corrections can be also computed. Such a subtraction term is very important to make the theoretical masses down close to experimental values. This is particularly so for the first $1/2^+$ excited state (the Roper resonance).

Coupling the orbital angular momentum l with the intrinsic spin s of three quarks, take for example $s = 1/2$, we can predict theoretical mass spectra of the $N = 2$ and $N = 1$ rotational bands. If the masses of the constituent quarks are the same for u, d and s quarks, the formula applies to all the baryons that we are considering now.

The theoretical predictions should be compared with excitation energies of baryons, in which relative values are relevant:

$$\Delta E(N,l) = E_{Nl} - E_{00}\,. \tag{8.13}$$

Hence, there is only one parameter, which is the oscillator parameter ω. It is determined here by the average mass splitting between the first excited states of $1/2^+$ and the corresponding ground states. The resulting value is $\omega = 644$ MeV [Takayama et al. 1999]. In Fig. 8.3, the dashed lines are the excitation energies of such predictions.

Considering the simplicity of the present method, it is remarkable that many observed states fit very well to the theoretical rotational band. In particular we note that the first $1/2^+$ excited state, the Roper resonance, emerges as the band head of the $N = 2$ rotational band. If we look at the figures in more detail, the third level of the theoretical rotational band ($l = 4$) appear slightly higher then experimental masses. In the framework of the deformed oscillator model, a slight reduction of energy for the $l = 4$ state is obtained if we take into account the intruder states. This is the topic of problem 8.3.

One of the long standing and still unresolved problems is the nature of the Roper resonance, the first excited state of $j^P = 1/2^+$. In conventional quark models, this state is identified with the $2\hbar\omega$ state, whose energy is twice as large as that of the first negative parity $1/2^-$ state of $1\hbar\omega$, an apparent contradiction with experiments where $1/2^\pm$ states appear almost degenerate. Considerable efforts have been made to explain the ordering of the $1/2^{\pm}$ states. In the non-relativistic quark model, various types of two-body residual interactions need to be introduced. Some of them have an origin in gluons and others in chiral mesons [Isgur and Karl 1978a; Isgur and Karl 1979b]. In a recent study, Glozman and collaborators have elaborated the quark model with chiral meson exchanges and have achieved a reasonably good agreement with experiments [Glozman and Riska 1996]. In other approaches, the Roper resonance is regarded as a collective state of monopole vibration of the nucleon [Brown et al. 1983; Hatsuda 1986], and is also regarded as a meson-nucleon resonance state [Breit and Nappi 1984; Hayashi et al. 1984; Zahed et al. 1984; Mattis and Peskin 1985; Hosaka and Toki 1988a]. In any descriptions so far, however, many parameters or dynamical degrees of freedom are necessary in order to reproduce the whole baryon spectrum including the Roper state. In this respect, it should be emphasized that the model of deformed baryons can describe, at least qualitatively, the systematics in baryon spectrum including the Roper resonance just by a single parameter ω.

8.3 Electromagnetic Transitions

Introduction

In order to obtain further information of baryon structure, it is very common to investigate various form factors. From experiences of atomic and nuclear physics, electromagnetic transitions are expected to be useful for the study of intrinsic structure, especially related to shape deformation. One of distinctive aspects in baryon physics is that the relevant momentum transfer q is much higher than the region where the long wave length approximation can apply. For instance, typical nuclear transition occurs at the length scale

$$\frac{1}{q(\text{Nuclear transition} \sim \text{few MeV})} \sim 100\,\text{fm}\,, \tag{8.14}$$

which is much longer than typical nuclear size of few fm. In contrast, in baryon transitions,

$$\frac{1}{q(\text{Baryon transition} \sim \text{few hundred MeV})} \sim 0.5\,\text{fm}\,, \tag{8.15}$$

which is comparative to or even less than baryon size.

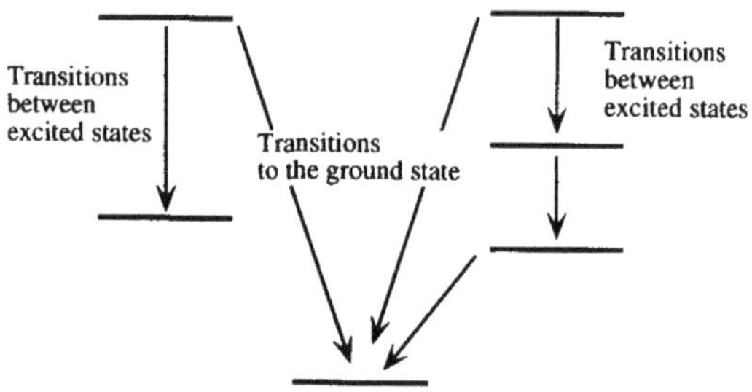

Fig. 8.5 Two types of transitions.

We can classify the transitions into two categories; one is the transitions between an excited state and the ground state, and the other is the transition between excited states (Fig. 8.5). In atomic and nuclear physics, both transitions have been observed extensively, which has established our present knowledge of atoms and nuclei. As compared to this, baryon data are rather poor and limited only to transitions from or to the ground state. This is so because the excited baryons decay fast through strong interactions by emitting mesons, mostly the pions. For these reasons, the present knowledge on baryon physics is much less than atomic and nuclear physics. This situation has now motivated further studies of baryon structure in recent projects of hadron physics (see, for instance, web pages of fascilities such as JLab, BATES, MAMI, COSY, CELCIUS, KEK and SPring8).

Experimentally, the relevant transition amplitudes are extracted from single pion photoproductions (Fig. 8.6), assuming resonance dominance. A production amplitude is then a sum of the background Born terms and resonance (excited state) contributions. A distinguished feature of this process is that not only the magnitude but also the sign of the resonance terms can be determined relative to the sign of the Born terms [Walker 1969; Moorhouse 1978; PDG 1976]. More precisely, the sign ϵ in the following combination can be determined experimentally:

$$\langle N|H_\pi|N^*\rangle \cdot \langle N^*|H_\gamma|N\rangle \equiv \epsilon|\langle N|H_\pi|N^*\rangle \cdot \langle N^*|H_\gamma|N\rangle| , \qquad (8.16)$$

where H_π and H_γ are the interaction Hamiltonian for the pion and photon couplings. The sign ϵ is then included in the electromagnetic couplings $\langle N^*|H_\gamma|N\rangle$. Therefore, in theoretical calculations, both the electromagnetic and pion couplings have to be calculated simultaneously.

In this section we develop how to compute the matrix elements appear-

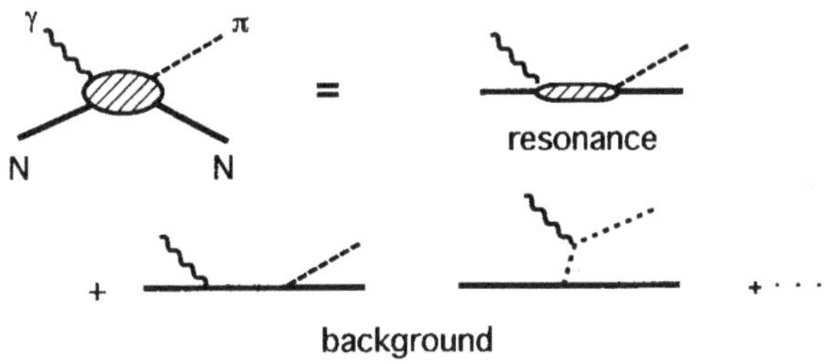

Fig. 8.6 Pion photoproduction which is decomposed into resonance and background terms.

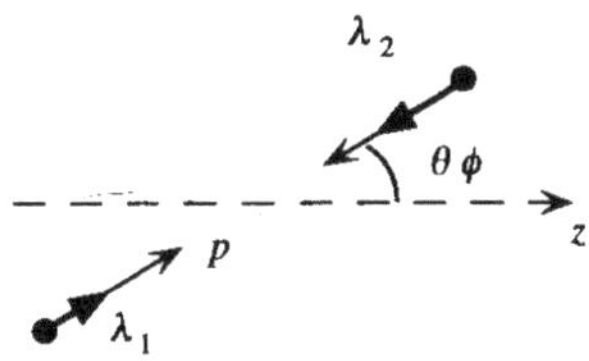

Fig. 8.7 Two particles of helicities of λ_1 and λ_2 with relative momentum p.

ing in Eq. (8.16) using the non-relativistic quark model. Our description is general and can apply to other processes such as strong interaction processes. Since conventional experimental data are presented in the helicity basis, we mostly work with this basis. However, we also derive matrix elements in the multipole basis which is rather familiar in atomic and nuclear physics.

Helicity formalism

Experimental amplitudes (T-matrix) are usually presented in the helicity basis [Walker 1969; Moorhouse 1978]. For this reason, we begin with a discussion on the helicity formalism developed by Jacob and Wick [Jacob and Wick 1959].

Let us consider two initial particles with helicities λ_1 and λ_2, and their relative momentum $\boldsymbol{p}$ which is oriented to $\theta\phi$ from the z-axis (see Fig. 8.7). The total helicity is then $\lambda_1 - \lambda_2 \equiv \lambda$. We denote this helicity state as

$$|p, \theta\phi, \lambda\rangle \,. \tag{8.17}$$

If $\boldsymbol{p}$ is along the z-axis ($\theta\phi = 00$), λ equals the z-component of the total angular momentum.

The helicity states (8.17) can be related to the eigenstates of angular momentum through

$$|p, \theta\phi, \lambda\rangle = \sum_{jm} \sqrt{\frac{2j+1}{4\pi}} D^j_{m\lambda}(\phi, \theta, -\phi)|p, jm, \lambda\rangle\,, \tag{8.18}$$

and the inverse relation

$$|p, jm, \lambda\rangle = \sqrt{\frac{2j+1}{4\pi}} \int d\Omega\, D^{j*}_{m\lambda}(\phi, \theta, -\phi)|p, \theta\phi, \lambda\rangle\,. \tag{8.19}$$

In these equations, $D^j_{mn}(\alpha, \beta, \gamma)$ are the Wigner's D-function [Varshalovich et al. 1988]. For rotation of a vector, the angle γ is redundant and can be fixed as $\alpha = -\gamma \equiv \phi$. The integral measure is then given by $d\Omega = d(\cos\theta)d\phi$, and so $\int d\Omega = 4\pi$.

It is instructive to rewrite the above relations using spherical harmonics. For this we consider

$$\begin{aligned} |p, 00, \lambda = 1/2\rangle &\sim e^{i\boldsymbol{p}\cdot\boldsymbol{x}}\,\chi_{1/2} \\ &= 4\pi \sum_{lm} i^l\, Y_{lm}(\hat{x}) Y^*_{lm}(\hat{p})\, j_l(pr)\, \chi_{1/2}\,, \end{aligned} \tag{8.20}$$

where $\chi_{1/2}$ is the spin up state for the nucleon. Using $Y^*_{lm}(\hat{p} = \hat{z}) = \sqrt{(2l+1)/4\pi}\,\delta_{m0}$ and recoupling the angular momentum, we find

$$|p, 00, 1/2\rangle \sim \sqrt{2\pi} \sum_j \sqrt{2j+1}\left(i^{j-1/2}\tilde{\mathcal{Y}}^{j-1/2}_{j\,1/2} - i^{j+1/2}\tilde{\mathcal{Y}}^{j+1/2}_{j\,1/2}\right)\,, \tag{8.21}$$

where $\tilde{\mathcal{Y}}^l_{j\,m}$ is the spinor harmonics (see Eq. (5.11)) including the spherical Bessel functions $j_l(kr)$:

$$\tilde{\mathcal{Y}}^l_{j\,m} = \sum_\mu (l\, m{-}\mu\, 1/2\, \mu\, |j\, m) Y_{l\, m-\mu}\, \chi_\mu\, j_l\,. \tag{8.22}$$

Similarly, we find

$$|p, 00, -1/2\rangle \sim \sqrt{2\pi} \sum_j \sqrt{2j+1}\left(i^{j-1/2}\tilde{\mathcal{Y}}^{j-1/2}_{j\,1/2} + i^{j+1/2}\tilde{\mathcal{Y}}^{j+1/2}_{j\,1/2}\right)\,. \tag{8.23}$$

The states with a definite angular momentum can be obtained by applying the projection operator:

$$\begin{aligned} &|p, jm, \pm 1/2\rangle \\ &\sim \sqrt{\frac{2j+1}{4\pi}} \int d\Omega\, D^{j*}_{m\,\pm 1/2}(\phi,\theta,-\phi) R(\phi,\theta,-\phi)\, |k, 00, \pm 1/2\rangle \\ &= \frac{4\pi}{\sqrt{2}} \left(i^{j-1/2} \tilde{\mathcal{Y}}^{j-1/2}_{j\,1/2} \mp i^{j+1/2} \tilde{\mathcal{Y}}^{j+1/2}_{j\,1/2} \right) . \end{aligned} \tag{8.24}$$

Here $R(\phi,\theta,-\phi)$ is the rotational operator as defined in Eq. (6.18), and in deriving the second equation we have used the definition of the D-functions. Obviously, in these states (8.24) parities are mixed. States with a definite parity can be obtained by forming the combinations:

$$\begin{aligned} |p, jm, \pm\rangle &= \frac{1}{\sqrt{2}} \left(|p, jm, +1/2\rangle \pm |p, jm, -1/2\rangle \right) \\ &= \pm 4\pi\, i^{j \mp 1/2}\, \mathcal{Y}^{j \mp 1/2}_{jm} . \end{aligned} \tag{8.25}$$

These are the eigenstates of parity $P = \eta(-)^{l_\pi} = \eta(-)^{j \mp 1/2}$, where η is the intrinsic parity of the system.

Helicity amplitude

Now we consider the process

$$\gamma N \to \pi N . \tag{8.26}$$

As shown in Fig. 8.8, we denote the initial and final momenta by k and q, respectively. λ_γ, λ_i, $\lambda_\pi (= 0)$ and λ_f are the helicities of the photon, of the initial nucleon, of the pion and of the final nucleon, respectively. Also we define $\lambda = \lambda_\gamma - \lambda_i$ and $\mu = \lambda_\pi - \lambda_f = -\lambda_f$. Therefore, the initial and final helicity states are denoted as

$$\begin{aligned} \text{initial} &\sim |k, 00, \lambda\rangle , \\ \text{final} &\sim |q, \theta\phi, \mu\rangle . \end{aligned} \tag{8.27}$$

Expanding these helicity states in the eigenstates of j as in Eq. (8.18),

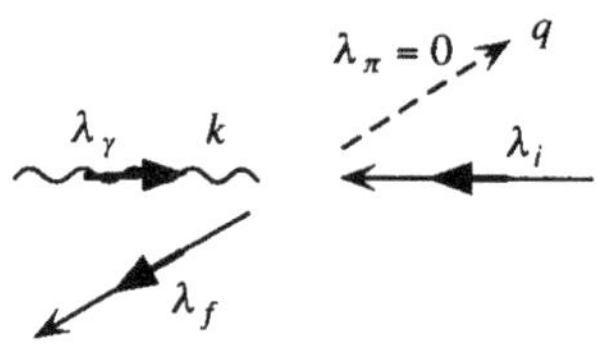

Fig. 8.8 Kinematics of the pion photoproduction.

the production amplitude may be written as

$$\begin{aligned}\langle q, \theta\phi, \mu|\mathcal{A}|k, 00, \lambda,\rangle &= \sum_{jj'm} \sqrt{\frac{2j'+1}{4\pi}}\sqrt{\frac{2j+1}{4\pi}} \\ &\times \ D^{j*}_{m\mu}(\phi, \theta, -\phi)\langle q, jm, \mu|\mathcal{A}|k, j'\lambda, \lambda\rangle \,. \quad (8.28)\end{aligned}$$

Due to the conservation of angular momentum, $j' = j$ and $m = \lambda$, we obtain

$$\begin{aligned}&\langle q, \theta\phi, \mu|\mathcal{A}|k, 00, \lambda\rangle \\ &= \frac{1}{4\pi}\sum_j (2j+1) D^{j*}_{\lambda\mu}(\phi, \theta, -\phi)\langle q, j\lambda, \mu|\mathcal{A}|k, j\lambda, \lambda\rangle \\ &= \frac{1}{4\pi}\sum_j (2j+1) A^j_{\mu\lambda}\, d^j_{\lambda\mu}(\theta)\, e^{i(\lambda-\mu)\phi} \,. \quad (8.29)\end{aligned}$$

This is the multipole expansion of the helicity amplitude. The expansion coefficients $A^j_{\mu\lambda}$ which were called the helicity elements in the literature [Walker 1969] contain dynamical information of the excited states of spin j. In the pion photoproduction, λ takes four values of $-3/2, -1/2, +1/2$ and $+3/2$, while $\mu = -1/2$ and $+1/2$. For a given j, there are eight components of $A^j_{\mu\lambda}$; four of them are independent due to time reversal symmetry.

The helicity elements $A^j_{\mu\lambda}$ in its form, however, do not have a definite parity. Using Eq. (8.25), parity eigenstates are projected out by the following linear combinations [Walker 1969; Moorhouse 1978; Jacob and Wick 1959]:

$$C^{l_\pi +}_\lambda = \frac{1}{\sqrt{2}}\left(A^j_{1/2\,\lambda} + A^j_{-1/2\,\lambda}\right) , \quad (8.30)$$

$$C^{(l_\pi+1)-}_\lambda = \frac{1}{\sqrt{2}}\left(A^j_{1/2\,\lambda} - A^j_{-1/2\,\lambda}\right) , \quad (8.31)$$

where the total spin j is given by $j = l_\pi + 1/2$, and the parity is $P = -(-)^{l_\pi}$ for $C_\lambda^{l_\pi +}$ and $P = (-)^{l_\pi}$ for $C_\lambda^{(l_\pi+1)-}$. Here l_π is the orbital angular momentum of the pion in the final state.

Resonance parameters are determined by fitting the experimental amplitudes (8.30) and (8.31) by a simple Breit-Wigner form assuming resonance dominance [Gourdin and Salin 1963] *:

$$\begin{aligned} C_\lambda^{l_\pi \pm}(W) &= -\sum_{N^*} \epsilon \left(\frac{\Gamma_\gamma^\lambda \Gamma_\pi}{kq} \right)^{1/2} \frac{W}{W^2 - M^{*2} + iW\Gamma} + \text{background} \\ &\to i\,\epsilon \left(\frac{\Gamma_\gamma^\lambda \Gamma_\pi}{kq\Gamma^2} \right)^{1/2} , \end{aligned} \tag{8.32}$$

where Γ is the total decay width of the resonance N^* , Γ_γ^λ and Γ_π are the partial width for gamma (with the helicity λ) and pion emission. In the second line of (8.32) the initial energy W in the center of mass system is set to the resonance energy M^*, where the single resonance N^* is assumed to dominate the amplitude. The relative sign of each term is included in the parameter $\epsilon = \pm 1$. The helicity amplitudes for the electromagnetic couplings $A_\lambda^{j^P}$ are then defined by the imaginary part of the amplitude:

$$A_\lambda^{j^P} = -i \left(\frac{1}{(2j+1)\pi} \frac{k}{q} \frac{M}{M^*} \frac{\Gamma_\pi}{\Gamma^2} \right)^{-1/2} C_\lambda^{l_\pi \pm}(M^*)\, C_{\pi N} , \tag{8.33}$$

where

$$j = l_\pi \pm 1/2 , \; P = -(-)^{l_\pi} ,$$

and $C_{\pi N}$ is the isospin Clebsch-Gordan coefficient for the decay of N^* into the relevant πN charge state [PDG 1976; PDG 1996]. In this equation, the strength of the pion coupling in $C_\lambda^{l\pm}$ is removed by multiplying the factor $\Gamma_\pi^{-1/2}$.

Theoretically, the amplitude $C_\lambda^{l_\pi \pm}(W)$ may be computed as

$$\begin{aligned} C_\lambda^{l_\pi \pm}(W) &= i \langle N|H_\pi|N^*\rangle \frac{W}{W^2 - M^{*2} + iW\Gamma} \langle N^*|H_\gamma|N\rangle \\ &\to \langle N|H_\pi|N^*\rangle \langle N^*|H_\gamma|N\rangle \frac{1}{\Gamma} , \quad (W \to M^*) . \end{aligned} \tag{8.34}$$

*Note that the expression the resonance propagator (8.32) differs from that given in Refs. [Moorhouse 1978; PDG 1976; PDG 1996]. However, Eq. (8.32) agrees with, for instance, Eq. (2.18) of Ref. [Moorhouse 1978].

Table 8.4 Relation between various amplitudes and quark model states.

Excited states		Photon couplings		Pion photoproduction		
j^P	$[l,\ s]^j$	Multipole	Helicity	l_π	Multipole	Helicity
$1/2^+$	$[0,\ 1/2]^{1/2}$	$M1$	$A_{1/2}$	1	M_{1-}	$C^{1-}_{1/2}$
	$[2,\ 3/2]^{1/2}$					
$3/2^+$	$[2,\ 1/2]^{3/2}$	$M1, E2$	$A_{1/2,3/2}$	1	M_{1+}, E_{1+}	$C^{1+}_{1/2,3/2}$
	$[0,\ 3/2]^{3/2}$	$M1$				
$5/2^+$	$[2,\ 1/2]^{5/2}$	$M3, E2$	$A_{1/2,3/2}$	3	M_{3-}, E_{3-}	$C^{3-}_{1/2,3/2}$
	$[4,\ 3/2]^{3/2}$	$M3$				
$1/2^-$	$[1,\ 1/2]^{1/2}$	$E1$	$A_{1/2}$	0	E_{0+}	$C^{0+}_{1/2}$
	$[1,\ 3/2]^{1/2}$					
$3/2^-$	$[1,\ 1/2]^{3/2}$	$M2, E1$	$A_{1/2,3/2}$	2	M_{2-}, E_{2-}	$C^{2-}_{1/2,3/2}$
	$[3,\ 3/2]^{3/2}$	$M2$				
$5/2^-$	$[3,\ 1/2]^{5/2}$	$M2, E3$	$A_{1/2,3/2}$	2	M_{2+}, E_{2+}	$C^{2+}_{1/2,3/2}$
	$[1,\ 3/2]^{3/2}$	$M2$				

From this and Eq. (8.33), one obtains an alternative expression

$$A_\lambda^{j_P} = -\epsilon\,(KF)\,(-i)^l\,\langle N^*|H_\gamma|N\rangle\,,\quad (KF) = \sqrt{\frac{E_N(k)}{2\omega M}}\,, \tag{8.35}$$

where l is an orbital angular momentum of an excited quark in N^*, and ϵ is the sign extracted from, together with another phase factor $(-i)^l$, the pion matrix element as given in the next subsection (see Eq. (8.43)). At the end of this subsection, we summarize selection rules for various amplitudes and excited states in Table 8.4.

Pion coupling

The main subject in this section is the electromagnetic coupling. However, as we have seen in Eq. (8.35), the pion coupling is also needed to determine the sign of the photoproduction amplitudes, when comparing theoretical results with experimental data.

Let us start with the relation (8.16):

$$A_\lambda^{j^P} \;=\; -i\left(\frac{1}{(2j+1)\pi}\frac{k}{q}\frac{M}{M^*}\frac{\Gamma_\pi}{\Gamma^2}\right)^{-1/2} C_\lambda^{l_\pi\pm}(M^*)C_{\pi N}$$

$$
\begin{aligned}
&= (\cdots)^{-1/2} \langle q, j\lambda, \pm | H_\pi | N^*(j\lambda)\rangle \frac{W}{W^2 - M^{*2} + iW\Gamma} \\
&\quad \times \langle N^*(j\lambda)|H_\gamma|k, j\lambda, \lambda\rangle C_{\pi N}\,. \qquad (8.36)
\end{aligned}
$$

In order to compute the matrix element $\langle q, j\lambda, \pm | H_\pi | N^*(j\lambda)\rangle$, it is convenient to adopt an explicit form for the final state as given in (8.25):

$$
\begin{aligned}
\langle q, j\lambda, \pm| &= \left(\pm i^{l_\pi} |\tilde{\mathcal{Y}}^{l_\pi}_{j\lambda}\rangle\right)^\dagger \\
&= \left(\pm i^{l_\pi} \sum_m (l_\pi\ m\ 1/2\ \lambda\,|-m\ j) Y_{l_\pi m} j_{l_\pi} |\chi_{\lambda - m}\rangle\right)^\dagger \qquad (8.37) \\
&= \pm(-i)^{l_\pi} \sum_m (l_\pi\ -m\ 1/2\ \lambda{+}m\,|\,j\ \lambda)\,(-)^m \langle \chi_{\lambda+m}| Y_{l_\pi m} j_{l_\pi}\,,
\end{aligned}
$$

where in the last step we have used

$$
Y^*_{l_\pi m} = (-1)^{l_\pi} Y_{l_\pi\ -m}\,. \qquad (8.38)
$$

In Eq. (8.37), $l_\pi \equiv j \mp 1/2$ is the orbital angular momentum for the pion relative to the nucleon.

The interaction hamiltonian for the pion-quark coupling is given by

$$
H_\pi = -\frac{g}{2m} \boldsymbol{\sigma} \cdot \boldsymbol{\nabla}_\pi\,, \qquad (8.39)
$$

where the derivative ∇_π operates to the pion wave function. Sandwiching this hamiltonian by the final state of Eq. (8.37) and the intermediate resonance state $|N^*(j\lambda)\rangle$, we find after some algebra

$$
\begin{aligned}
&\langle q, j\lambda, \pm | H_\pi | N^*(j\lambda)\rangle \\
&\quad = \mp 4\pi \frac{gq}{2m} i^{l_\pi} \left(\sqrt{\frac{l_\pi + 1}{2l_\pi + 1}} \langle 1/2 || [Y_{l_\pi+1}\sigma]^{l_\pi} j_{l_\pi+1} || j\rangle \right. \\
&\qquad \left. + \sqrt{\frac{l_\pi}{2l_\pi + 1}} \langle 1/2 || [Y_{l_\pi-1}\sigma]^{l_\pi} j_{l_\pi-1} || j\rangle \right)\,. \qquad (8.40)
\end{aligned}
$$

In fact, one of the two terms is relevant depending on the orbital angular momentum $l_q \equiv l$ of an excited quark in N^*. Due to the pseudoscalar nature of the pion-quark coupling, it follows that $l = l_\pi \pm 1$. When $j = l - 1/2$ and when $j = l - 3/2$, the first term of Eq. (8.40) survives, while when $j = l + 3/2$ and when $j = l + 1/2$, the second term does.

For later convenience when performing actual computations, we summarize relevant pion matrix elements according to $j = l + 3/2,\ l + 1/2,\ l - 1/2,\ l - 3/2$:

(1) When $j = l + 3/2,\ l + 1/2 \xrightarrow{l_\pi = l+1} j = l_\pi \pm 1/2$,

$$\begin{aligned} &\langle q, j\lambda \pm |H_\pi| N^*(j\lambda)\rangle \\ &= \mp 4\pi \frac{gq}{2m} i^{l+1} \sqrt{\frac{l+1}{2l+3}} \langle 1/2 \| [Y_l \sigma]^{l+1} j_l \| j\rangle . \end{aligned} \tag{8.41}$$

(2) When $j = l - 1/2, l - 3/2 \xrightarrow{l_\pi = l-1} j = l_\pi \pm 1/2$,

$$\begin{aligned} &\langle q, j\lambda \pm |H_\pi| N^*(j\lambda)\rangle \\ &= \pm 4\pi \frac{gq}{2m} i^{l+1} \sqrt{\frac{l}{2l-1}} \langle 1/2 \| [Y_l \sigma]^{l-1} j_l \| j\rangle . \end{aligned} \tag{8.42}$$

Writing the pion matrix element as †

$$\langle q, j\lambda \pm |H_\pi| N^*(j\lambda)\rangle \equiv -i(-i)^l\ \epsilon\ |\langle q, j\lambda \pm |H_\pi| N^*(j\lambda)\rangle| , \tag{8.43}$$

the sign ϵ is given as summarized in Table 8.5. In the table, η is the sign of the radial integral $(nl|j_l|00)$ which is defined in Eq. (8.83).

Table 8.5 The sign ϵ extracted from the pion matrix elements.

Nucleon			Delta		
	j	ϵ		j	ϵ
$S = 1/2$	$l + 1/2$	$-\eta$	$S = 1/2$	$l + 1/2$	$+\eta$
	$l - 1/2$	$+\eta$		$l - 1/2$	$-\eta$
$S = 3/2$	$l + 3/2$	$+\eta$	$S = 3/2$	$l + 3/2$	$+\eta$
	$l + 1/2$	$-\eta$		$l + 1/2$	$-\eta$
	$l - 1/2$	$-\eta$		$l - 1/2$	$-\eta$
	$l - 3/2$	$+\eta$		$l - 3/2$	$+\eta$

$\eta \equiv \mathrm{sgn}\,[(00|j_l|nl)]$

†In Eq. (8.43), the phase factor $(-i)^{l+1}$ has been extracted. This arises in Eqs. (8.41) and (8.42) from the reduced matrix elements.

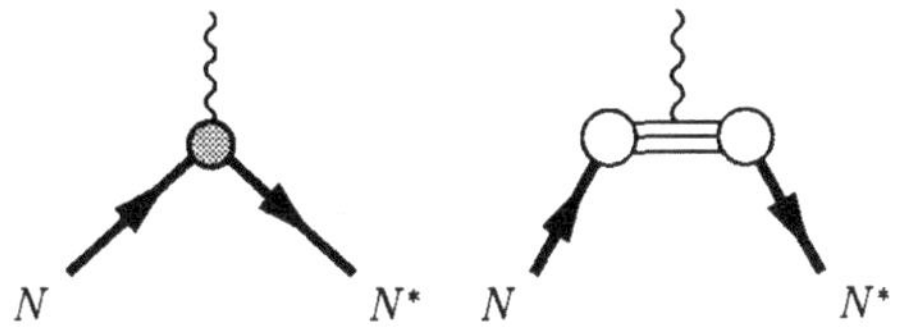

Fig. 8.9 A photon coupling to a quark in baryons.

Photon coupling

It is convenient to work in the transverse or radiation gauge [Jackson 1974] (Ch.6.5), where

$$A_0 = 0, , \quad \nabla \cdot \boldsymbol{A} = 0 \,. \tag{8.44}$$

In this gauge, the electromagnetic coupling is given by the interaction hamiltonian

$$H_\gamma = -e \int d^3x\, \boldsymbol{J} \cdot \boldsymbol{A} \,, \tag{8.45}$$

where $\boldsymbol{A}$ is the photon field, and for the current $\boldsymbol{J}$ we adopt the non-relativistic form

$$\begin{aligned} \boldsymbol{J} &= \sum_{j=1}^{3} \boldsymbol{J}^{(j)} = 3\boldsymbol{J}^{(3)} \,, \\ \boldsymbol{J}^{(3)} &= \frac{1}{2m} \left(u_f^\dagger (i \overleftarrow{\nabla} - i \overrightarrow{\nabla}) u_i + \overrightarrow{\nabla} \times (u_f^\dagger \boldsymbol{\sigma} u_i) \right) \tau_\mu \,, \end{aligned} \tag{8.46}$$

where u_i and u_f are the two component Pauli spinors for the initial and final state quarks, respectively (see Fig. 8.9). In the first equation we have used overall permutation symmetry of quarks in the baryons. The isospin matrix τ_μ is either $\tau_0 = 1$ for the isoscalar current and $\tau_i = \tau_3$ for the isovector current.

Helicity amplitudes $A_{1/2,3/2}$ are then computed as given in (8.35):

$$\begin{aligned} A_{1/2} &= +\epsilon(-i)^l\, (KF)\, \langle N^*(j, 1/2) | \boldsymbol{J} \cdot \boldsymbol{A}_{+1} | N(1/2, -1/2) \rangle \,, \\ A_{3/2} &= +\epsilon(-i)^l\, (KF)\, \langle N^*(j, 3/2) | \boldsymbol{J} \cdot \boldsymbol{A}_{+1} | N(1/2, 1/2) \rangle \,. \end{aligned} \tag{8.47}$$

In order to relate the helicity amplitudes with the multipole amplitudes, we consider the multipole expansion for the photon field [‡]

$$\boldsymbol{A}_\lambda = \epsilon_\lambda e^{i\boldsymbol{k}\cdot\boldsymbol{x}} \,. \tag{8.48}$$

Here ϵ_λ is the polarization vector of helicity $\lambda = \pm 1$. The momentum $\boldsymbol{k}$ is related to the physical momentum $\boldsymbol{k}_{phys}$ by

$$\boldsymbol{k} = \sqrt{\frac{2}{3}}\boldsymbol{k}_{phys} \,, \tag{8.49}$$

in the convention of coordinate transformation (8.3). If we choose the z-axis along the direction of photon propagation $\boldsymbol{k}$, the plane wave of a definite helicity (8.48) can be expanded by multipoles as [T. deForest and Walecka 1966; Rose 1957]

$$\boldsymbol{A}_\lambda = -\sqrt{2\pi}\sum_{l=1}^{\infty} i^l\sqrt{2l+1}\left(\lambda \boldsymbol{A}_{l\lambda}^{(M)} + i\boldsymbol{A}_{l\lambda}^{(E)}\right) \,. \tag{8.50}$$

The magnetic (M) and electric (E) multipole fields are given by

$$\begin{aligned} \boldsymbol{A}_{lm}^{(M)}(\boldsymbol{r}) &= j_l(kr)\boldsymbol{Y}^l_{lm}(\hat{r}) \,, \qquad \boldsymbol{Y}^l_{lm} = \frac{1}{\sqrt{l(l+1)}}\boldsymbol{L}Y_{lm}(\hat{r}) \,, \\ \boldsymbol{A}_{lm}^{(E)}(\boldsymbol{r}) &= -\frac{i}{k}\boldsymbol{\nabla}\times \boldsymbol{A}_{lm}^{(M)}(\boldsymbol{r}) \,. \end{aligned} \tag{8.51}$$

Here we have adopted the standard convention for the vector spherical harmonics [Varshalovich et al. 1988],

$$\boldsymbol{Y}^l_{jm} = \sum_{\mu\nu}(l\ \mu\ 1\ \nu|jm)Y_{l\mu}\boldsymbol{e}_\nu \tag{8.52}$$

where $\boldsymbol{e}$ is the rank one unit vector. Substituting Eq. (8.50) in Eq. (8.45) and applying the Wigner-Eckart theorem (see Appendix H) to each multipole amplitude, we find the following formulae relating the helicity and multipole amplitudes ($\mathcal{M}(M\,l)$ for magnetic and $\mathcal{M}(W\,l)$ for electric) [Hosaka et al. 2000]:

[‡] The same symbol A is used for the photon field and helicity amplitudes, and ϵ for the photon polarization vector and the sign of the amplitudes. Both notations are standard and there will be no confusion in the following discussions.

For $j = l + 3/2$:

$$
\begin{aligned}
A_{1/2} &= -\epsilon\,(KF)\,\sqrt{2\pi}\sqrt{\frac{l+1}{2l+4}}\,\mathcal{M}(M\,l+1) \\
A_{3/2} &= -\epsilon\,(KF)\,\sqrt{2\pi}\sqrt{\frac{l+3}{2l+4}}\,\mathcal{M}(M\,l+1)
\end{aligned}
\tag{8.53}
$$

For $j = l + 1/2$:

$$
\begin{aligned}
A_{1/2} &= \epsilon\,(KF)\,\sqrt{2\pi} \\
&\times \left(+\sqrt{\frac{l+2}{2l+2}}\,\mathcal{M}(M\,l+1) - \sqrt{\frac{l}{2l+2}}\,\mathcal{M}(E\,l)\right) \\
A_{3/2} &= \epsilon\,(KF)\,\sqrt{2\pi} \\
&\times \left(-\sqrt{\frac{l}{2l+2}}\,\mathcal{M}(M\,l+1) - \sqrt{\frac{l+2}{2l+2}}\,\mathcal{M}(E\,l)\right)
\end{aligned}
\tag{8.54}
$$

For $j = l - 1/2$:

$$
\begin{aligned}
A_{1/2} &= \epsilon\,(KF)\,\sqrt{2\pi} \\
&\times \left(+\sqrt{\frac{l-1}{2l}}\,\mathcal{M}(M\,l-1) + \sqrt{\frac{l+1}{2l}}\,\mathcal{M}(E\,l)\right) \\
A_{3/2} &= \epsilon\,(KF)\,\sqrt{2\pi} \\
&\times \left(+\sqrt{\frac{l+1}{2l}}\,\mathcal{M}(M\,l-1) - \sqrt{\frac{l-1}{2l}}\,\mathcal{M}(E\,l)\right)
\end{aligned}
\tag{8.55}
$$

For $j = l - 3/2$:

$$
\begin{aligned}
A_{1/2} &= -\epsilon\,(KF)\,\sqrt{2\pi}\sqrt{\frac{l}{2l-2}}\,\mathcal{M}(M\,l-1) \\
A_{3/2} &= +\epsilon\,(KF)\,\sqrt{2\pi}\sqrt{\frac{l-2}{2l-2}}\,\mathcal{M}(M\,l-1)
\end{aligned}
\tag{8.56}
$$

Writing $|N\rangle \to |i\rangle$ and $|N^*\rangle \to |f\rangle$, the reduced matrix elements are defined by (see Appendix H)

$$
\begin{aligned}
\mathcal{M}(M\,l) &\equiv i\langle f||\boldsymbol{J}\cdot\boldsymbol{A}_l^{(M)}||i\rangle \cdot \langle f|\tau_\mu|i\rangle\,, \\
\mathcal{M}(E\,l) &\equiv i\langle f||\boldsymbol{J}\cdot\boldsymbol{A}_l^{(E)}||i\rangle \cdot \langle f|\tau_\mu|i\rangle\,,
\end{aligned}
\tag{8.57}
$$

where the isospin part is the ordinary matrix elements. In Eqs. (8.57), the phase i is introduced to make the matrix elements $\mathcal{M}$ pure real. Furthermore, in Eqs. (8.53) – (8.56), the phase factor $(-i)^l$ in Eq. (8.47) is canceled by another factor i^l coming from the multipole expansion (8.50).

Let us briefly outline how to compute the multipole amplitudes (8.57). For illustration, we consider magnetic transitions $M\,l$. Let us write the reduced matrix element as [§]

$$\langle f \,||\, \boldsymbol{J} \cdot \boldsymbol{A}_l^{(M)} \,||\, i \rangle \to \frac{1}{2m} \left(i\, u_f^\dagger (\overleftarrow{\nabla} - \overrightarrow{\nabla}) u_i + \overrightarrow{\nabla} \times (u_f^\dagger \boldsymbol{\sigma} u_i) \right) \cdot \boldsymbol{A}_l^{(M)} \,, \quad (8.58)$$

where on the right hand side it is understood that the integral and sum are taken over configuration and spin spaces. The initial and final states are shown explicitly by u_i and u_f in order to indicate the operation of derivatives. The isospin part can be incorporated separately in a trivial manner and is not shown here. Now in Eq. (8.58) integrating the left derivative $(\overleftarrow{\nabla})$ term by parts and using the relations $\boldsymbol{\nabla} \cdot \boldsymbol{A} = 0$ and $\hat{x} \cdot \boldsymbol{A}^{(M)} = 0$ (the radial vector $\hat{x}$ appears when the derivative hits the initial state nucleon which is spherical), one finds that the convection $(\overleftarrow{\nabla} - \overrightarrow{\nabla})$ term vanishes. Thus only the spin term survives:

$$\mathcal{M}(M\,l) = \frac{i}{2m}\, u_f^\dagger \boldsymbol{\sigma} u_i \cdot \boldsymbol{\nabla} \times \boldsymbol{A}_l^{(M)} \,. \quad (8.59)$$

Using the relation between the magnetic and electric fields in Eq. (8.51), one arrives at the expression

$$\begin{aligned} \mathcal{M}(M\,l) &= \frac{i}{2m} \left[-\sqrt{\frac{l}{2l+1}}\, \langle f || [Y_{l+1} \sigma]^l j_{l+1} || i \rangle \right. \\ &\qquad \left. + \sqrt{\frac{l+1}{2l+1}}\, \langle f || [Y_{l-1} \sigma]^l j_{l-1} || i \rangle \right] \langle f | \tau_\mu | i \rangle \,. \end{aligned} \quad (8.60)$$

For electric transitions $E\,l$, again integrating by parts, one finds

$$\mathcal{M}(E\,l) = \frac{1}{m} u_f^\dagger \,\overrightarrow{\nabla}\, u_i \cdot \boldsymbol{A}_l^{(E)} + \frac{k}{2m} u_f^\dagger \,\boldsymbol{\sigma}\, u_i \cdot \boldsymbol{A}_l^{(M)} \,. \quad (8.61)$$

Since the initial state is spherical, the first term can be written as

$$(u_f^\dagger \,\overrightarrow{\nabla}\, u_i) \cdot \boldsymbol{A}_l^{(E)} = u_f^\dagger u_i' \,\hat{x} \cdot \boldsymbol{A}_l^{(E)} \,, \quad (8.62)$$

[§]We have not included the factor 3 in Eq. (8.46) in the following analytic expressions. However, numerical values in Tables 8.6 and 8.7 include this factor 3.

where prime $'$ denotes a radial derivative. Then using the following property of the spherical harmonics,

$$\hat{x} \cdot \boldsymbol{A}_{lm}^{(E)} = \sqrt{l(l+1)}\, \frac{j_l(kr)}{kr}\, Y_{lm} \,, \tag{8.63}$$

for the first term and

$$\boldsymbol{\sigma} \cdot \boldsymbol{A}_{lm}^{(M)} = \boldsymbol{\sigma} \cdot \boldsymbol{Y}_{lm}^{l}\, j_l(kr) = [Y_l, \sigma]^{lm}\, j_l(kr) \,, \tag{8.64}$$

for the second term of (8.61), one arrives at the expression

$$\begin{aligned} \mathcal{M}(El) \;=\; & -i \left[\frac{\sqrt{l(l+1)}}{m} \langle f \,||\, Y_l \frac{j_l(kr)}{kr} ||i' \rangle \right. \\ & \left. + \frac{k}{2m} \langle f \,||\, [Y_l \sigma]^l j_l(kr) ||\, i \rangle \right] \langle f | \tau_\mu | i \rangle \,, \end{aligned} \tag{8.65}$$

where $|i'\rangle = \partial/\partial r |i\rangle$. Eqs. (8.60) and (8.65) are the general expressions for electromagnetic transitions $N\gamma \to N^*$.

Wave functions

In order to compute the matrix elements of Eqs. (8.60) and (8.65), we use in this section the conventional non-relativistic quark model. An extension to the deformed case has been discussed in detail in Ref. [Hosaka et al. 2000].

We can apply immediately the formulae (A.26) – (A.31) in Appendix A to write down the wave functions of excited states. The harmonic oscillator functions in the spherical basis f_{nlm} for ρ and λ variables are used for the single particle states. We denote the three-particle (internal) states as $\psi_{lm}^{N\sigma}(\boldsymbol{\rho}, \boldsymbol{\lambda})$. The results are:

$$\psi_{00}^{0S} = f_{000}(\boldsymbol{\lambda}) f_{000}(\boldsymbol{\rho}) \,, \tag{8.66}$$

$$\psi_{1m}^{1\lambda} = f_{01m}(\boldsymbol{\lambda}) f_{000}(\boldsymbol{\rho}) \,, \tag{8.67}$$

$$\psi_{1m}^{1\rho} = f_{000}(\boldsymbol{\lambda}) f_{01m}(\boldsymbol{\rho}) \,, \tag{8.68}$$

$$\psi_{00}^{2S} = \frac{1}{\sqrt{2}} \left(f_{100}(\boldsymbol{\lambda}) f_{000}(\boldsymbol{\rho}) + f_{000}(\boldsymbol{\lambda}) f_{100}(\boldsymbol{\rho}) \right) \,, \tag{8.69}$$

$$\psi_{00}^{2\lambda} = \frac{1}{\sqrt{2}} \left(f_{100}(\boldsymbol{\lambda}) f_{000}(\boldsymbol{\rho}) - f_{000}(\boldsymbol{\lambda}) f_{100}(\boldsymbol{\rho}) \right) \,, \tag{8.70}$$

$$\psi_{2m}^{2S} = \frac{1}{\sqrt{2}} \left(f_{02m}(\boldsymbol{\lambda}) f_{000}(\boldsymbol{\rho}) + f_{000}(\boldsymbol{\lambda}) f_{02m}(\boldsymbol{\rho}) \right) \,, \tag{8.71}$$

$$\psi^{2\lambda}_{2m} = \frac{1}{\sqrt{2}}\left(f_{02m}(\boldsymbol{\lambda})f_{000}(\boldsymbol{\rho}) - f_{000}(\boldsymbol{\lambda})f_{02m}(\boldsymbol{\rho})\right) , \tag{8.72}$$

$$\psi^{2\rho}_{00} = [f_{l=1}(\boldsymbol{\lambda}), f_{l=1}(\boldsymbol{\rho})]^0_0 , \tag{8.73}$$

$$\psi^{2\rho}_{2m} = [f_{l=1}(\boldsymbol{\lambda}), f_{l=1}(\boldsymbol{\rho})]^2_m . \tag{8.74}$$

In the last two equations we have introduced the coupling of angular momenta, $[l_1, l_2]^l_m$.

Using these orbital wave functions, the total wave functions for nucleon excitations are given by

$$\begin{aligned}
&\text{N1} \quad |N; [l_S,\, 1/2]^j\rangle = \frac{1}{\sqrt{2}}\left([\psi^{NS}_l,\, \chi^\rho]^j\phi^\rho + [\psi^{NS}_l,\, \chi^\lambda]^j\phi^\lambda\right) , \\
&\text{N2} \quad |N; [l_{MS},\, 1/2]^j\rangle = \frac{1}{2}\Big([\psi^{N\rho}_l,\, \chi^\rho]^j\phi^\lambda + [\psi^{N\rho}_l,\, \chi^\lambda]^j\phi^\rho \\
&\qquad + [\psi^{N\lambda}_l,\, \chi^\rho]^j\phi^\rho - [\psi^{N\lambda}_l,\, \chi^\lambda]^j\phi^\lambda\Big) , \\
&\text{N3} \quad |N; [l_{MS},\, 3/2]^j\rangle \\
&\qquad = \frac{1}{\sqrt{2}}\left([\psi^{N\rho}_l,\, \chi^S]^j\phi^\rho + [\psi^{N\lambda}_l,\, \chi^S]^j\phi^\lambda\right) ,
\end{aligned} \tag{8.75}$$

where we have used the same notation for the spin (χ) and isospin (ϕ) wave functions as in Chapter 2. Similarly the total wave functions for delta excitations are given by

$$\mathbf{\Delta 1} \quad |\Delta; [l_S,\, 3/2]^j\rangle = [\psi^{NS}_l,\, \chi^S]^j\phi^S , \tag{8.76}$$

$$\mathbf{\Delta 2} \quad |\Delta; [l_{MS},\, 1/2]^j\rangle = \frac{1}{\sqrt{2}}\left([\psi^{N\rho}_l,\, \chi^\rho]^j + [\psi^{N\lambda}_l,\, \chi^\lambda]^j\right)\phi^S . \tag{8.77}$$

Multipole amplitudes

Using the wave functions for N^* as given in Eqs. (8.75) – (8.77) one can compute multipole amplitudes in a straightforward manner. For a given excited baryon, there are three possible amplitudes:

$$\mathcal{M}(M\,l+1) ,\quad \mathcal{M}(M\,l-1) ,\quad \mathcal{M}(E\,l) . \tag{8.78}$$

In order to illustrate actual computations, we consider $\mathcal{M}(M\,l+1)$ for the transition $|N(\text{groundstate})\rangle \to |N^*; [l_S,\, 1/2]^j\rangle$.

Replacing $l \to l+1$ in Eq. (8.60), only the second term survives due to the matching of orbital angular momentum. Since the initial ground state for the nucleon is spherical, $l = 0$, the orbital angular momentum of the absorbed photon and that of the excited baryon must be the same. Thus we have

$$\begin{aligned}\mathcal{M}(M\,l+1) &= -\frac{k}{4m}\sqrt{\frac{l+2}{2l+3}}\,\Big(\,\langle[\psi_l^{NS}\chi^\rho]^j||[Y_l\sigma]^l j_l||\psi_{l=0}^{0S}\chi^\rho\rangle\tau_\mu(\rho)\\ &\quad+\ \langle[\psi_l^{NS}\chi^\lambda]^j||[Y_l\sigma]^l j_l||\psi_{l=0}^{0S}\chi^\lambda\rangle\tau_\mu(\lambda)\,\Big)\,. \qquad (8.79)\end{aligned}$$

Here we have included the isospin matrix elements which are defined by

$$\tau_\mu(\rho) = \langle\phi^\rho|\tau_\mu|\phi^\rho\rangle\,,\quad \tau_\mu(\lambda) = \langle\phi^\lambda|\tau_\mu|\phi^\lambda\rangle\,, \qquad (8.80)$$

and the radial integral $\int r^2dr$ is implicit. Now the reduced matrix elements can be computed in a straightforward manner using the decomposition theorems(see Appendix H). The result is:

$$\begin{aligned}\mathcal{M}(M\,l+1) &= -\frac{k}{4m}\sqrt{\frac{l+2}{2l+3}}\,\hat{j}\,\widehat{l+1}\left\{\begin{array}{ccc}1/2 & j & l\\ l+1 & 1 & 1/2\end{array}\right\}\\ &\times\ \frac{1}{\sqrt{4\pi}}(nl|j_l|00)\,\sqrt{3}\,(\tau_\mu(\rho)-\frac{1}{3}\tau_\mu(\lambda))\,, \qquad (8.81)\end{aligned}$$

where $\hat{l}=\sqrt{2l+1}$, and n and l satisfy

$$2n+l=N\,. \qquad (8.82)$$

In Eq. (8.81) we have introduced the standard 6-j symbols and also the radial matrix elements:

$$\begin{aligned}(nl|j_l|00) &= \int r^2dr\ R_{nl}\ j_l\ R_{00}\,,\\ (nl'|j_l|00) &= \int r^2dr\ R'_{nl}\ j_l\ R_{00}\,, \qquad (8.83)\end{aligned}$$

where R_{nl} are the radial functions of the harmonic oscillator and R'_{nl} their radial derivatives.

All other matrix elements can be computed in similar ways. Here we summarize the results for all amplitudes. Let us introduce the following notation for the combinations of isospin matrix elements

$$\begin{aligned}T_{\pm1} &= \tau_\mu(\rho)\pm\tau_\mu(\lambda)\,,\\ T_{\pm1/3} &= \tau_\mu(\rho)\pm\frac{1}{3}\tau_\mu(\lambda)\,.\end{aligned}$$

Then transition matrix elements for various excited states N^* are summarized as follows:

N1 $N^* = |[l_S, 1/2]^j\rangle$:

$$\mathcal{M}(M\,l+1) = -\frac{k}{4m}\sqrt{\frac{l+2}{2l+3}}\,\hat{j}\,\widehat{l+1}\left\{\begin{array}{ccc} 1/2 & j & l \\ l+1 & 1 & 1/2 \end{array}\right\} \times (nl|j_l|00)\,\sqrt{3}\,T_{-1/3}\,, \quad (8.84)$$

$$\mathcal{M}(M\,l-1) = \frac{k}{4m}\sqrt{\frac{l-1}{2l-1}}\,\hat{j}\,\widehat{l-1}\left\{\begin{array}{ccc} 1/2 & j & l \\ l-1 & 1 & 1/2 \end{array}\right\} \times (nl|j_l|00)\,\sqrt{3}\,T_{-1/3} \quad (8.85)$$

$$\mathcal{M}(El) = \frac{\sqrt{l(l+1)}}{2m}(-1)^l\,\hat{j}\,(nl'|j_l|00)\,T_{+1} \quad (8.86)$$
$$+ \frac{k}{4m}\,\hat{j}\,\hat{l}\left\{\begin{array}{ccc} 1/2 & j & l \\ l & 1 & 1/2 \end{array}\right\}(nl|j_l|00)\,\sqrt{3}\,T_{-1/3}\,.$$

N2 $N^* = |[l_{MS}, 1/2]^j\rangle$:

This case is obtained simply by performing the following replacements in the results of (N1).

- Multiply $1/\sqrt{2}$ as an overall factor.
- Change the sign of one of the isospin matrices: $+\tau_\mu(\lambda) \to -\tau_\mu(\lambda)$.

N3 $N^* = |[l_{MS}, 3/2]^j\rangle$:

$$\mathcal{M}(M\,l+1) = -\frac{k}{4m}\sqrt{\frac{l+2}{2l+3}}\,\hat{j}\,\widehat{l+1}\left\{\begin{array}{ccc} 3/2 & j & l \\ l+1 & 1 & 1/2 \end{array}\right\} \times (nl|j_l|00)\frac{4}{\sqrt{6}}\,\tau_\mu(\lambda)\,, \quad (8.87)$$

$$\mathcal{M}(M\,l-1) = \frac{k}{4m}\sqrt{\frac{l-1}{2l-1}}(-1)^{j+3/2}\,\hat{j}\,\widehat{l-1}\left\{\begin{array}{ccc} 3/2 & j & l \\ l-1 & 1 & 1/2 \end{array}\right\} \times (nl|j_l|00)\,\frac{4}{\sqrt{6}}\,\tau_\mu(\lambda)\,, \quad (8.88)$$

$$\mathcal{M}(El) = -\frac{k}{4m}\,\hat{j}\,\hat{l}\left\{\begin{array}{ccc} 3/2 & j & l \\ l & 1 & 1/2 \end{array}\right\} \times (nl|j_l|00)\,\frac{4}{\sqrt{6}}\,\tau_\mu(\lambda)\,. \quad (8.89)$$

The decays of delta excited states $\Delta 1$ and $\Delta 2$ are obtained by the following manipulations.

Δ1 $\Delta^* = |[l_{MS}, 1/2]^j\rangle$: In the results of **N1**

- Pick up the $\tau_\mu(\lambda)$-term and replace $\tau_\mu(\lambda) \to \langle\phi^\lambda|\tau_\mu|\phi^S\rangle$.

Δ2 $\Delta^* = |[l_S, 3/2]^j\rangle$: In the results of **N3**

- Replace $\tau_\mu(\lambda)$ by $\langle\phi^\lambda|\tau_\mu|\phi^S\rangle$.
- Multiply $\sqrt{2}$ as an overall factor.

For practical purposes, it is convenient to write the spin of excited baryons j by l ($j = l \pm 1/2$ or $l \pm 3/2$), so that the 6-j symbols are computed explicitly. Results are:

N1.a $N^* = |[l_S, 1/2]^{l+1/2}\rangle$, allowed transitions: $M\ l+1$ and $E\ l$,

$$\mathcal{M}(M\ l+1) = -\frac{k}{4m}\sqrt{(l+2)}\ (nl|j_l|00)\ T_{-1/3} \tag{8.90}$$

$$\begin{aligned}\mathcal{M}(E\ l) &= \frac{\sqrt{l}(l+1)}{2m}\ (nl'|j_l|00)\ T_{+1} \\ &+ \frac{k\sqrt{l}}{4m}\ (nl|j_l|00)\ T_{-1/3}\end{aligned} \tag{8.91}$$

N1.b $N^* = |[l_S, 1/2]^{l-1/2}\rangle$, allowed transitions: $M\ l-1$ and $E\ l$,

$$\mathcal{M}(M\ l-1) = \frac{k}{4m}\sqrt{(l-1)}\ (nl|j_l|00)\ T_{-1/3} \tag{8.92}$$

$$\begin{aligned}\mathcal{M}(E\ l) &= -\frac{\sqrt{(l+1)}l}{2m}\ (nl'|j_l|00)\ T_{+1} \\ &+ \frac{k\sqrt{(l+1)}}{4m}\ (nl|j_l|00)\ T_{-1/3}\end{aligned} \tag{8.93}$$

N2 The results for this case, where $N^* = |[l_S, 1/2]^{l\pm1/2}\rangle$ are obtained by making the replacements as described before.

N3.a $N^* = |[l_{MS}, 3/2]^{l+3/2}\rangle$, allowed transitions: $M\ l+1$,

$$\mathcal{M}(M\ l+1) = \frac{k}{4m}\sqrt{\frac{1}{3}}\frac{l+2}{\sqrt{2l+3}}\ (nl|j_l|00)\ \tau_\mu(\lambda) \tag{8.94}$$

N3.b $N^* = |[l_{MS}, 3/2]^{l+1/2}\rangle$, allowed transitions: $M\ l+1$ and $E\ l$,

$$\mathcal{M}(M\ l+1) = -\frac{k}{12m}\sqrt{\frac{l(l+2)}{(2l+3)}}\ (nl|j_l|00)\ \tau_\mu(\lambda) \tag{8.95}$$

$$\mathcal{M}(E\,l) = -\frac{k}{12m}\sqrt{2l+3}\,(nl|j_l|00)\;\tau_\mu(\lambda) \tag{8.96}$$

N3.c $N^* = |[l_{MS}, 3/2]^{l-1/2}\rangle$, allowed transitions: $M\,l-1$ and $E\,l$,

$$\mathcal{M}(M\,l-1) = -\frac{k}{12m}\sqrt{\frac{(l-1)(l+1)}{(2l-1)}}\,(nl|j_l|00)\;\tau_\mu(\lambda) \tag{8.97}$$

$$\mathcal{M}(E\,l) = \frac{k}{12m}\sqrt{2l-1}\,(nl|j_l|00)\;\tau_\mu(\lambda) \tag{8.98}$$

N3.d $N^* = |[l_{MS}, 3/2]^{l-3/2}\rangle$, allowed transitions: $M\,l-1$,

$$\mathcal{M}(M\,l-1) = \frac{k}{4m}\sqrt{\frac{1}{3}}\frac{l-1}{\sqrt{2l-1}}\,(nl|j_l|00)\;\tau_\mu(\lambda) \tag{8.99}$$

Similarly, we can derive the formulae for deltas.

Numerical values

It is possible to compute matrix elements of Eqs. (8.90) – (8.99) for various excited states. Here we show some numerical numbers for those states of $N = 1$ (negative parity) and $N = 2$ (positive parity). We use the oscillator parameter as determined in the previous section: $\omega = 644$ MeV. For the photon momentum k we use the values computed from the theoretical mass differences between excited states and the ground state nucleon. The use of the experimental values do not change the essential features in the following discussions.

We have computed the amplitudes for all the quark model states which are more than experimentally observed. For a more realistic analysis, we should consider configuration mixings due to residual interactions. They could be induced by gluon and meson exchange interactions. For instance, a tensor force between quarks strongly mixes the two states of $^2P_{MS}$ and $^4P_{MS}$ for $1/2^-$ $N(1535)$ and $N(1650)$ [Isgur and Karl 1977; Isgur and Karl 1978b; Arima et al. 1992]. In Tables 8.6-8.7, theoretical helicity amplitudes for photon couplings (first column) are compared with experimental data (second column). Observed states are identified naively with the quark model states (see also Table 8.1 and discussions there) as shown in the last columns.

Table 8.6 Helicity amplitudes of nucleon excitations.

Positive parity			$A_{1/2}$ Theor.	Exp.	$A_{3/2}$ Theor.	Exp.	
$1/2^+$	$^2S'_S$	p	+25	-68 ± 5	–	–	$P_{11}(1440)$
		n	−16	$+39 \pm 15$	–	–	
	$^2S_{MS}$	p	+17	$+5 \pm 16$	–	–	$P_{11}(1710)$
		n	−6	-5 ± 23	–	–	
	$^4D_{MS}$	p	0		–	–	
		n	+4	-5 ± 23	–	–	
$3/2^+$	^2D_S	p	+113	$+52 \pm 39$	−36	-35 ± 24	$P_{13}(1720)$
		n	−34	-2 ± 26	0	-43 ± 94	
	$^2D_{MS}$	p	+80		−25		
		n	−56		+25		
	$^4D_{MS}$	p	0		0		
		n	−8.5		+15		
	$^4S_{MS}$	p	0		0		
		n	−14		−23		
$5/2^+$	^2D_S	p	+12	-17 ± 10	−71	$+127 \pm 12$	$F_{15}(1680)$
		n	−42	$+31 \pm 13$	0	-30 ± 14	
	$^2D_{MS}$	p	+9		−50		
		n	+21		+50		
	$^4D_{MS}$	p	0		0		
		n	−6		−24		
Negative parity			$A_{1/2}$ Theor.	Exp.	$A_{3/2}$ Theor.	Exp.	
$1/2^-$	$^2P_{MS}$	p	+154	74 ± 11	–	–	$S_{11}(1535)$
		n	−126	-72 ± 25	–	–	
	$^4P_{MS}$	p	0	48 ± 16	–	–	$S_{11}(1650)$
		n	+14	-17 ± 37	–	–	
$3/2^-$	$^2P_{MS}$	p	−19	-23 ± 9	−137	163 ± 8	$D_{13}(1520)$
		n	+59	-64 ± 8	+137	-141 ± 11	
	$^4P_{MS}$	p	0	-22 ± 13	0	0 ± 19	$D_{13}(1700)$
		n	−6	0 ± 56	−33	-2 ± 44	
$5/2^-$	$^4P_{MS}$	p	0	$+19 \pm 12$	0	$+19 \pm 12$	$D_{15}(1675)$
		n	−28	-47 ± 23	−39	-69 ± 19	

As explained in the preceding subsection, we have computed both the photon and pion couplings simultaneously. Therefore, there is no ambiguity in relative signs of the photon couplings. In the nucleon channels of $5/2^+(^2D_S)$ and $3/2^-(^2P_{MS})$, however, the signs of the present result do not agree with those of, for instance, the pioneering work by Feynman, Kislinger and Ravndal [Feynman et al. 1971], and with the experimental data. However, if we included the sign convention as tabulated by Moor-

Table 8.7 Helicity amplitudes of delta excitations

		$A^{1/2}$		$A^{3/2}$	
Positive parity	Theor.	Exp.	Theor.	Exp.	
$1/2^+$ $^2S_{MS}$	+14		–	–	
^4D_S	+18	-12 ± 30	–	–	$P_{31}(1910)$
$3/2^+$ $^2D_{MS}$	−14		+18		
^4D_S	−24		+41		$P_{33}(1920)$
$^4S'_S$	−37	-20 ± 29	−65	$+1 \pm 22$	$P_{33}(1600)$
$5/2^+$ $^2D_{MS}$	+46		+36		
^4D_S	−16	27 ± 13	−65	-47 ± 19	$F_{35}(1905)$
Negative parity					
$1/2^-$ $^2P_{MS}$	−40	19 ± 16	–	–	$S_{31}(1620)$
$3/2^-$ $^2P_{MS}$	78	116 ± 17	+78	77 ± 28	$D_{33}(1700)$

house and Parsons [Moorhouse and Parsons 1973], and an additional phase factor $i^{l=2} = -1$ for the D-wave component in the multipole expansion Eq. (8.50), the present results agree with the results of FKR. Sometimes the convention of the sign has not been considered seriously, even in recent works. In the work of Koniuk and Isgur [Koniuk and Isgur 1980], their convention was taken to reproduce empirical one, which has been followed by several literatures [Murthy et al. 1984; Dong et al. 1999; Meyer et al. 1997]. This situation was partly discussed by Capstick in Ref. [Capstick 1992].

To illustrate the importance of the sign, let us look at the amplitudes for the $N(1440)$ and $\Delta(1600)$. The latter is interpreted as the Roper resonance of the delta. The relevant formulae are Eq. (8.90) and the corresponding one for Δ. We have

$$\begin{aligned}\mathcal{M}(M1; N; {}^2S'_S) &= \frac{3}{4\sqrt{2}}\mathcal{M}(M1; \Delta; {}^4S'_S) \\ &= -\frac{k}{2m}\frac{1}{\sqrt{4\pi}}(00|j_0|10)\cdot(\text{isospin part})\,. \quad (8.100)\end{aligned}$$

From this, after taking into account the sign of the pion matrix element (see Table 8.5), we observe that the signs of the transitions for $N(1440)$ and $\Delta(1600)$ are different. This fact is correctly presented in Tables 8.6 and 8.7. In literatures, however, these signs are set the same [Capstick 1992; Close and Li 1990].

Taking into account the sign conventions as mentioned above, the nonrelativistic quark model reproduces qualitatively experimental numbers including both signs and absolute values. This is particularly so for channels where the amplitudes are large. This justifies crudely the use of the non-relativistic quark model for the description of excited baryons. However, if we look at more details, we can point out that (1) the data are not precise enough in many channels; even the signs are not fixed, and (2) theoretical predictions are not so accurate either; the predictions can change by varying the model parameters within reasonable range.

The most serious discrepancy between the theory and experiment is in $1/2^+$ excited state for the Roper resonance; not only the magnitude but also the sign do not agree with data. The relevant amplitude is $A_{1/2}$ or $M1$. Indeed the discrepancy in the sign is puzzling, if the $1/2^+$ excitation is simply considered as a radial excitation of the ground state nucleon. It does not matter whether it is a single particle or collective excitation. Due to an S-wave nature of the orbital motion, spin structure and hence the magnetic moment of the excited state should be the same as the ground state. Now we know that the magnetic moments of the nucleon, the delta [Nefkens et al. 1978] and the M1 transition of the delta [Sato and Lee 1996] are well explained by the quark model. Therefore, it is not easy to understand that the magnetic transition to the $1/2^+$ excited state can not be explained by the same framework.

The discrepancy in the sign for the Roper resonance is known for a long time and various ideas to overcome the problem have been proposed. In Refs. [Murthy and Bhaduri 1985] and [Hosaka et al. 2000], deformation effects was considered. However, that effect increases the amplitude but with the sign unchanged. The problem in the Roper resonance is in itself an interesting question. We have not yet, however, reached the right understanding of this problem and hence the nature of this resonance $N(1440)$ [Krehl et al. 2000].

Problems

8.1

Let us add a volume type energy to the hamiltonian of the harmonic oscillator,

$$H = \sum_{i=1}^{3} \left(\frac{p_i^2}{2m} + \frac{m}{2}(\omega_x^2 x_i^2 + \omega_y^2 y_i^2 + \omega_z^2 z_i^2) \right) + \frac{B}{\omega_x \omega_y \omega_z}, \tag{8.101}$$

where B is a positive constant corresponding to the bag constant in the bag model. Perform variation with respect to $\omega_x, \omega_y, \omega_z$, and find the minimum for $N = 0, 1, 2$.

8.2

Let us evaluate the moment of inertia $\mathcal{I}$ and the angular momentum fluctuation $\langle l^2 \rangle$ in Eqs. (8.10) and (8.11).

(1) Moment of inertia of the state $|N\rangle$ may be computed by the Inglis formula

$$\mathcal{I} = 2\sum_{N'} \frac{|\langle N'|\hat{l}_x|N\rangle|^2}{E_{N'} - E_N}, \tag{8.102}$$

where $\hat{l}_x = yp_z - zp_y$ is the angular momentum operator (compare this formula with Eq. (6.64)). For simplicity, first consider the case where there is just one degree of freedom and write the state $|N\rangle$ as $|n_x n_y n_z\rangle$. Using the creation and annihilation operators for the harmonic oscillator,

$$\begin{aligned} \hat{x} &= \sqrt{\frac{\hbar}{2m\omega_x}} \left(a_x + a_x^\dagger\right) , \\ \hat{p}_x &= i\sqrt{\frac{\hbar m \omega_x}{2}} \left(-a_x + a_x^\dagger\right) , \quad \text{etc.}, \end{aligned}$$

show the following expression

$$\begin{aligned} \mathcal{I} &= \frac{\hbar^2}{2\omega_y \omega_z} \left\{ \frac{(\omega_z - \omega_y)^2}{\omega_y + \omega_z}(n_y + n_z + 1) \right. \\ &\quad \left. + \frac{(\omega_z + \omega_y)^2}{\omega_y - \omega_z}(n_y - n_z) \right\} . \end{aligned} \tag{8.103}$$

When there are two degrees of freedom for $\boldsymbol{\rho}$ and $\boldsymbol{\lambda}$, show

$$\begin{aligned} \mathcal{I} &= \frac{\hbar^2}{2\omega_y\omega_z}\left\{\frac{(\omega_z-\omega_y)^2}{\omega_y+\omega_z}(N_y+N_z+2)\right. \\ &\qquad\left. + \frac{(\omega_z+\omega_y)^2}{\omega_y-\omega_z}(N_y-N_z)\right\} \\ &= \hbar^2\left(\frac{N_y+1}{\omega_y}+\frac{N_z+1}{\omega_z}\right) \end{aligned} \tag{8.104}$$

where $N_y = n_y^\rho + n_y^\lambda$ and $N_z = n_z^\rho + n_z^\lambda$.
(2) Following similar calculational procedure, derive the average of the angular momentum square of Eq. (8.11).

8.3
The theoretical predictions in Fig. 8.3 (dashed lines) are obtained under the condition that the deformation of the intrinsic state is not changed when it starts to rotate (rigid rotation). If, however, the system is soft, one should take into account the effect of stretching due to the centrifugal force. This can effectively be incorporated by considering rotational bands of higher N.
(1) Construct rotational bands of various N's, namely, find out the energy spectrum $E(N, l)$.
(2) For a given l, choose the minimum energy $E(N, l)$ as a function of N,

$$E_{min}(\tilde{N}, l) \le E(N, l)\,, \quad \text{for all } N\,. \tag{8.105}$$

This construction provides an improved rotational band due to intruder states.

8.4
Verify the multipole expansions of Eqs. (8.18) and (8.19).

8.5
In the non-relativistic quark model, the $M1$ amplitude is responsible for the transition from the nucleon ($1/2^+$) to the Roper resonance ($1/2^+$). Using the non-relativistic form of the electromagnetic current and the $\boldsymbol{\sigma}\cdot\boldsymbol{q}$ form of the pion-quark coupling, verify that both photon and pion matrix elements are proportional to $\langle\sigma\rangle$ and hence it is not possible to flip the sign of the photoproduction amplitude when the sign of the pion matrix element is included.

Appendix

A Harmonic Oscillator Wave Function

We consider three-particle wave functions for the harmonic oscillator, whose hamiltonian is

$$H = \sum_{i=1}^{3} \left(\frac{p_i^2}{2m} + \frac{m\omega^2}{2}(x_i^2 + y_i^2 + z_i^2) \right) . \tag{A.1}$$

The eigenstates are denoted by ψ_σ^N, where N and σ specify oscillator quantum numbers and symmetry properties, respectively. They are linear combinations of the products of single-particle wave functions $f_{n_x^i n_y^i n_z^i}(\boldsymbol{x}_i)$ $(i = 1, 2, 3)$:

$$f_{n_x^1 n_y^1 n_z^1}(\boldsymbol{x}_1) f_{n_x^2 n_y^2 n_z^2}(\boldsymbol{x}_2) f_{n_x^3 n_y^3 n_z^3}(\boldsymbol{x}_3) . \tag{A.2}$$

The function $f_{n_x n_y n_z}(\boldsymbol{x})$ satisfy the eigenvalue equation

$$\begin{aligned} \left(\frac{1}{2m}(p_x^2 + p_y^2 + p_z^2) + \frac{m\omega^2}{2}(x^2 + y^2 + z^2) \right) f_{n_x n_y n_z}(\boldsymbol{x}) \\ = \epsilon_{n_x n_y n_z} f_{n_x n_y n_z}(\boldsymbol{x}) , \end{aligned} \tag{A.3}$$

where the eigenenergy is given by

$$\epsilon_{n_x n_y n_z} = \left(n_x + \frac{1}{2} + n_y + \frac{1}{2} + n_z + \frac{1}{2} \right) \omega . \tag{A.4}$$

$f_{n_x n_y n_z}(\boldsymbol{x})$ is further expressed as a product of one-dimensional harmonic oscillator wave functions as

$$f_{n_x n_y n_z}(\boldsymbol{x}) = \phi_{n_x}(x)\phi_{n_y}(y)\phi_{n_z}(z)\,, \tag{A.5}$$

where

$$\phi_n(x) = \left(\frac{m\omega}{\pi}\right)^{1/4} \frac{1}{\sqrt{2^n n}} \exp(-\frac{m\omega}{2}x^2) H_n(\sqrt{m\omega}x) \tag{A.6}$$

with the Hermite polynomials of low orders given as

$$\begin{aligned} H_0(z) &= 1\,, \\ H_1(z) &= 2z\,, \\ H_2(z) &= 4z^2 - 2\,, \\ &\cdots \end{aligned} \tag{A.7}$$

Ground state

The ground state wave function which is totally symmetric contains just a single term:

$$\begin{aligned} \psi_S^{N=0}(\boldsymbol{x}_1, \boldsymbol{x}_2, \boldsymbol{x}_3) &= f_{000}(\boldsymbol{x}_1) f_{000}(\boldsymbol{x}_2) f_{000}(\boldsymbol{x}_3) \\ &= \left(\frac{m\omega}{\pi}\right)^{3/4} \exp\left(-\frac{m\omega}{2}(\boldsymbol{x}_1^2 + \boldsymbol{x}_2^2 + \boldsymbol{x}_3^2)\right)\,. \end{aligned} \tag{A.8}$$

Excited state: N = 1

We consider an excitation by one quantum in the z-direction. Single-particle states for three quarks are

$$(f_{000}\,,\ f_{000}\,,\ f_{001})\,.$$

The symmetric combination

$$\begin{aligned} & f_{000}(\boldsymbol{x}_1) f_{000}(\boldsymbol{x}_2) f_{001}(\boldsymbol{x}_3) + (123 \to 231) + (123 \to 312) \\ & \sim\ (z_1 + z_2 + z_3)\psi_S^{N=0} \end{aligned} \tag{A.9}$$

corresponds to the excitation of the center of mass coordinate along the z-direction. Hence we consider a mixed symmetric combination for an excitation of internal motion. It is easy to verify that the λ and ρ combination yield the following functions (see Chapter 2, section 2.5):

$$\begin{aligned}\psi_\lambda^{N=1} &= \frac{1}{\sqrt{6}}(f_{000}(\boldsymbol{x}_1)f_{001}(\boldsymbol{x}_2)f_{000}(\boldsymbol{x}_3)+f_{001}(\boldsymbol{x}_1)f_{000}(\boldsymbol{x}_2)f_{000}(\boldsymbol{x}_3)\\ &\qquad -2f_{000}(\boldsymbol{x}_1)f_{000}(\boldsymbol{x}_2)f_{001}(\boldsymbol{x}_3))\\ &= \lambda_z\,\psi_S^{N=0}\,, \qquad (A.10)\\ \psi_\rho^{N=1} &= \frac{1}{\sqrt{2}}(f_{000}(\boldsymbol{x}_1)f_{001}(\boldsymbol{x}_2)-f_{001}(\boldsymbol{x}_1)f_{000}(\boldsymbol{x}_2))f_{000}(\boldsymbol{x}_3)\\ &= \rho_z\,\psi_S^{N=0}\,. \qquad (A.11)\end{aligned}$$

Excited state: $N = 2$

Let us illustrate again excitations along the z-direction. There are two possibilities to describe the excitations:

$$(f_{000}, f_{000}, f_{002}) \text{ and } (f_{000}, f_{001}, f_{001})\,. \qquad (A.12)$$

Using Eqs. (A.5) and (A.6), we have

$$\begin{aligned} f_{000}f_{000}f_{002} &= \sqrt{2}\left(m\omega z_3^2-\frac{1}{2}\right)\psi_S^{N=0}\,, \qquad (A.13)\\ f_{000}f_{001}f_{001} &= 2m\omega z_2 z_3\,\psi_S^{N=0}\,. \qquad (A.14)\end{aligned}$$

Here and in what follows we omit the arguments $\boldsymbol{x}_i$ for simplicity.

Let us form a linear combination of

$$a f_{000}f_{000}f_{002}+b f_{000}f_{001}f_{001} \qquad (A.15)$$

for the totally symmetric combination. Explicitly, it is

$$\Big(\sqrt{2}am\omega(z_1^2+z_2^2+z_3^2) + 2bm\omega(z_1z_2+z_2z_3+z_3z_1)-\frac{3\sqrt{2}a}{2}\Big)\,\psi_S^{N=0}\,. \qquad (A.16)$$

In terms of the Jacobi coordinates defined by

$$\begin{aligned} R &= \frac{1}{\sqrt{3}}(x_1+x_2+x_3)\\ \rho &= \frac{1}{\sqrt{2}}(x_1-x_2) \qquad (A.17)\end{aligned}$$

$$\lambda = \frac{1}{\sqrt{6}}(x_1 + x_2 - 2x_3)$$

and

$$\begin{aligned} x_1 &= \frac{R}{\sqrt{3}} + \frac{\lambda}{\sqrt{6}} + \frac{\rho}{\sqrt{2}} \\ x_2 &= \frac{R}{\sqrt{3}} + \frac{\lambda}{\sqrt{6}} - \frac{\rho}{\sqrt{2}} \\ x_1 &= \frac{R}{\sqrt{3}} - \frac{2\lambda}{\sqrt{6}} \end{aligned} \tag{A.18}$$

Eq. (A.16) can be written as

$$\sim (\sqrt{2}a - b)(\lambda_z^2 + \rho_z^2) + (\sqrt{2}a + 2b)R_z^2 \,. \tag{A.19}$$

From this, we see that both $f_{000}f_{000}f_{002}$ and $f_{000}f_{001}f_{001}$ contain a spurious excitation of the center of mass coordinate. This can be eliminated by choosing

$$a = \sqrt{\frac{2}{3}}\,, \quad b = -\sqrt{\frac{1}{3}}\,. \tag{A.20}$$

Therefore, the appropriate combination for internal excitation for the symmetric $N = 2$ state is

$$\begin{aligned} &\text{Symm.}\left(\sqrt{\frac{2}{3}}\, f_{000}f_{000}f_{002} - \sqrt{\frac{1}{3}}\, f_{000}f_{001}f_{001}\right) \\ &= \left(m\omega(\lambda_z^2 + \rho_z^2) - 1\right)\, \psi_S^{N=0} \,. \end{aligned} \tag{A.21}$$

Here Symm. denotes the totally symmetrizing operation.

Mixed symmetry states can be constructed in an analogous way. We illustrate it here only for the ρ symmetric case. We find for the ρ symmetric combination of Eq. (A.15)

$$\begin{aligned} &\frac{a}{\sqrt{2}}(f_{000}f_{002} - f_{002}f_{000})f_{000} + \frac{b}{\sqrt{2}}(f_{000}f_{001} - f_{001}f_{000})f_{001} \\ &= m\omega(z_2 - z_1)\left(\frac{1}{3}(a + \sqrt{2}b)(z_1 + z_2 - 2z_3)\right. \\ &\qquad \left. + \frac{1}{3}(2a - \sqrt{2}b)(z_1 + z_2 + z_3)\right)\psi_S^{N=0} \,. \end{aligned} \tag{A.22}$$

Therefore, by choosing

$$a = \sqrt{\frac{1}{3}}\,, \quad b = \sqrt{\frac{2}{3}}\,, \tag{A.23}$$

we can eliminate the dependence on the center of mass coordinate. Hence

$$\psi_{\rho}^{N=2} = 2m\omega\lambda_z\rho_z\,\psi_S^{N=0}\,. \tag{A.24}$$

Similarly, we obtain for the λ symmetric state

$$\psi_{\lambda}^{N=2} = m\omega(\lambda_z^2 - \rho_z^2)\,\psi_S^{N=0}\,. \tag{A.25}$$

In the end, we provide alternative expressions for the wave functions in terms of $\boldsymbol{\lambda}$ and $\boldsymbol{\rho}$ variables. They are

$$\begin{aligned}
\psi_S^{N=0} &= f_{000}(\boldsymbol{\lambda})f_{000}(\boldsymbol{\rho})\,, && \text{(A.26)}\\
\psi_{\lambda}^{N=1} &= f_{001}(\boldsymbol{\lambda})f_{000}(\boldsymbol{\rho})\,, && \text{(A.27)}\\
\psi_{\rho}^{N=1} &= f_{000}(\boldsymbol{\lambda})f_{001}(\boldsymbol{\rho})\,, && \text{(A.28)}\\
\psi_S^{N=2} &= \frac{1}{\sqrt{2}}\left(f_{002}(\boldsymbol{\lambda})f_{000}(\boldsymbol{\rho}) + f_{000}(\boldsymbol{\lambda})f_{002}(\boldsymbol{\rho})\right)\,, && \text{(A.29)}\\
\psi_{\lambda}^{N=2} &= \frac{1}{\sqrt{2}}\left(f_{002}(\boldsymbol{\lambda})f_{000}(\boldsymbol{\rho}) - f_{000}(\boldsymbol{\lambda})f_{002}(\boldsymbol{\rho})\right)\,, && \text{(A.30)}\\
\psi_{\rho}^{N=2} &= f_{001}(\boldsymbol{\lambda})f_{001}(\boldsymbol{\rho})\,. && \text{(A.31)}
\end{aligned}$$

B Quantization of Collective Motion

When there are symmetries in a system, collective zero energy motions associated with the symmetries dominate the dynamics at low energies. They are "force-free" motions in the parameter space of symmetry transformations. In this appendix we first present how to construct wave functions by a group theoretical method rather than by solving Schrödinger equations. A classic reference to such a method is the textbook by Wigner [Wigner 1959].

Free motion along a straight line

This is the simplest example but is useful to illustrate the idea. The position of a particle is labeled by one real parameter q. The hamiltonian for a free particle in one dimensional space is given by

$$H = \frac{\hat{p}^2}{2m} \,. \tag{B.1}$$

We should notice that the momentum $\hat{p}$ is the generator of displacement in q-space. In the q-representation it is $\hat{p} = -i\partial/\partial q$. We should recall that in the p-representation, the position operator q is expressed as $\hat{q} = i\partial/\partial p$. The two variables q and p, one of which is expressed as the derivative with respect to the other, are conjugate variables.

An eigenstate of the momentum is labeled by a parameter p and is defined by

$$\hat{p}|p\rangle = p|p\rangle \,. \tag{B.2}$$

In the q-representation it is expressed as

$$\langle q|p\rangle = \exp(ipq) \,. \tag{B.3}$$

This is also the eigenstate of the hamiltonian of Eq. (B.1),

$$H|p\rangle = E(p)|p\rangle \,, \qquad E(p) = \frac{p^2}{2m} \,. \tag{B.4}$$

We can construct the wave function of Eq. (B.3) in the following way. The state $|q\rangle$ is generated by shifting a reference state $|0\rangle$ by q. This can be achieved by applying the shifting operator as

$$|q\rangle = \exp(-i\hat{p}q)|0\rangle \,. \tag{B.5}$$

Note that we need $\exp(-i\hat{p}q)$ rather than $\exp(+i\hat{p}x)$. In an elementary example, the operation $\exp(-i\hat{p}q) = \exp(-q\partial/\partial x)$ on a function $f(x)$ yields the shifted function $f(x-q)$ (Fig. B.1 (a)). The wave function of Eq. (B.3) is therefore given by

$$\begin{aligned} \langle q|p\rangle &= \langle 0|\exp(i\hat{p}q)|p\rangle \\ &= \exp(ipq)\langle 0|p\rangle \,. \end{aligned} \tag{B.6}$$

If we normalize the reference value $\langle 0|p\rangle = 1$, we obtain the desired result

$$\langle q|p\rangle = \exp(ipq)\,. \tag{B.7}$$

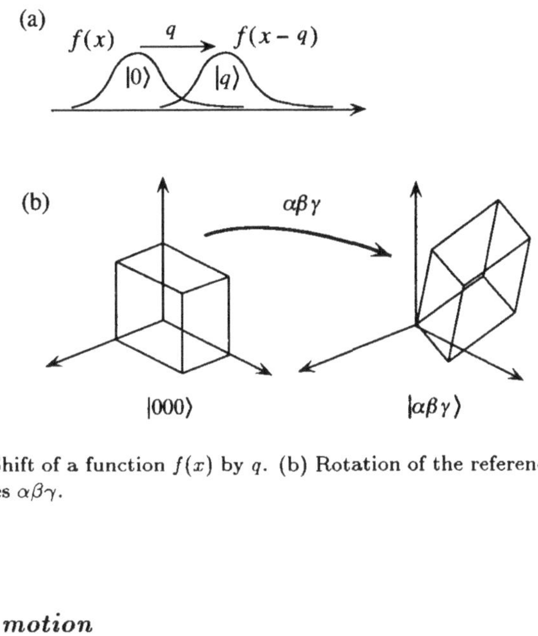

Fig. B.1 (a) Shift of a function $f(x)$ by q. (b) Rotation of the reference state $|000\rangle$ by the Euler angles $\alpha\beta\gamma$.

Rotational motion

Rotational motion occurs in a parameter space of $SU(2)$. The hamiltonian for symmetric free motion is then given by

$$H = \frac{1}{2\Lambda}\left(t_x^2 + t_y^2 + t_z^2\right)\,, \tag{B.8}$$

where t_i are generators of $SU(2)$. Here we have assumed the rotational symmetry and therefore, the hamiltonian is a function of a single Casimir operator of $SU(2)$, $C(2) \equiv t_x^2 + t_y^2 + t_z^2$. As usual, irreducible representations of $SU(2)$ are labeled by J and M, $|JM\rangle$. A particularly useful way to represent a position in the $SU(2)$ parameter space is to use the Euler angles, α, β and γ. They are defined in the region

$$0 \leq \alpha, \gamma \leq 2\pi\,, \quad 0 \leq \beta \leq \pi \tag{B.9}$$

and the integral measure is given by

$$d[\alpha\beta\gamma] = d\alpha d(\cos\beta) d\gamma\,. \tag{B.10}$$

Wave functions of $SU(2)$ are constructed in a way similar to the previous example. First we generate a state in the position space $|\alpha\beta\gamma\rangle$ from a reference state $|000\rangle$. A classical counter example is a rigid body placed at the standard orientation where the Euler angles between the body-fixed and laboratory frames are zero (Fig. B.1 (b) left). If there is an additional axial symmetry in the system, the reference state should be chosen such that the angular momentum projection K along the body fixed z-axis becomes zero. Such an example is a point-like particle on a sphere, whose reference state is the point on the z-axis. If there is no axial symmetry, K can be finite:

$$t_z|000\rangle = K|000\rangle\,. \tag{B.11}$$

As shown in Fig. B.1 (b), the state $|\alpha\beta\gamma\rangle$ can be generated by rotating the reference state $|000\rangle$,

$$|\alpha\beta\gamma\rangle = R(\alpha\beta\gamma)^\dagger|000\rangle\,. \tag{B.12}$$

where the rotational operator is defined by

$$R(\alpha\beta\gamma) = \exp(it_z\alpha)\exp(it_y\beta)\exp(it_z\gamma)|000\rangle\,. \tag{B.13}$$

Wave functions are then expressed as

$$\begin{aligned} \langle\alpha\beta\gamma|JM\rangle &= \langle 000|R(\alpha\beta\gamma)|JM\rangle \\ &= \sum_m \langle 000|Jm\rangle\langle Jm|R(\alpha\beta\gamma)|JM\rangle\,. \end{aligned} \tag{B.14}$$

Using Eq. (B.11) we find ¶

$$\langle JK|R(\alpha\beta\gamma)|JM\rangle = D^J_{KM}(\alpha\beta\gamma) \sim \Psi^{J\,*}_{K\,-M}(\alpha\beta\gamma)\,. \tag{B.15}$$

¶Complex conjugation is taken in order to be consistent with the commutation relation for collective coordinates of Eq. (B.25)

Here we have introduced the rotational D-functions. Using the integral formula

$$\int d[\alpha\beta\gamma] D^{J*}_{KM}(\alpha\beta\gamma) D^{J'}_{K'M'}(\alpha\beta\gamma) = \frac{8\pi^2}{2J+1}\delta_{JJ'}\delta_{KK'}\delta_{MM'}\,, \qquad \text{(B.16)}$$

the properly normalized wave functions are

$$\Psi^J_{KM}(\alpha\beta\gamma) = \sqrt{\frac{2J+1}{8\pi^2}} D^{J\,*}_{K\,-M}(\alpha\beta\gamma)\,. \qquad \text{(B.17)}$$

The reason that we have taken a complex conjugate in Eq. (B.15) and the physical meaning of the subscripts M and K in the context of the hedgehog quantization will be discussed in the next section (Eqs. (C.40)-(C.43), see also Eq. (7.59)).

Finally we prove that the wave function $\Psi^J_{MK}(\alpha\beta\gamma)$ satisfies the desired eigenvalue equation. Consider a matrix element $\langle\alpha\beta\gamma|H|JM\rangle$. By operating the hamiltonian H to the right state $|JM\rangle$, it brings the eigenvalue $J(J+1)/2\Lambda \equiv E_J$. By operating the hamiltonian H to the left state $|\alpha,\beta,\gamma\rangle$, we find a differential operator for the hamiltonian. Therefore, we find

$$H(\alpha\beta\gamma)\Psi^J_{KM} = E_J \Psi^J_{KM}\,, \qquad \text{(B.18)}$$

where the differential form of the hamiltonian is given by ([Wigner 1959], see also Appendix C)

$$\begin{aligned} H(\alpha\beta\gamma) &= \frac{1}{2\Lambda}\left[\frac{1}{\sin\beta}\frac{\partial}{\partial\beta}\left(\sin\beta\frac{\partial}{\partial\beta}\right)\right. \\ &+ \left.\frac{1}{\sin^2\beta}\left(\frac{\partial}{\partial\alpha} - \cos\beta\frac{\partial}{\partial\gamma}\right)^2 + \frac{\partial^2}{\partial\gamma^2}\right]. \end{aligned} \qquad \text{(B.19)}$$

Commutation relations for collective variables

Finally, we establish commutation relations for collective variables which differ in sign from ordinary ones. We demonstrate this for the simple translational motion in one dimension.

Consider an extended object centered at a point q. If there is a translational invariance, it is expressed by a single-variable function

$$f(x-q)\,. \qquad \text{(B.20)}$$

For collective motion, the position parameter q is regarded as collective dynamical variable. The increase in q by a is generated by

$$e^{-iap} f(x-q) e^{iap} = f(x-(q+a)), \tag{B.21}$$

where p is the momentum conjugate to x,

$$p = -i\frac{\partial}{\partial x}. \tag{B.22}$$

We would like to do this by using the momentum π which is conjugate to q,

$$e^{-ia\pi} f(x-q) e^{ia\pi} = f(x-(q+a)). \tag{B.23}$$

This is accomplished by defining

$$\pi = +i\frac{\partial}{\partial q} \tag{B.24}$$

that differs from the ordinary definition in sign. Thus the commutation relation for the collective variables should be

$$[q, \pi] = -i. \tag{B.25}$$

C Quantization of Hedgehog Rotation

In this appendix we apply the method developed in the previous section to the quantization of the rotating hedgehog. The ansatz we consider is

$$U(t,\boldsymbol{x}) = A(t)U_h(\boldsymbol{x})A(t)^\dagger\,, \tag{C.1}$$
$$\psi(t,\boldsymbol{x}) = A(t)\psi_h(\boldsymbol{x})\,. \tag{C.2}$$

The 2×2 unitary matrix $A(t)$ is the relevant dynamical variable. From Eq. (6.65), the lagrangian can be written as

$$L = \frac{1}{2}\Lambda\,\boldsymbol{\Omega}^\sigma\cdot\boldsymbol{\Omega}^\sigma = \frac{1}{2}\Lambda\,\boldsymbol{\Omega}^\tau\cdot\boldsymbol{\Omega}^\tau\,, \tag{C.3}$$

where we have omitted the subscript $\Lambda_{tot}\to\Lambda$. Here we have ignored the constant term E_h, and have introduced the spin and isospin angular velocities which are defined by

$$\boldsymbol{\Omega}^\sigma = i\,\mathrm{tr}\,(\boldsymbol{\tau}A^\dagger\dot{A}) = -i\,\mathrm{tr}\,(\boldsymbol{\tau}\dot{A}^\dagger A)\,, \tag{C.4}$$
$$\boldsymbol{\Omega}^\tau = -i\,\mathrm{tr}\,(\boldsymbol{\tau}\dot{A}A^\dagger) = i\,\mathrm{tr}\,(\boldsymbol{\tau}A\dot{A}^\dagger)\,. \tag{C.5}$$

In deriving the second equation we have used the identity

$$AA^\dagger = 1 \;\to\; \dot{A}A^\dagger + A\dot{A}^\dagger = 0\,. \tag{C.6}$$

The two definitions of Eqs. (C.4) and (C.5) are associated with right and left multiplications of matrices. They are related to spin and isospin transformations as discussed below. We should note that the dynamics of the rotating hedgehog is precisely the same as that of the symmetric top [Wigner 1959]. Interpretation of quantum numbers in these two different dynamical systems is our concern.

In Eq. (C.3), the square of the angular velocity is equivalent to

$$\boldsymbol{\Omega}^2 = -2\,\mathrm{tr}\,(\dot{A}^\dagger A\,\dot{A}^\dagger A)\,. \tag{C.7}$$

This follows from the fact that $\boldsymbol{\Omega}$ is nothing but the Maurer-Cartan one-form of the A-matrix:

$$A^\dagger\dot{A} = -i\frac{\boldsymbol{\tau}}{2}\cdot\boldsymbol{\Omega}^\sigma\,,\quad \dot{A}A^\dagger = i\frac{\boldsymbol{\tau}}{2}\cdot\boldsymbol{\Omega}^\tau\,. \tag{C.8}$$

Therefore, another equivalent form of the lagrangian is

$$L = -\Lambda\,\mathrm{tr}\,(A^\dagger\dot{A}A^\dagger\dot{A})$$

$$= \quad +\Lambda \operatorname{tr}(\dot{A}^\dagger \dot{A}) . \tag{C.9}$$

As demonstrated in section 6.2, the pion field is subject to the following configuration space (spin S) and isospace (isospin T) rotations,

$$U(t, \boldsymbol{x}) \quad \rightarrow \quad TU(t, D(S^\dagger)\boldsymbol{x})T^\dagger , \tag{C.10}$$

where $D(B)$ is an adjoint 3×3 matrix for a rotation and is related to the fundamental 2×2 matrix B by

$$D_{ij}(B) = \frac{1}{2} \operatorname{tr}(\tau_i B \tau_j B^\dagger) . \tag{C.11}$$

Note also that D_{ij} is the D-function of rank one (see Eq. (C.46).

Now for the rotating hedgehog ansatz, Eq. (C.10) reads

$$\begin{aligned} & TA \exp(i\tau_a D_{ai}(S^\dagger)\hat{x}_i F(r))A^\dagger T^\dagger \\ = \quad & TA \exp(iS^\dagger \tau_i S \hat{x}_i F(r))A^\dagger T^\dagger \\ = \quad & TAS^\dagger \exp(i\boldsymbol{\tau} \cdot \hat{\boldsymbol{x}} F(r)) S A^\dagger T^\dagger . \end{aligned} \tag{C.12}$$

This implies that the dynamical variable $A(t)$ is subject to the transformation

$$A(t) \rightarrow TA(t)S^\dagger . \tag{C.13}$$

It is apparent that the lagrangian of Eq. (C.9) is invariant under this transformation.

Upon canonical quantization, we need to define momentum variables. For this, we can take a derivative of the lagrangian with respect to either the spin or the isospin angular velocity. If we do so for the spin angular velocity, one obtains the spin momentum (charge) $\boldsymbol{J}$ as

$$\boldsymbol{J} = \frac{\partial L}{\partial \Omega^\sigma} = \Lambda \Omega^\sigma . \tag{C.14}$$

Inferred by Eq. (B.23) and being careful with the sign, the following commutation relations generate the spin transformation $A \rightarrow AS^\dagger$ as

$$-i[J_i, A] = A(-it_i) . \tag{C.15}$$

Similarly, the isospin transformation is generated by the isospin momentum $\boldsymbol{I}$,

$$\boldsymbol{I} \quad = \quad \frac{\partial L}{\partial \Omega^\tau} ,$$

$$-i[I_i, A] = it_i A\,. \tag{C.16}$$

For practical purposes, we need to express matrix equations by some parameters. For fundamental representations ($J = I = 1/2$), four-vector parameterization is useful for A:

$$A = a_0 + i\boldsymbol{\tau}\cdot\boldsymbol{a}\,, \quad \dot{A} = \dot{a}_0 + i\boldsymbol{\tau}\cdot\dot{\boldsymbol{a}}\,. \tag{C.17}$$

The lagrangian is then written as

$$L = 2\Lambda \sum_{\mu=0,\ldots 3} \dot{a}_\mu^2\,. \tag{C.18}$$

Hence the canonical momentum and the hamiltonian are

$$\begin{aligned} \pi_\mu &= 4\Lambda\dot{a}_\mu \to i\frac{\partial}{\partial a_\mu}\,, \\ H &= \frac{1}{8\Lambda}\pi_\mu^2\,. \end{aligned} \tag{C.19}$$

Note the sign in the first equation for the momentum, for a and π are regarded as collective variables, leading to the commutation relations

$$[a_\mu\, \pi_\nu] = -i\delta_{\mu\nu}\,. \tag{C.20}$$

In this parameterization, the spin and isospin charges are expressed as

$$J_i = \frac{i}{2}\left(-a_0\partial_i + a_i\partial_0 - \epsilon_{ijk}a_j\partial_k\right)\,, \tag{C.21}$$

$$I_i = \frac{i}{2}\left(a_0\partial_i - a_i\partial_0 - \epsilon_{ijk}a_j\partial_k\right)\,. \tag{C.22}$$

We can verify explicitly the following commutation relations from Eq. (C.20),

$$[J_i,\, J_j] = i\epsilon_{ijk}J_k\,, \tag{C.23}$$

$$[I_i,\, I_j] = i\epsilon_{ijk}I_k\,. \tag{C.24}$$

In some circumstances, it is convenient to adopt the Euler angle representation. The matrix A is regarded as the fundamental representation of the rotational operator $R(\alpha\beta\gamma)$:

$$\begin{aligned} A &= \exp(\frac{i}{2}\tau_z\alpha)\exp(\frac{i}{2}\tau_y\beta)\exp(\frac{i}{2}\tau_z\gamma) \\ &= \begin{pmatrix} \exp\left(\frac{i}{2}(\alpha+\gamma)\right)\cos\frac{\beta}{2} & \exp\left(\frac{i}{2}(\alpha-\gamma)\right)\sin\frac{\beta}{2} \\ -\exp\left(-\frac{i}{2}(\alpha-\gamma)\right)\sin\frac{\beta}{2} & \exp\left(-\frac{i}{2}(\alpha+\gamma)\right)\cos\frac{\beta}{2} \end{pmatrix}. \end{aligned} \tag{C.25}$$

Each component of this matrix is then related to Wigner's D-function by

$$\begin{aligned} A_{mn}(\alpha\beta\gamma) &= D^{1/2}_{mn}(\alpha\beta\gamma) \\ &= \langle m|R(\alpha\beta\gamma)|n\rangle . \end{aligned} \tag{C.26}$$

From these, one can derive

$$\begin{aligned} J_i &= i\Lambda \,\mathrm{tr}\, \tau_i A^\dagger \dot{A} \;=\; \Lambda \Omega_i^\sigma \\ &= \Lambda \begin{cases} x: & -\dot\alpha \cos\gamma \sin\beta + \dot\beta \sin\gamma \\ y: & -\dot\alpha \sin\gamma \sin\beta - \dot\beta \cos\gamma \\ z: & -\dot\alpha \cos\beta - \dot\gamma \end{cases} \end{aligned} \tag{C.27}$$

$$\begin{aligned} I_a &= -i\Lambda \,\mathrm{tr}\, \tau_a \dot{A} A^\dagger \;=\; \Lambda \Omega_i^\tau \\ &= \Lambda \begin{cases} x: & -\dot\gamma \cos\alpha \sin\beta + \dot\beta \sin\alpha \\ y: & \dot\gamma \sin\alpha \sin\beta + \dot\beta \cos\alpha \\ z: & \dot\gamma \cos\beta + \dot\alpha \end{cases} \end{aligned} \tag{C.28}$$

which are familiar in classical mechanics. Note that they are symmetric (modulo the overall sign) under

$$\begin{aligned} (\alpha,\ \beta,\ \gamma) &\leftrightarrow (-\gamma,\ -\beta,\ -\alpha) \\ A &\leftrightarrow A^\dagger \end{aligned}$$

which is consistent with the fact that the left (isospin) and right (spin) are exchanged under the parity transformation.

Using these expressions for $\boldsymbol{J} = \Lambda\boldsymbol{\Omega}^\sigma$ or $\boldsymbol{I} = \Lambda\boldsymbol{\Omega}^\tau$, we see that the lagrangian can be written as

$$L = \frac{\Lambda}{2}(\dot\alpha^2 + 2\cos\beta\,\dot\alpha\,\dot\gamma + \dot\gamma^2 + \dot\beta^2) . \tag{C.29}$$

Hence the canonical momenta conjugate to α, β and γ are

$$\pi_\alpha = \frac{\partial L}{\partial \dot\alpha} = \Lambda(\dot\alpha + \cos\beta\,\dot\gamma) \to i\frac{\partial}{\partial\alpha} , \tag{C.30}$$

$$\pi_\beta = \frac{\partial L}{\partial \dot\beta} = \Lambda\dot\beta \to i\frac{\partial}{\partial\beta} , \tag{C.31}$$

$$\pi_\gamma = \frac{\partial L}{\partial \dot\gamma} = \Lambda(\dot\gamma + \cos\beta\,\dot\alpha) \to i\frac{\partial}{\partial\gamma} . \tag{C.32}$$

Therefore, Euler angle representations for angular momenta are,

$$J_x = i\left(-\frac{\cos\gamma}{\sin\beta}\partial_\alpha + \sin\gamma\,\partial_\beta + \cot\beta\,\cos\gamma\,\partial_\gamma\right) , \tag{C.33}$$

$$J_y = i\left(-\frac{\sin\gamma}{\sin\beta}\partial_\alpha - \cos\gamma\,\partial_\beta + \cot\beta\,\sin\gamma\,\partial_\gamma\right), \tag{C.34}$$

$$J_z = -i\,\partial_\gamma\,. \tag{C.35}$$

Similarly we obtain

$$I_x = i\left(-\cos\alpha\,\cot\beta\,\partial_\alpha + \sin\alpha\,\partial_\beta - \frac{\sin\alpha}{\sin\beta}\partial_\gamma\right), \tag{C.36}$$

$$I_y = i\left(-\sin\alpha\,\cot\beta\,\partial_\alpha + \cos\alpha\,\partial_\beta + \frac{\sin\alpha}{\sin\beta}\partial_\gamma\right), \tag{C.37}$$

$$I_z = i\,\partial_\alpha\,. \tag{C.38}$$

The wave function for the nucleon state with spin (J, m) and isospin (I, t) are given by the D-functions (*cf.* Eq. (B.17))

$$\psi^N_{tm}(\alpha\beta\gamma) = \sqrt{\frac{2I+1}{8\pi^2}}\,D^{I=J\,*}_{t\,-m}(\alpha\beta\gamma)\,. \tag{C.39}$$

Explicitly, for proton and neutron states, we have

$$|p\uparrow\rangle \sim \begin{cases} A^*_{12} = -ia_1 + a_2 \\ D^*_{1/2,-1/2} = \exp\left(-\frac{i}{2}(\alpha-\gamma)\right)\sin\frac{\beta}{2} \end{cases} \tag{C.40}$$

$$|p\downarrow\rangle \sim \begin{cases} A^*_{11} = a_0 - ia_3 \\ D^*_{1/2,1/2} = \exp\left(-\frac{i}{2}(\alpha+\gamma)\right)\cos\frac{\beta}{2} \end{cases} \tag{C.41}$$

$$|n\uparrow\rangle \sim \begin{cases} A^*_{22} = a_0 + ia_3 \\ D^*_{-1/2,-1/2} = \exp\left(\frac{i}{2}(\alpha+\gamma)\right)\cos\frac{\beta}{2} \end{cases} \tag{C.42}$$

$$|n\downarrow\rangle \sim \begin{cases} A^*_{21} = -ia_1 - a_2 \\ D^*_{-1/2,1/2} = -\exp\left(\frac{i}{2}(\alpha-\gamma)\right)\sin\frac{\beta}{2} \end{cases} \tag{C.43}$$

Quantum numbers of these states are verified explicitly by a simple computation, either by the A or Euler angle representations:

$$I_3|p\uparrow\rangle = \frac{1}{2}|p\uparrow\rangle\,,$$

$$J_3|p\uparrow\rangle = \frac{1}{2}|p\uparrow\rangle\,,$$

and etc.

It often happens that we need to compute matrix elements of collective operators. The one which frequently appears is

$$\frac{1}{2}\,\mathrm{tr}\,(\tau_i A \tau_j A^\dagger) = 2\,\mathrm{tr}\,(t_i R t_j R^\dagger)\,. \tag{C.44}$$

The expression on the right hand side is more general than the left hand side in 2×2 matrix form. Since the generators t_i are rank one operators, we see that

$$R t_j R^\dagger = \sum_k t_k\, D^1_{kj}\,, \tag{C.45}$$

where D^1_{kj} is a D-function of rank one. Therefore,

$$\frac{1}{2}\,\mathrm{tr}\,(\tau_i A \tau_j A^\dagger) = 2 D^1_{kj}\,\mathrm{tr}\,(t_i t_k) = D^1_{ij}\,, \tag{C.46}$$

where we have used the normalization of the generators

$$\mathrm{tr}\,(t_i t_j) = \frac{1}{2}\delta_{ij}\,. \tag{C.47}$$

Now the nucleon matrix elements are easily computed as, for instance,

$$\begin{aligned} &\langle I\,tm | D^1_{00} | I\,tm\rangle \\ &= \frac{2I+1}{8\pi^2}\int d[\alpha\beta\gamma] D^I_{t\,-m}(\alpha\beta\gamma) D^1_{00}(\alpha\beta\gamma) D^{I*}_{t\,-m}(\alpha\beta\gamma) \\ &= (It\,10|It)(I\,-m\,10|I\,-m) = -\frac{tm}{I(I+1)}\,. \end{aligned} \tag{C.48}$$

This justifies the replacement which has been made in the text many times, e.g., in Eq. (6.104),

$$D_{ij} \to -\frac{1}{3}\sigma_i\,\tau^j\,. \tag{C.49}$$

D Moment of Inertia

Consider the rotation energy of the quarks:

$$E = \frac{N_c}{4} \sum_{mn} \frac{\langle m|\boldsymbol{\tau} \cdot \boldsymbol{\Omega}|n\rangle\langle n|\boldsymbol{\tau} \cdot \boldsymbol{\Omega}|m\rangle}{E_n - E_m}, \tag{D.1}$$

where we have factored out the trivial sum over N_c quarks. Choosing the axis of the rotation along the z–axis, $\boldsymbol{\tau} \cdot \boldsymbol{\Omega} = \tau_3 \Omega \equiv \tau_0 \Omega$ and so

$$E = \frac{N_c}{4} \sum_{mn} \frac{\langle m|\tau_0|n\rangle\langle n|\tau_0|m\rangle}{E_n - E_m} \Omega^2, \tag{D.2}$$

We apply the Wigner-Eckart theorem to $\langle n|\tau_0|m\rangle$ and factor out the Clebsch-Gordan coefficient:

$$\begin{aligned} \langle n|\tau_0|m\rangle &= \langle K'M'|\tau_0|KM\rangle \\ &= \frac{\delta_{MM'}}{\sqrt{2K+1}}(K\,M\,1\,0|K'\,M')\langle K'||\tau||K\rangle\,. \end{aligned} \tag{D.3}$$

The sum over M in Eq. (D.2) is performed and the square of the CG-coefficients becomes

$$\sum_M |(K\,M\,1\,0|K'\,M)|^2 = \frac{2K+1}{3}. \tag{D.4}$$

It follows then that

$$E = \frac{N_C}{12} \sum_{K,K'} \frac{|\langle K||\tau||K'\rangle|^2}{E_n - E_m} \Omega^2 \tag{D.5}$$

and

$$\Lambda_q = \frac{N_C}{6} \sum_{K,K'} \frac{|\langle K||\tau||K'\rangle|^2}{E_n - E_m}. \tag{D.6}$$

Since τ is a rank one operator, the allowed transitions are for $|K' - K| \leq 1$. Therefore,

$$\begin{aligned} \Lambda_q &= \frac{N_C}{6} \sum_{K\geq 0} \left\{ \frac{|\langle K||\tau||\widetilde{K+1}\rangle|^2}{\tilde{E}_{K+1} - E_K} \right. \\ &+ \left. \frac{|\langle K+1||\tau||\tilde{K}\rangle|^2}{\tilde{E}_K - E_{K+1}} + \frac{|\langle K+1||\tau||\widetilde{K+1}\rangle|^2}{\tilde{E}_{K+1} - E_{K+1}} \right\}, \end{aligned} \tag{D.7}$$

where tilded (untilded) quantities are for unoccupied (occupied) states.

Computation of the reduced matrix elements is straightforward but lengthy. We present here the final results,

$$\begin{aligned}\langle K\|\tau\|\widetilde{K+1}\rangle &= -\langle\widetilde{K+1}\|\tau\|K\rangle \\ &= -ic_1\tilde{c}_2\sqrt{\frac{(2K+1)(2K+3)}{K+1}} \\ &\times \int_0^R r^2dr\,\left(j_K\tilde{j}_K + j_{K+1}\tilde{j}_{K+1}\right) \qquad \text{(D.8)} \\ \langle K\|\tau\|\tilde{K}\rangle &= -\sqrt{2K+1} \\ &\times \int_0^R r^2dr\,\left[X\,c_1\tilde{c}_1\left\{j_K\tilde{j}_K + j_{K+1}\tilde{j}_{K+1}\right\}\right. \\ &\quad \left. -\; X^{-1}\,c_2\tilde{c}_2\left\{j_K\tilde{j}_K + j_{K-1}\tilde{j}_{K-1}\right\}\right], \qquad \text{(D.9)}\end{aligned}$$

where we have introduced the notation $X = \sqrt{K/(K+1)}$. Furthermore, the tilded (untilded) spherical bessel functions have the argument Er where E is the eigenenergy of the unoccupied (occupied) states of angular momentum K or $K+1$ as indicated in the bra or ket vectors on left hand side. The coefficients c_1 and c_2 are for the quark wave function as given in (5.110). The radial integrals of two spherical Bessel functions in the same order can be carried out analytically using the formulae:

$$\begin{aligned}&\int r^2dr\,j_L(\alpha r)j_L(\beta r) \\ &= \frac{r^2}{\alpha^2-\beta^2}\left[\beta j_L(\alpha r)j_{L-1}(\beta r) - \alpha j_{L-1}(\alpha r)j_L(\beta r)\right] \\ &= \frac{r^2}{\alpha^2-\beta^2}\left[\alpha j_{L+1}(\alpha r)j_L(\beta r) - \beta j_L(\alpha r)j_{L+1}(\beta r)\right]. \qquad \text{(D.10)}\end{aligned}$$

E Quark Contributions to g_A and $g_A^{(0)}$ of Order Ω^1

We compute quark contribution to g_A of order Ω^1 by evaluating the matrix element of $\sigma_3\tau_3$ (see Eq. (5-32)),

$$
\begin{aligned}
&\langle \int d^3x\, \psi^\dagger \sigma_3\tau_3\psi\rangle_\Omega \\
&= -\frac{1}{2}\sum_{mn} \frac{\langle m|\sigma_3\tau_a|n\rangle\langle n|\tau_b|m\rangle}{E_n - E_m}\left(D_{a3}\Omega_b - \Omega_b D_{a3}\right) \\
&= \frac{1}{6\Lambda_{\rm tot}}\sum_{mn} \frac{\langle m|\sigma_3\tau_a|n\rangle\langle n|\tau_b|m\rangle}{E_n - E_m} i\,\epsilon_{abc}\sigma_c\tau_3 \,.
\end{aligned}
\tag{E.1}
$$

Here we have quantized the rotational matrix $D_{a3} \to -1/3\,\sigma_a\tau_3$, and the angular velocity $\Omega_b \to \sigma_b/2\Lambda_{\rm tot}$. From a simple consideration, one can conclude that the index c in this equation must be 3. For example, suppose that $(a,b,c) = (3,1,2)$. Then the operator $\sigma_3\tau_a \to \sigma_3\tau_3$ has the spherical component $\mu = 0$, while $\tau_b \to \tau_1$ and τ_2 have the components 1 and -1. This means that the product $\langle m|\sigma_3\tau_3|n\rangle\langle n|\tau_{1\text{ or }2}|m\rangle$ vanishes since one of the two matrix elements vanishes. Thus, in the spherical representation

$$
\begin{aligned}
&\langle \int d^3x\, \psi^\dagger \sigma_3\tau_3\psi\rangle_\Omega \\
&= \frac{1}{6\Lambda_{\rm tot}}\sum_{\substack{mn\\ \mu=\pm1}} \left\{\frac{\langle m|\sigma_0\tau_{-\mu}|n\rangle\langle n|\tau_\mu|m\rangle}{E_n - E_m}\right\}\cdot\sigma_3\tau_3 \\
&= \frac{1}{6\Lambda_{\rm tot}}\sum_{\substack{KMM'\\ \mu=\pm1}} \left\{\frac{\langle K\,M|\sigma_0\tau_{-\mu}|\widetilde{K+1}\,M'\rangle\langle \widetilde{K+1}\,M'|\tau_\mu|K\,M\rangle}{\tilde{E}_{K+1} - E_K}\right. \\
&+ \frac{\langle K+1\,M|\sigma_0\tau_{-\mu}|\widetilde{K}\,M'\rangle\langle \widetilde{K}\,M'|\tau_\mu|K+1\,M\rangle}{\tilde{E}_K - E_{K+1}} \\
&+ \left.\frac{\langle K+1\,M|\sigma_0\tau_{-\mu}|\widetilde{K+1}\,M'\rangle\langle \widetilde{K+1}\,M'|\tau_\mu|K+1\,M\rangle}{\tilde{E}_{K+1} - E_{K+1}}\right\}\cdot\sigma_3\tau_3\,.
\end{aligned}
\tag{E.2}
$$

By virtue of the identity

$$
\sigma_3\tau_\mu = \frac{1}{\sqrt{2}}\left([\sigma\,\tau]^2_\mu + (-)^{\frac{\mu+1}{2}}[\sigma\,\tau]^1_\mu\right)\,, \tag{E.3}
$$

only the rank one part, the second term of this equation, contributes to the matrix elements. Then it follows that

$$
\begin{aligned}
&\langle \int d^3x\, \psi^\dagger \sigma_3 \tau_3 \psi \rangle_\Omega \\
= \; & \frac{\sigma_3 \tau_3}{9\sqrt{2}\Lambda_{\rm tot}} \sum_{K\geq 1} \left\{ \frac{\langle K||[\sigma\,\tau]^1||\widetilde{K+1}\rangle\langle \widetilde{K+1}||\tau_0||K\rangle}{\tilde{E}_{K+1} - E_K} \right. \\
+ \; & \frac{\langle K+1||[\sigma\,\tau]^1||\tilde{K}\rangle\langle \tilde{K}||\tau_0||K+1\rangle}{\tilde{E}_K - E_{K+1}} \\
- \; & \left. \frac{\langle K+1||[\sigma\,\tau]^1||\widetilde{K+1}\rangle\langle \widetilde{K+1}||\tau_0||K+1\rangle}{\tilde{E}_{K+1} - E_{K+1}} \right\} . \qquad \text{(E.4)}
\end{aligned}
$$

The reduced matrix elements are given by

$$
\begin{aligned}
& \langle K||[\sigma\,\tau]^1||K+1\rangle = \langle K+1||[\sigma\,\tau]^1||K\rangle \\
= \; & -i\sqrt{2}\left\{ \alpha\tilde{c}_2\sqrt{2K+3}j_K\,\tilde{j}_K - c_1\tilde{\alpha}\sqrt{2K+1}j_{K+1}\,\tilde{j}_{K+1} \right\} , \qquad \text{(E.5)} \\
& \langle K||[\sigma\,\tau]||K\rangle = \sqrt{2(2K+1)}\,(c_1\tilde{c}_2 - c_2\tilde{c}_1)\, j_K\,\tilde{j}_K \, , \qquad \text{(E.6)}
\end{aligned}
$$

where

$$
\alpha = c_1 P + c_2 Q\,, \quad \beta = c_1 Q - c_2 P\,, \qquad \text{(E.7)}
$$

with P and Q being given in Eq. (5.98). In Eqs. (E.5) and (E.6), the radial integral $\int_0^R r^2 dr$ is implicit.

The isoscalar matrix element $g_A^{(0)} \sim \langle \sigma \rangle$ is computed similarly, except for differences in some kinematical factors. We present only the final results here:

$$
\begin{aligned}
&\langle \int d^3x\, \psi^\dagger \sigma_3 \psi \rangle_\Omega \\
= \; & \frac{-1}{6\Lambda_{\rm tot}} \sum_{K\geq 0} \left\{ \frac{\langle K||\sigma||\widetilde{K+1}\rangle\langle \widetilde{K+1}||\tau_0||K\rangle}{\tilde{E}_{K+1} - E_K} \right. \\
+ \; & \frac{\langle K+1||\sigma||\tilde{K}\rangle\langle \tilde{K}||\tau_0||K+1\rangle}{\tilde{E}_K - E_{K+1}} \\
- \; & \left. \frac{\langle K+1||\sigma||\widetilde{K+1}\rangle\langle \widetilde{K+1}||\tau_0||K+1\rangle}{\tilde{E}_{K+1} - E_{K+1}} \right\} \cdot \sigma_3 \, , \qquad \text{(E.8)}
\end{aligned}
$$

where the reduced matrix elements are

$$
\begin{aligned}
&\langle K||\sigma||K+1\rangle = -\langle K+1||\sigma||K\rangle \\
=& -i\tilde{c}_2\, j_K\tilde{j}_K\left(-\sqrt{2K+3}\,\alpha+\sqrt{\frac{K(2K+3)}{K+1}}\,\beta\right) \\
+& \ ic_1\, j_{K+1}\tilde{j}_{K+1}\left(\sqrt{2K+1}\,\tilde{\alpha}+\sqrt{\frac{(K+2)(2K+1)}{K+1}}\,\tilde{\beta}\right), \qquad \text{(E.9)}\\
&\langle K||\sigma||K\rangle \\
=& \sqrt{2K+1}\left[\left(\alpha\tilde{\beta}-\tilde{\alpha}\beta+\frac{\beta\tilde{\beta}}{\sqrt{K(K+1)}}\right)j_K\tilde{j}_K\right. \qquad \text{(E.10)}\\
+& \left.\left(-\sqrt{\frac{K}{K+1}}c_1\tilde{c}_1 j_{K+1}\tilde{j}_{K+1}+\sqrt{\frac{K+1}{K}}c_2\tilde{c}_2 j_{K-1}\tilde{j}_{K-1}\right)\right].
\end{aligned}
$$

Here again the radial integral $\int_0^R r^2 dr$ is implicit.

F Magnetic Moment

Isoscalar magnetic moment

We compute the quark contribution to the nucleon matrix element of

$$\bar{\psi}(\boldsymbol{x} \times \boldsymbol{\gamma})_3 \psi \,. \tag{F.1}$$

A non-vanishing term appears at order Ω^1:

$$\begin{aligned}
m &= \frac{1}{2}\sum_{mn} \frac{\langle m|(\boldsymbol{x}\times\boldsymbol{\gamma})_3|n\rangle\langle n|\tau_a|m\rangle}{E_n - E_m}\Omega_a \\
&+ \frac{1}{2}\sum_{mn} \frac{\langle m|\tau_a|n\rangle\langle n|(\boldsymbol{x}\times\boldsymbol{\gamma})_3|m\rangle}{E_n - E_m}\Omega_a \\
&= \sum_{mn} \frac{\langle m|(\boldsymbol{x}\times\boldsymbol{\gamma})_3|n\rangle\langle n|\tau_a|m\rangle}{E_n - E_m}\Omega_a \\
&= -i\sqrt{2}\sum_{K\geq 1,M}\left\{\frac{\langle K\,M|\hat{m}_0|\widetilde{K+1}\,M\rangle\langle \widetilde{K+1}\,M|\tau_0|K\,M\rangle}{\tilde{E}_{K+1} - E_K}\right. \\
&+ \frac{\langle K+1\,M|\hat{m}_0|\widetilde{K\,M}\rangle\langle \widetilde{K\,M}|\tau_0|K+1\,M\rangle}{\tilde{E}_K - E_{K+1}} \\
&+ \left.\frac{\langle K+1\,M|\hat{m}_0|\widetilde{K+1}\,M\rangle\langle \widetilde{K+1}\,M|\tau_0|K+1\,M\rangle}{\tilde{E}_{K+1} - E_{K+1}}\right\}\Omega_3\,,
\end{aligned} \tag{F.2}$$

where the rank one tensor operator for the isoscalar magnetic moment is defined by

$$\hat{m}_\mu = [\boldsymbol{x}\,\boldsymbol{\alpha}]^1_\mu \,. \tag{F.3}$$

In Eq. (F.2) we factor out Clebsch-Gordan coefficients by the Wigner-Eckart theorem and perform the sum over M. Using the identity

$$\begin{aligned}
&\frac{1}{\sqrt{(2K+1)(2K'+1)}}\sum_M (K'\,M\,1\,0|K\,M)(K\,M\,1\,0|K'\,M) \\
&= \frac{(-)^{K-K'+1}}{3}
\end{aligned} \tag{F.4}$$

we get

$$
\begin{aligned}
m &= -i\frac{\sqrt{2}}{3}\sum_{K\geq 1}\left\{\frac{\langle K||\hat{m}||\widetilde{K+1}\rangle\langle\widetilde{K+1}||\tau||K\rangle}{\tilde{E}_{K+1}-E_K}\right. \\
&\quad + \frac{\langle K+1||\hat{m}||\tilde{K}\rangle\langle\tilde{K}||\tau||K+1\rangle}{\tilde{E}_K - E_{K+1}} \\
&\quad - \left.\frac{\langle K+1||\hat{m}||\widetilde{K+1}\rangle\langle\widetilde{K+1}||\tau||K+1\rangle}{\tilde{E}_{K+1}-E_{K+1}}\right\}\Omega_3 \, .
\end{aligned}
\tag{F.5}
$$

The computation of the reduced matrix elements is lengthy. Again we present the final result for $\langle\hat{m}\rangle$:

$$
\begin{aligned}
\langle K||\hat{m}||\widetilde{K+1}\rangle &= \langle K+1||\hat{m}||\tilde{K}\rangle \\
&= \frac{1}{\sqrt{2}}\left\{\sqrt{\frac{(K+2)(2K+1)}{2K+3}}c_1\tilde{c}_1\left(j_K\,\tilde{j}_{K+1} - j_{K+1}\,\tilde{j}_{K+2}\right)\right. \\
&\quad - 2\sqrt{\frac{K+1}{(2K+1)(2K+3)}}c_1\tilde{c}_2\left(j_K\,\tilde{j}_{K+1} + j_{K+1}\,\tilde{j}_K\right) \\
&\quad + \sqrt{\frac{K(2K+3)}{2K+1}}c_2\tilde{c}_2\left(j_{K-1}\,\tilde{j}_K - j_K\,\tilde{j}_{K+1}\right)
\end{aligned}
\tag{F.6}
$$

$$
\begin{aligned}
&\langle K||\hat{m}||\tilde{K}\rangle \\
&= \frac{i}{2(2K+1)}\left(A\tilde{c}_1 j_K\tilde{j}_{K+1} + \tilde{A}c_1 j_{K+1}\tilde{j}_K\right. \\
&\quad + \left. B\tilde{c}_2 j_K\tilde{j}_{K-1} + \tilde{B}c_2 j_{K-1}\tilde{j}_K\right) ,
\end{aligned}
\tag{F.7}
$$

where

$$
A = 2c_1\sqrt{K(K+1)} - c_2, \quad B = 2c_2\sqrt{K(K+1)} + c_1 ,
$$

and the radial integral $\int_0^R r^3 dr$ is implicit (note the weight r^3 for magnetic moments). In the following formulae for magnetic moments, the same integral is always assumed.

Isovector magnetic moment

Zeroth order

We compute the quark contribution to the nucleon matrix element of

$$\bar{\psi}(\boldsymbol{x} \times \boldsymbol{\gamma})_3 \tau_3 \psi \,. \tag{F.8}$$

The expectation value in the rotating quark wave function to the leading order of Ω^0 is

$$\begin{aligned} m &= \int_0^R d^3x \, \langle \bar{\psi} | (\boldsymbol{x} \times \boldsymbol{\gamma})_3 \tau_3 | \psi \rangle \\ &= \int_0^R d^3x \, \langle h | A^\dagger (\boldsymbol{x} \times \boldsymbol{\gamma})_3 \tau_3 A | h \rangle \\ &= \sum_a \int_0^R d^3x \, \langle h | (\boldsymbol{x} \times \boldsymbol{\gamma})_3 \tau_a | h \rangle \, D_{3a} \,, \end{aligned} \tag{F.9}$$

where $D_{3a} = \frac{1}{2} \operatorname{tr}(\tau_a A^\dagger \tau_3 A)$. The matrix element of Eq. (F.9) is written in the form of the mode sum:

$$\begin{aligned} m = -\frac{1}{2} \sum_n \operatorname{sgn}(E_n) \sum_a \int_0^R d^3x \, \bar{\psi}_n (\boldsymbol{x} \times \boldsymbol{\gamma})_3 \tau_a \psi_n \, e^{-\eta^2 |E_n|^2} \, D_{3a} \\ + \, \theta(\pi/2 - F) \int_0^R d^3x \bar{\psi}_{0+} (\boldsymbol{x} \times \boldsymbol{\gamma})_3 \tau_a \psi_{0+} \, D_{3a} \,, \end{aligned} \tag{F.10}$$

where we have adopted the gaussian regulator for the vacuum part, and the valence contribution is separately shown in the second term. Writing the quark state ψ_n as ψ_{KM}, the Clebsch-Gordan coefficient is factored out by the Wigner-Eckart theorem:

$$\bar{\psi}_{KM} (\boldsymbol{x} \times \boldsymbol{\gamma})_3 \tau_a \psi_{KM} \sim (K\,M\,T\,a | K\,M) \langle \bar{\psi}_{KM} || (\boldsymbol{x} \times \boldsymbol{\gamma}) \tau || \psi_{KM} \rangle \tag{F.11}$$

where possible values of T, the rank of the operator $(\boldsymbol{x} \times \boldsymbol{\gamma})\tau$, are $T = 0$, 1, 2, and non-vanishing contributions are only when $a = 3$.

Now we write the operator in the spherical representation

$$\begin{aligned} (\boldsymbol{x} \times \boldsymbol{\gamma})_3 \tau_3 &= -i\sqrt{2} \, [\boldsymbol{x} \, \boldsymbol{\gamma}]^1_0 \, \tau^1_0 \\ &= -i\sqrt{2} \left(-\sqrt{\frac{1}{3}} \left[[\boldsymbol{x} \, \boldsymbol{\gamma}]^1 \, \tau^1 \right]^0_0 + \sqrt{\frac{2}{3}} \left[[\boldsymbol{x} \, \boldsymbol{\gamma}]^1 \, \tau^1 \right]^2_0 \right) . \end{aligned} \tag{F.12}$$

Using the identity for the CG coefficients,

$$\sum_M (K\,M\,2\,0\,|K\,M) = 0\,, \tag{F.13}$$

the rank two part in Eq. (F.12) disappears. Thus we get m_0 for the vacuum part

$$\begin{aligned}
m_0 &= -\frac{1}{2}\sum_{KM} \mathrm{sgn}(E_K) i\sqrt{\frac{2}{3}} (KM| \left[[\boldsymbol{x}\,\gamma]^0\, \tau^1 \right]^0_0 |KM\rangle e^{-\eta^2|E_K|^2} D_{33} \\
&= -i\sqrt{\frac{1}{6}}\sum_{KM} \mathrm{sgn}(E_K) \frac{1}{\sqrt{2K+1}} (K\,M\,0\,0|K\,M) \\
&\qquad \times \langle K|| \left[[\boldsymbol{x}\,\gamma]^0\, \tau^1 \right]^0 ||K\rangle e^{-\eta^2|E_K|^2} D_{33} \\
&= -i\sqrt{\frac{1}{6}}\sum_{K} \mathrm{sgn}(E_K) \sqrt{2K+1} \langle K|| \left[[\boldsymbol{x}\,\gamma]^0\, \tau^1 \right]^0 ||K\rangle \\
&\qquad \times e^{-\eta^2|E_K|^2} D_{33}\,,
\end{aligned} \tag{F.14}$$

where we have used the relation

$$\sum_M (K\,M\,0\,0|K\,M) = \sum_M 1 = 2K+1 \tag{F.15}$$

Computation of the reduced matrix element $\langle K|| \left[[\boldsymbol{x}\,\gamma]^0\, \tau^1 \right]^0 ||K\rangle$ is straightforward but lengthy. We briefly sketch how it is performed. Since the quark state $|K\rangle$ is constructed by the coupling $K = [L\ [\sigma\ \tau]^G]^K$, the operator $\left[[\boldsymbol{x}\,\gamma]^0\, \tau^1 \right]^0$ is recoupled,

$$\langle [L\ [\sigma\ \tau]^G]^K || \left[\boldsymbol{x}\, [\gamma\ \tau]^1 \right]^0 || [L\ [\sigma\ \tau]^G]^K \rangle\,. \tag{F.16}$$

It is now straightforward to apply the formulae for coupled tensor operators. After a lengthy calculation, we get a rather compact expression:

$$\begin{aligned}
m_0 &= \frac{2}{3}\sum_K \sqrt{2K+1}(c_1 P + c_2 Q) e^{-\eta^2|E_K|^2} \\
&\times\ j_K(E_K r) \Big(c_1\sqrt{K+1}\, j_{K+1}(E_K r) \\
&\qquad - c_2\sqrt{K}\, j_{K-1}(E_K r) \Big) D_{33}\,,
\end{aligned} \tag{F.17}$$

where c_1, c_2, P and Q are given in Eqs. (5.110) and (5.98).

First order

Let us start with the expression analogous to Eq. (E.4):

$$
\begin{aligned}
&\langle \int d^3x\, \bar{\psi} \left[[\boldsymbol{x}\, \boldsymbol{\alpha}]^1 \tau\right]^1 \psi \rangle_\Omega \\
= & -i\frac{\sigma_3 \tau_3}{9\Lambda_{\mathrm{tot}}} \sum_{K\geq 0} \left\{ \frac{\langle K || \left[[\boldsymbol{x}\, \boldsymbol{\alpha}]^1 \tau\right]^1 || \widetilde{K+1}\rangle \langle \widetilde{K+1} || \tau || K\rangle}{\tilde{E}_{K+1} - E_K} \right. \\
& + \frac{\langle K+1 || \left[[\boldsymbol{x}\, \boldsymbol{\alpha}]^1 \tau\right]^1 || \tilde{K}\rangle \langle \tilde{K} || \tau || K+1\rangle}{\tilde{E}_K - E_{K+1}} \\
& \left. - \frac{\langle K+1 || \left[[\boldsymbol{x}\, \boldsymbol{\alpha}]^1 \tau\right]^1 || \widetilde{K+1}\rangle \langle \widetilde{K+1} || \tau || K+1\rangle}{\tilde{E}_{K+1} - E_{K+1}} \right\}.
\end{aligned}
\tag{F.18}
$$

The computation of the reduced matrix elements is again lengthy, since first we need to recouple the operator as

$$
\left[[\boldsymbol{x}\, \boldsymbol{\alpha}]^1 \tau\right]^1 = \sum_{t=0,1,2} \sqrt{3}\,\hat{t} \left\{ \begin{array}{ccc} t & 1 & 1 \\ 1 & 1 & 1 \end{array} \right\} \left[\boldsymbol{x}\, [\boldsymbol{\alpha}\, \boldsymbol{\tau}]^t\right]^1 . \tag{F.19}
$$

For each t, matrix elements are computed and they are summed up. Altogether, the result is

$$
\begin{aligned}
\langle K || \left[[\boldsymbol{x}\, \boldsymbol{\alpha}]^1 \tau\right]^1 || \widetilde{K+1}\rangle &= -\langle \widetilde{K+1} || \left[[\boldsymbol{x}\, \boldsymbol{\alpha}]^1 \tau\right]^1 || K\rangle \\
&= \frac{1}{2}\left[\sqrt{\frac{(K+2)(2K+1)}{2K+3}} c_1 \tilde{c}_1 \left(j_K \tilde{j}_{K+1} - j_{K+1}\tilde{j}_{K+2}\right) \right. \\
&+ \sqrt{\frac{K(2K+3)}{2K+1}} c_2 \tilde{c}_2 \left(j_K \tilde{j}_{K+1} - j_{K-1}\tilde{j}_K\right) \\
&\left. + 4\sqrt{\frac{(K+1)^3}{(2K+1)(2K+3)}} c_1 \tilde{c}_2 \left(j_K \tilde{j}_{K+1} + j_{K+1}\tilde{j}_K\right) \right],
\end{aligned}
\tag{F.20}
$$

$$
\begin{aligned}
\langle K || \left[[\boldsymbol{x}\, \boldsymbol{\alpha}]^1 \tau\right]^1 || K\rangle &= -\frac{i}{2}\sqrt{2K+1} \\
\times \quad & \left(c_2 \tilde{c}_1 j_K \tilde{j}_{K+1} + c_1 \tilde{c}_2 j_K j_{K-1} - \tilde{c}_2 c_1 \tilde{j}_K j_{K+1} - \tilde{c}_1 c_2 \tilde{j}_K j_{K-1}\right)
\end{aligned}
\tag{F.21}
$$

G Equations of Motions for Vector Mesons

In this appendix, the hedgehog energies, moment of inertia and the resulting equations of motion for the hedgehog and the excitation fields for the mesons in the chiral bag model with vector mesons are summarized.

Substituting (6.214) in the lagrangian, we obtain the meson energy for the MYM scheme

$$\begin{aligned}
E_{\rm mes}^{\rm MYM} &= 4\pi \int_R^\infty r^2 dr \\
&\times \left\{ \frac{1}{g^2}\left[\left(AB - \frac{G'}{r}\right)^2 + \frac{1}{2}\left(A^2 + \frac{G(G-2)}{r^2}\right)^2 \right.\right. \\
&+ \left.\left(A' + \frac{A-B}{r} + \frac{BG}{r}\right)^2 \right] \\
&+ \frac{m_v^2}{g^2}\left[\left(\frac{G}{r}\right)^2 + A^2 + \frac{1}{2}B^2 \right] \\
&+ f_\pi^2 \left[2\left(A\cos F - \frac{G-1}{r}\sin F\right)^2 + (B+F')^2 \right] \\
&- \left. \frac{1}{2}\omega'^2 - \frac{1}{2}m_v^2\omega^2 + \frac{N_C}{2} g\,\omega \frac{\sin^2 F}{2\pi^2 r^2} F' \right\} , \qquad \text{(G.1)}
\end{aligned}$$

and that for the HLS scheme

$$\begin{aligned}
E_{\rm mes}^{\rm HLS} &= 4\pi \int_R^\infty r^2 dr \\
&\times \left\{ \frac{1}{g^2}\left[\left(\tilde{A}\tilde{B} - \frac{\tilde{G}'}{r}\right)^2 + \frac{1}{2}\left(\tilde{A}^2 + \frac{\tilde{G}(\tilde{G}-2)}{r^2}\right)^2 \right.\right. \\
&+ \left.\left(\tilde{A}' + \frac{\tilde{A}-\tilde{B}}{r} + \frac{\tilde{B}\tilde{G}}{r}\right)^2 \right] \\
&+ \frac{m_v^2}{g^2}\left[\left(\frac{\tilde{G}}{r} - \frac{1-\cos F}{r}\right)^2 + \left(\tilde{A} - \frac{\sin F}{r}\right)^2 + \frac{1}{2}\left(\tilde{B} - F'\right)^2 \right] \\
&+ f_\pi^2 \left(2\tilde{A}^2 + \tilde{B}^2 \right)
\end{aligned}$$

$$- \quad \frac{1}{2}\omega'^2 - \frac{1}{2}m_v^2\omega^2 + \frac{N_C}{2}g\,\omega\frac{\sin^2 F}{2\pi^2 r^2}F'\Bigg\} \,. \tag{G.2}$$

The corresponding equations of motion are for the MYM scheme

$$\begin{aligned}
F'' &= -\frac{2}{r}(F'+B) - B' + \sin 2F\left[\left(\frac{G-1}{r}\right)^2 - A^2\right] \\
&- 2A\frac{G-1}{r}\cos 2F - \frac{1}{4}N_C g\,\omega'\frac{\sin^2 F}{2\pi^2 f_\pi^2 r^2}\,, &\text{(G.3)}\\
G'' &= (AB'+BA')r + B(2A-B) \\
&+ \left(A^2 + \frac{G(G-2)}{r^2}\right)(G-1) + B(A'r+BG) \\
&+ m_v^2 G - m_v^2(rA\cos F - (G-1)\sin F)\sin F\,, &\text{(G.4)}\\
A'' &= -\frac{2}{r}A' + \frac{B'}{r} - \frac{B'}{r}G - \frac{2}{r}BG' \\
&+ A(A^2+B^2) + A\frac{G(G-2)}{r^2} \\
&+ m_v^2 A + m_v^2\left(A\cos F - \frac{G-1}{r}\sin F\right)\cos F\,, &\text{(G.5)}\\
0 &= \left(AB - \frac{G'}{r}\right)A\left(A' + \frac{A-B}{r} + \frac{BG}{r}\right)\frac{G-1}{r} \\
&+ \frac{1}{2}m_v^2 B + \frac{1}{2}m_v^2(B+F^2)\,, &\text{(G.6)}\\
\omega'' &= -\frac{2}{r}\omega' + m_v^2\omega - \frac{1}{2}N_C g\frac{\sin^2 F}{2\pi^2 r^2}F'\,, &\text{(G.7)}
\end{aligned}$$

These are solved under the boundary conditions at $r = R$:

$$\begin{aligned}
2f_\pi^2(B+F') + \frac{N_C\,g}{4\pi^2}\frac{\sin^2 F}{2\pi^2 r^2}F' &= A_q(F)\,, &\text{(G.8)}\\
G' - RAB &= 0\,, &\text{(G.9)}\\
\omega' &= 0\,, &\text{(G.10)}\\
RA' + A - B + GB &= 0\,, &\text{(G.11)}\\
2B + F' &= 0\,, &\text{(G.12)}
\end{aligned}$$

and those at $r \to \infty$

$$F(r) \;\to\; \frac{1}{r^2}\,, \tag{G.13}$$

$$\begin{aligned} G(r),\ \omega(r) &\to 0\,, &\text{(G.14)}\\ A(r),\ B(r) &\to \frac{1}{r^3}\,. &\text{(G.15)}\end{aligned}$$

These differential equations have been efficiently solved using the computer code COLSYS [U. Ascher and Russell 1981]. Eq. (G.6) does not contain second derivative terms and is a constraint equation for B. Numerically, the constraint equation is solved by adding a small kinetic term $\epsilon B''$ with a small parameter ϵ. The boundary condition (G.8) is the continuity equation of the axial flux at the bag surface; the right hand side is the quark axial flux $\hat{x}_a \hat{x}_i \langle \bar{\psi}\gamma^i \gamma_5 \frac{1}{2}\tau^a \rangle$, and the left hand side is the corresponding one of the meson sector as verified by using Eq. (G.12).

Similarly, the equations of motion in the HLS scheme is given as

$$\begin{aligned}
F'' &= -\frac{2}{r}\left(F' - \tilde{B}\right) + \tilde{B}' - \frac{2}{r^2}((\tilde{G}-1)\sin F + \tilde{A}\cos F)\\
&- \frac{1}{4}N_C g\,\omega' \frac{\sin^2 F}{2\pi^2 f_\pi^2 r^2}\,, &\text{(G.16)}\\
\tilde{G}'' &= (\tilde{A}\tilde{B}' + \tilde{B}\tilde{A}')r + \tilde{A}\tilde{B} + \left(\tilde{A}^2 + \frac{\tilde{G}(\tilde{G}-2)}{r^2}\right)(\tilde{G}-1)\\
&+ \tilde{B}(\tilde{A}'r - \tilde{B} + \tilde{B}\tilde{G}) + m_v^2(\tilde{G} - 1\cos F)\,, &\text{(G.17)}\\
\tilde{A}'' &= -\frac{2}{r}\tilde{A}' + \frac{\tilde{B}'}{r} - \frac{\tilde{B}'}{r}G - \frac{2}{r}\tilde{B}\tilde{G}' + \tilde{A}(\tilde{A}^2 + \tilde{B}^2)\\
&+ \tilde{A}\frac{\tilde{G}(\tilde{G}-2)}{r^2} + m_v^2\left(\tilde{A} - \frac{\sin F}{r}\right) + m_v^2\tilde{A}\,, &\text{(G.18)}\\
0 &= \left(\tilde{A}\tilde{B} - \frac{\tilde{G}'}{r}\right)\tilde{A}\left(\tilde{A}' + \frac{\tilde{A}-\tilde{B}}{r} + \frac{\tilde{B}\tilde{G}}{r}\right)\frac{\tilde{G}-1}{r}\\
&+ \frac{1}{2}m_v^2(\tilde{B} - F') + \frac{1}{2}m_v^2\tilde{B}\,, &\text{(G.19)}\\
\omega'' &= -\frac{2}{r}\omega' + m_v^2\omega - \frac{1}{2}N_C g\frac{\sin^2 F}{2\pi^2 r^2}F'\,. &\text{(G.20)}
\end{aligned}$$

The boundary conditions are

$$\begin{aligned}
f_\pi^2(F' - \tilde{B}) + \frac{N_C\, g}{4\pi^2}\frac{\sin^2 F}{2\pi^2 r^2}F' &= A_q(F)\,, &\text{(G.21)}\\
\tilde{G}' - R\tilde{A}\tilde{B} &= 0\,, &\text{(G.22)}\\
\omega' &= 0\,, &\text{(G.23)}
\end{aligned}$$

$$R\tilde{A}' + \tilde{A} - \tilde{B} + \tilde{G}\tilde{B} = 0\,, \tag{G.24}$$

$$\tilde{B} - F' = 0\,. \tag{G.25}$$

at $r = R$ and Eqs. (G.13) - (G.15) at $r \to \infty$.

The excitation fields $u_1(r)$, $u_2(r)$, $\varphi(r)$ and $\Omega_A(r)$ are determined by minimizing the classical rotational energy $\frac{1}{2}\Omega^2\Lambda_{\rm mes}$, or equivalently minimizing the moment of inertia $\Lambda_{\rm mes}$. The moment of inertia $\Lambda_{\rm mes}$ takes the form in the MYM scheme as

$$\begin{aligned}
\Lambda_{\rm mes}^{\rm MYM} &= 4\pi \int_R^\infty r^2 dr \\
&\times \left\{ \frac{1}{6g^2} \left[\frac{v_2'}{2} + \left(v_1\Gamma - A\varphi - \frac{v_2}{r} \right)^2 \right.\right. \\
&+ (v_1' - B\varphi)^2 + \left(v_2\Gamma - \frac{v_1}{r} \right)^2 \\
&+ \left.(Av_1 - \varphi\Gamma)^2 + (Bv_1 + \varphi')^2 + \left(Av_2 + \frac{\varphi}{r} \right)^2 \right] \\
&+ \frac{f_\pi^2}{3}(\varphi\cos F - v_1 \sin F)^2 \\
&+ \left. \frac{m_v^2}{6g^2}[(v_1+2)^2 + \frac{1}{2}(v_2+2)^2 + \varphi^2] + \lambda_\omega \right\}\,,
\end{aligned} \tag{G.26}$$

with

$$\begin{aligned}
\lambda_\omega &= -\frac{N_C}{6} g\,\omega_A \frac{\sin^2 F}{\pi^2 r} F' \\
&- \frac{1}{6}\left(\omega_A'^2 + \frac{3}{r^2}\omega_A^2 + \frac{2}{r}\omega_A\omega_A' + m_v^2\omega_A^2 \right)\,.
\end{aligned} \tag{G.27}$$

The corresponding expression for the HLS scheme is

$$\begin{aligned}
\Lambda_{\rm mes}^{\rm HLS} &= 4\pi \int_R^\infty r^2 dr \\
&\times \left\{ \frac{1}{6g^2} \left[\frac{\tilde{v}_2'}{2} + \left(\tilde{v}_1\tilde{\Gamma} - \tilde{A}\tilde{\varphi} - \frac{\tilde{v}_2}{r} \right)^2 \right.\right. \\
&+ (\tilde{v}_1' - \tilde{B}\tilde{\varphi})^2 + \left(\tilde{v}_2\tilde{\Gamma} - \frac{\tilde{v}_1}{r} \right)^2 \\
&+ \left.(\tilde{A}\tilde{v}_1 - \tilde{\varphi}\tilde{\Gamma})^2 + (\tilde{B}\tilde{v}_1 + \tilde{\varphi}')^2 + \left(\tilde{A}\tilde{v}_2 + \frac{\tilde{\varphi}}{r} \right)^2 \right] + \frac{f_\pi^2}{3}\tilde{\varphi}^2
\end{aligned}$$

$$\begin{aligned}
&+ \quad \frac{m_v^2}{6g^2}[(\tilde{v}_1 \cos F - \tilde{\varphi} \sin F + 2)^2 + \frac{1}{2}(\tilde{v}_2 + 2)^2 \\
&+ \quad (\tilde{v}_1 \sin F + \tilde{\varphi} \cos F)^2] + \lambda_\omega\} \,, \qquad \text{(G.28)}
\end{aligned}$$

with the same λ_ω as in Eq. (G.27). In Eqs. (G.26) and (G.28), we have introduced the notations

$$v_1 = u_1 - 2\,, \quad v_2 = u_2 - 2\,, \quad \Gamma = \frac{G-1}{r}\,. \qquad \text{(G.29)}$$

Equations of motion for the excitation fields are for the MYM scheme

$$\begin{aligned}
u_1'' &= -\frac{2}{r}u_1' + \frac{2}{r}\varphi B + 2\varphi' B + \varphi B' \\
&+ m_v^2(u_1 \sin^2 F - 2\sin^2 F - \varphi \sin F \cos F) + m_v^2 u_1 \\
&+ \left(\frac{G}{r}\right)^2 (u_1 - 2) + 2\frac{G-1}{r^2}(u_2 - u_1) \\
&+ (A^2 + B^2)(u_1 - 2)\,, \qquad \text{(G.30)} \\
u_2'' &= -\frac{2}{r}u_2' + m_v^2 u_2 + 2\left(\frac{G}{r}\right)^2 (u_2 - 2) \\
&+ \frac{4}{r^2}(u_2 - u_1)(1 - G) + \frac{4}{r}\varphi A + 2u_2 A^2 - 4A^2\,, \qquad \text{(G.31)} \\
\varphi'' &= -\frac{2}{r}\varphi' + \frac{4}{r}B - \frac{2}{r}Bu_1 + 2B' - B'u_1 - 2Bu_1' \\
&+ m_v^2\varphi + m_v^2(\varphi \cos^2 F + 2\sin F \cos F - u_1 \sin F \cos F) \\
&+ (A^2 + B^2)\varphi + \frac{2}{r^2}\varphi + \left(\frac{G}{r}\right)^2 \varphi \\
&- \frac{4}{r}A + \frac{2}{r}Au_2 - \frac{2}{r^2}\varphi G\,, \qquad \text{(G.32)} \\
\omega_A'' &= -\frac{2}{r}\omega_A' + \frac{2}{r}\omega_A m_v^2 \omega_A + N_C g \frac{\sin^2 F}{2\pi^2 r}F'\,. \qquad \text{(G.33)}
\end{aligned}$$

The corresponding equations for the HLS scheme are

$$\begin{aligned}
\tilde{u}_1'' &= -\frac{2}{r}\tilde{u}_1' + \frac{2}{r}\tilde{\varphi}\tilde{B} + 2\tilde{\varphi}'\tilde{B} + \tilde{\varphi}\tilde{B}' + m_v^2\left(\tilde{u}_1 - 4\sin^2\frac{F}{2}\right) \\
&+ \left(\frac{\tilde{G}}{r}\right)^2 (\tilde{u}_1 - 2) + 2\frac{\tilde{G}-1}{r^2}(\tilde{u}_2 - \tilde{u}_1) \\
&+ (\tilde{A}^2 + \tilde{B}^2)(\tilde{u}_1 - 2)\,, \qquad \text{(G.34)}
\end{aligned}$$

$$\begin{aligned}
\tilde{u}_2'' &= -\frac{2}{r}\tilde{u}_2' + m_v^2\tilde{u}_2 + 2\left(\frac{\tilde{G}}{r}\right)^2(\tilde{u}_2 - 2) \\
&+ \frac{4}{r^2}(\tilde{u}_2 - \tilde{u}_1)(1 - \tilde{G}) + \frac{4}{r}\tilde{\varphi}\tilde{A} + 2\tilde{u}_2\tilde{A}^2 - 4\tilde{A}^2\,, \qquad \text{(G.35)} \\
\tilde{\varphi}'' &= -\frac{2}{r}\tilde{\varphi}' + \frac{4}{r}\tilde{B} - \frac{2}{r}\tilde{B}\tilde{u}_1 + 2\tilde{B}' - \tilde{B}'\tilde{u}_1 - 2\tilde{B}\tilde{u}_1' \\
&+ m_v^2(\tilde{\varphi} - 2\sin F) + m_v^2\tilde{\varphi} \\
&+ (\tilde{A}^2 + \tilde{B}^2)\varphi + \frac{2}{r^2}\tilde{\varphi} + \left(\frac{\tilde{G}}{r}\right)^2\varphi \\
&- \frac{4}{r}\tilde{A} + \frac{2}{r}\tilde{A}\tilde{u}_2 - \frac{2}{r^2}\tilde{\varphi}\tilde{G}\,, \qquad \text{(G.36)} \\
\omega_A'' &= -\frac{2}{r}\omega_A' + \frac{2}{r}\omega_A m_v^2\omega_A + N_C g\frac{\sin^2 F}{2\pi^2 r}F'\,. \qquad \text{(G.37)}
\end{aligned}$$

The boundary conditions take the same form for both schemes:

$$R\omega_A' + \omega_A = 0\,, \qquad \text{(G.38)}$$

$$u_1' - B\varphi = 0\,, \qquad \text{(G.39)}$$

$$u_2' = 0\,, \qquad \text{(G.40)}$$

$$\varphi' + B(u_1 - 2) = 0\,, \qquad \text{(G.41)}$$

at $r = R$ and

$$u_1\,,\ u_2\,,\ \varphi \text{ and } \omega_A \to 0 \qquad \text{(G.42)}$$

at $r \to \infty$.

Finally, we outline how the isovector charge is related to the moment of inertia. We demonstrate this in the MYM scheme. The isovector charge in the meson sector is written as

$$Q_V^{mes} = \frac{4\pi}{\Lambda_{\text{tot}}}\int_R^\infty r^2 dr\left[\frac{m_v^2}{6g^2}\left(u_1 + \frac{1}{2}u_2\right) - \frac{N_C g}{24\pi^2 r}\omega_A F'\sin^2 F\right]\,. \qquad \text{(G.43)}$$

In the moment of inertia of Eq. (G.26), after partial integration, the following substitutions are made:

$$\begin{aligned}
\Phi'^2 &\to -\Phi\left(\Phi'' + \frac{2}{r}\Phi'\right)\,, \quad \Phi = u_1\,,\ u_2\,,\ \phi\,, \\
\phi'^2 &\to -\phi\phi''\,, \quad \phi = \frac{\omega_A}{r}
\end{aligned}$$

With the help of the equations of motion (G.30) – (G.33) together with the boundary conditions (G.38) – (G.41), all the derivatives in the moment of inertia, which originate from the kinetic terms ($1/g^2$ dependent terms) in Eq. (G.26), are eliminated. After some algebra, one finds the expression of the isovector charge as Eq. (G.43).

H Computation in the Spherical Basis

This appendix collects some formulae for the computation of various matrix elements in the spherical basis.

Spherical harmonics

Scalar:

$$Y_{lm}(\hat{x}) = Y_{lm}(\theta\phi) \tag{H.1}$$

Special cases:

$$Y_{l0}(\theta\ \phi = 0) = \sqrt{\frac{2l+1}{4\pi}} P_l(\cos\theta) \tag{H.2}$$

$$Y_{1\mu}(\hat{x}) = \sqrt{\frac{3}{4\pi}} \hat{x}_\mu \tag{H.3}$$

$$\hat{x}_{\pm 1} = \mp\frac{1}{\sqrt{2}}(\hat{x}_x \pm i\hat{x}_y), \quad \hat{x}_0 = \hat{x}_z$$

Spinor:

$$\mathcal{Y}^l_{jm} = \sum_\mu (l\mu\ 1/2 m{-}\mu|jm) Y_{l\mu}(\hat{x}) \chi_{m-\mu}, \tag{H.4}$$

$$\chi_{+1/2} = \begin{pmatrix} 1 \\ 0 \end{pmatrix}, \quad \chi_{-1/2} = \begin{pmatrix} 0 \\ 1 \end{pmatrix}$$

Vector:

$$\boldsymbol{Y}^l_{jm} = \sum_\mu (l\mu\ 1 m{-}\mu|jm) Y_{l\mu}(\hat{x}) \boldsymbol{\epsilon}_{m-\mu}, \tag{H.5}$$

$$\boldsymbol{\epsilon}_{\pm 1} = \mp\frac{1}{\sqrt{2}}(\boldsymbol{\epsilon}_x \pm i\boldsymbol{\epsilon}_y), \quad \boldsymbol{\epsilon}_0 = \boldsymbol{\epsilon}_z,$$

$$\boldsymbol{\epsilon}_x = (1,0,0), \quad \boldsymbol{\epsilon}_y = (0,1,0), \quad \boldsymbol{\epsilon}_z = (0,0,1),$$

Complex conjugation:

$$Y^*_{lm}(\theta\phi) = Y^*_{lm}(\theta\ -\phi) = (-1)^m Y_{l,-m}(\theta\phi), \tag{H.6}$$

$$\boldsymbol{Y}^{l\,*}_{jm}(\theta\phi) = (-1)^{j+l+m+1} \boldsymbol{Y}^l_{j,-m}(\theta\phi). \tag{H.7}$$

Multipole expansion

$$e^{i\boldsymbol{k}\cdot\boldsymbol{x}} = 4\pi\sum_{lm} i^l j_l(kr)Y^*_{lm}(\hat{k})Y_{lm}(\hat{x}) \tag{H.8}$$

$$\begin{aligned} e^{i\boldsymbol{k}\cdot\boldsymbol{x}}\boldsymbol{\epsilon}_\lambda &= 4\pi\sum_{jlm} i^l j_l(kr)\left(\boldsymbol{\epsilon}_\lambda\cdot\boldsymbol{Y}^{l\,*}_{jm}\right)\boldsymbol{Y}^l_{jm}(\hat{x}) \\ &= -\sqrt{2\pi}\sum_{l=1}^{\infty} i^l\sqrt{2l+1}\left(\lambda\boldsymbol{A}^{(M)}_{l\lambda}+i\boldsymbol{A}^{(E)}_{l\lambda}\right) \quad (\boldsymbol{k}\parallel\boldsymbol{z}) \end{aligned} \tag{H.9}$$

Here the magnetic (M) and electric (E) multipole fields are given by

$$\boldsymbol{A}^{(M)}_{lm}(\boldsymbol{x}) = j_l(kr)\boldsymbol{Y}^l_{lm}(\hat{x}), \qquad \boldsymbol{Y}^l_{lm} = \frac{1}{\sqrt{l(l+1)}}\boldsymbol{L}Y_{lm}(\hat{x}), \tag{H.10}$$

$$\begin{aligned} \boldsymbol{A}^{(E)}_{lm}(\boldsymbol{x}) &= -\frac{i}{k}\boldsymbol{\nabla}\times\boldsymbol{A}^{(M)}_{lm}(\boldsymbol{x}), \\ &= \left(\frac{j+1}{2j+1}\right)^{1/2} j_{j-1}\boldsymbol{Y}^{j-1}_{jm} - \left(\frac{j}{2j+1}\right)^{1/2} j_{j+1}\boldsymbol{Y}^{j+1}_{jm} \end{aligned} \tag{H.11}$$

Useful relations

Gradient:

$$\begin{aligned} &\boldsymbol{\nabla}(f(r)Y_{lm}(\hat{x})) \\ &= -\left(\frac{l+1}{2l+1}\right)^{1/2}\left(\frac{df}{dr}-\frac{l}{r}f\right)\boldsymbol{Y}^{l+1}_{lm}(\hat{x}) \\ &\quad + \left(\frac{l}{2l+1}\right)^{1/2}\left(\frac{df}{dr}+\frac{l+1}{r}f\right)\boldsymbol{Y}^{l-1}_{lm}(\hat{x}) \\ &\to k\left[\left(\frac{l+1}{2l+1}\right)^{1/2} j_{l+1}(kr)\boldsymbol{Y}^{l+1}_{lm}(\hat{x}) \right. \\ &\quad \left. + \left(\frac{l}{2l+1}\right)^{1/2} j_{l-1}(kr)\boldsymbol{Y}^{l-1}_{lm}(\hat{x})\right] \quad (f=j_l(kr)) \end{aligned} \tag{H.12}$$

$$\boldsymbol{\sigma}\cdot\boldsymbol{\nabla}\left(f(kr)\mathcal{Y}^l_{jm}(\hat{\boldsymbol{x}})\right) = \begin{cases} -\left(\dfrac{df}{dr}-\dfrac{l}{r}f\right)\mathcal{Y}^{l+1}_{jm}(\hat{\boldsymbol{x}}) & j=l+\frac{1}{2} \\ -\left(\dfrac{df}{dr}+\dfrac{l+1}{r}f\right)\mathcal{Y}^{l-1}_{jm}(\hat{\boldsymbol{x}}) & j=l-\frac{1}{2} \end{cases} \tag{H.13}$$

Electromagnetic field:

$$\nabla \cdot [f(r) \boldsymbol{L} Y_{lm}(\hat{x})] = 0 \tag{H.14}$$

$$\nabla \cdot \boldsymbol{A}_{jm}^{(M)} = \nabla \cdot \boldsymbol{A}_{jm}^{(E)} = 0 \tag{H.15}$$

$$\nabla \times \boldsymbol{A}_{jm}^{(E)} = \frac{-i}{k} \nabla \times (\nabla \times \boldsymbol{A}_{jm}^{(M)}) = -ik \boldsymbol{A}_{jm}^{(M)} \tag{H.16}$$

Products:

$$\boldsymbol{\sigma} \cdot \hat{\boldsymbol{x}}\, \mathcal{Y}_{jm}^{l}(\hat{\boldsymbol{x}}) = -\mathcal{Y}_{jm}^{l+\epsilon}(\hat{\boldsymbol{x}}) \quad \text{for } j = l + \frac{\epsilon}{2} \tag{H.17}$$

$$\boldsymbol{\sigma} \cdot \boldsymbol{Y}_{jm}^{l} = [Y_l \sigma]_m^j \tag{H.18}$$

$$\boldsymbol{x} \cdot \boldsymbol{Y}_{jm}^{j} = 0 \tag{H.19}$$

$$\boldsymbol{x} \cdot \boldsymbol{Y}_{jm}^{j+1} = -\left(\frac{j+1}{2j+1}\right)^{1/2} Y_{jm} \tag{H.20}$$

$$\boldsymbol{x} \cdot \boldsymbol{Y}_{jm}^{j-1} = \left(\frac{j}{2j+1}\right)^{1/2} Y_{jm} \tag{H.21}$$

Electromagnetic field:

$$\boldsymbol{\sigma} \cdot \boldsymbol{A}_{jm}^{(M)} = \boldsymbol{\sigma} \cdot \boldsymbol{Y}_{jm}^{j} j_j = [Y_j \sigma]_m^j j_j \tag{H.22}$$

$$\begin{aligned} \boldsymbol{\sigma} \cdot \boldsymbol{A}_{jm}^{(E)} &= \left(\frac{j+1}{2j+1}\right)^{1/2} j_{j-1} [Y_{j-1}, \sigma]_m^j \\ &- \left(\frac{j}{2j+1}\right)^{1/2} j_{j+1} [Y_{j+1}, \sigma]_m^j \end{aligned} \tag{H.23}$$

$$\boldsymbol{x} \cdot \boldsymbol{A}_{jm}^{(E)} = \sqrt{j(j+1)} \frac{j_j(kr)}{kr} Y_{jm} \tag{H.24}$$

$$\boldsymbol{x} \cdot \boldsymbol{A}_{jm}^{(M)} = 0 \tag{H.25}$$

Wigner-Eckart theorem

$$\begin{aligned} &\langle JM | \mathcal{O}_m^k | J'M' \rangle \\ &= (-1)^{J-M} \begin{pmatrix} J & k & J' \\ -M & m & M' \end{pmatrix} \langle J || \mathcal{O}^k || J' \rangle \\ &= \frac{1}{\sqrt{2J+1}} (J'M'\, km | JM) \langle J || \mathcal{O}^k || J' \rangle \end{aligned} \tag{H.26}$$

Some special reduced matrix elements:

$$\langle J||1||J'\rangle = \sqrt{2J+1}\delta_{JJ'} \tag{H.27}$$

$$\langle J||\boldsymbol{J}||J'\rangle = \sqrt{J(J+1)(2J+1)} \tag{H.28}$$

$$\begin{aligned}\langle l||Y_k||l'\rangle &= (-1)^l \frac{\hat{l}\hat{k}\hat{l}'}{\sqrt{4\pi}} \begin{pmatrix} l & k & l' \\ 0 & 0 & 0 \end{pmatrix} \\ &= (-1)^{2l-k} \frac{\hat{l}\hat{k}}{\sqrt{4\pi}} (l0\ k0|l'0)\end{aligned} \tag{H.29}$$

Here we introduced $\hat{L} = \sqrt{2L+1}$. Special cases are

$$\langle 1/2||1||1/2\rangle = \sqrt{2} \tag{H.30}$$

$$\langle 1/2||\sigma||1/2\rangle = \sqrt{6} \tag{H.31}$$

$$\langle l||\hat{x}||l'\rangle = \begin{cases} -\sqrt{l+1} & l' = l+1 \\ -\sqrt{l} & l' = l-1 \end{cases} \tag{H.32}$$

$$\langle l+1||\boldsymbol{\nabla}||l\rangle = \sqrt{l+1}\ \langle f_{l+1}(r)|\frac{\partial}{\partial r} - \frac{l}{r}|f_l(r)\rangle \tag{H.33}$$

$$\langle l-1||\boldsymbol{\nabla}||l\rangle = -\sqrt{l}\ \langle f_l(r)|\frac{\partial}{\partial r} + \frac{l+1}{r}|f_{l+1}(r)\rangle \tag{H.34}$$

Tensor products

$$\begin{aligned}\mathcal{O}^k_m &\equiv [\mathcal{O}_{k_1}\ \mathcal{O}_{k_2}]^k_m \\ &= \sum_{m_1,m_2} (k_1m_1\ k_2m_2|km)\ \mathcal{O}_{k_1m_1}\ \mathcal{O}_{k_2m_2}\end{aligned} \tag{H.35}$$

Examples:

$$\boldsymbol{A}\cdot\boldsymbol{B} = -\sqrt{3}\,[AB]^0_0 \tag{H.36}$$

$$(\boldsymbol{A}\times\boldsymbol{B})_\mu = -\sqrt{2}i\,[A\ B]^1_\mu \tag{H.37}$$

Recoupling:

$$\begin{aligned}\left[[l_1s_1]^{j_1}\ [l_2s_2]^{j_2}\right]^J &= \sum_{LS} \hat{j}_1\hat{j}_2\hat{L}\hat{S} \left\{ \begin{matrix} l_1 & s_1 & j_1 \\ l_2 & s_2 & j_2 \\ L & S & J \end{matrix} \right\} \\ &\times \left[[l_1l_2]^L\ [s_1s_2]^S\right]^J\end{aligned} \tag{H.38}$$

This equation can be understood to be a definition of the 9-j symbol.

Reduced matrix element:

$$\begin{aligned}&\langle [j_1j_2]^J||[k_1k_2]^k||[j_1'j_2']^{J'}\rangle\\ &= \hat{J}\hat{k}\hat{J}'\left\{\begin{array}{ccc} j_1 & j_2 & J\\ j_1' & j_2' & J'\\ k_1 & k_2 & k\end{array}\right\}\langle j_1||k_1||j_1'\rangle\,\langle j_2||k_2||j_2'\rangle \qquad \text{(H.39)}\end{aligned}$$

If either k_1 or k_2 is zero, the 9-j symbol reduces to a 6-j symbol through

$$\left\{\begin{array}{ccc} j_1 & j_2 & J\\ j_1' & j_2' & J\\ k & k & 0\end{array}\right\} = (-1)^{j_2+J+j_1'+k}\frac{1}{\hat{J}\hat{k}}\left\{\begin{array}{ccc} j_1 & j_2 & J\\ j_2' & j_1' & k\end{array}\right\} \qquad \text{(H.40)}$$

Clebsch-Gordan coefficients

Definition:

$$|[j_1j_2]_{jm}\rangle = \sum_{m_1m_2}(j_1m_1\; j_2m_2|jm)|j_1m_1\rangle|j_2m_2\rangle \qquad \text{(H.41)}$$

$$|j_1m_1\rangle|j_2m_2\rangle = \sum_j (j_1m_1\; j_2m_2|jm)|[j_1j_2]^j_m\rangle \qquad \text{(H.42)}$$

Integral of the D-functions:

$$\begin{aligned}&(j_1m_1\; j_2m_2|jm)(j_1m_1'\; j_2m_2'|jm')\\ &= \frac{2j+1}{8\pi^2}\int d[\omega]\; D^{*\,j}_{mm'}(\omega)\; D^{j_1}_{m_1m_1'}(\omega)\; D^{j_2}_{m_2m_2'}(\omega) \qquad \text{(H.43)}\end{aligned}$$

Symmetry properties:

$$(j_1m_1\; j_2m_2|jm) = (-1)^{j_1+j_2-j}(j_2m_2\; j_1m_1|jm) \qquad \text{(H.44)}$$

$$(j_1\,-m_1\; j_2\,-m_2|j\;\,-m) = (-1)^{j_1+j_2+j+2m}(j_1m_1\; j_2m_2|jm) \qquad \text{(H.45)}$$

$$\begin{aligned}(j_1m_1\; j_2m_2|j_3m_3) &= (-1)^{j_1+j_2-j_3}(j_2m_2\; j_1m_1|j_3m_3)\\ &= (-1)^{j_1-m_1}\frac{\hat{j}_3}{\hat{j}_2}(j_1m_1\; j_3\,-m_3|j_2\,-m_2) \qquad \text{(H.46)}\\ &= (-1)^{j_1-m_1}\frac{\hat{j}_3}{\hat{j}_2}(j_3m_3\; j_1\,-m_1|j_2m_2) \qquad \text{(H.47)}\\ &= (-1)^{j_2+m_2}\frac{\hat{j}_3}{\hat{j}_1}(j_3\,-m_3\; j_2m_2|j_1\,-m_1) \qquad \text{(H.48)}\end{aligned}$$

$$= (-1)^{j_2+m_2}\frac{\hat{j_3}}{\hat{j_2}}(j_2\, -m_2\ j_3m_3|j_1m_1) \qquad \text{(H.49)}$$

3-j symbols

Definition:

$$\begin{pmatrix} j_1 & j_2 & j_3 \\ m_1 & m_2 & m_3 \end{pmatrix} = \frac{(-1)^{j_1-j_2-m_3}}{\sqrt{2j_3+1}}(j_1m_1\ j_2m_2|j_3\, -m_3) \qquad \text{(H.50)}$$

Symmetry properties:

$$\begin{aligned} \begin{pmatrix} j_1 & j_2 & j_3 \\ m_1 & m_2 & m_3 \end{pmatrix} &= \begin{pmatrix} j_2 & j_3 & j_1 \\ m_2 & m_3 & m_1 \end{pmatrix} \\ &= (-1)^{j_1+j_2+j_3}\begin{pmatrix} j_1 & j_3 & j_2 \\ m_1 & m_3 & m_2 \end{pmatrix} \qquad \text{(H.51)} \\ &= (-1)^{-j_1-j_2-j_3}\begin{pmatrix} j_1 & j_2 & j_3 \\ -m_1 & -m_2 & -m_3 \end{pmatrix} \end{aligned}$$

Special values of CG coefficients and 3-j symbols

$$(jm\ j'\, -m'|00) = \frac{(-1)^{j-m}}{\sqrt{2j+1}}\delta_{jj'}\delta_{mm'} \qquad \text{(H.52)}$$

$$\begin{pmatrix} j & 1 & j \\ -m & 0 & m \end{pmatrix} = (-1)^{j-m}\frac{m}{\sqrt{j(2j+1)(j+1)}} \qquad \text{(H.53)}$$

$$\begin{aligned} \begin{pmatrix} j & 2 & j \\ -m & 0 & m \end{pmatrix} &= (-1)^{j-m} \\ &\times \frac{3m^2 - j(j+1)}{\sqrt{(2j-1)j(2j+1)(j+1)(2j+3)}} \qquad \text{(H.54)} \end{aligned}$$

6-j symbols

Definition by the 9-j symbol:

$$\begin{Bmatrix} j_1 & j_2 & j \\ j_2' & j_1' & k \end{Bmatrix} = (-1)^{j_2+j+j_1'+k}\,\hat{j}\hat{k}\begin{Bmatrix} j_1 & j_2 & j \\ j_1' & j_2' & j \\ k & k & 0 \end{Bmatrix} \qquad \text{(H.55)}$$

Table H.1 $(lm_l\ 1/2m_s|jm)$

$j \setminus m_s$	$1/2$	$-1/2$
$l+1/2$	$\sqrt{\dfrac{l+1/2+m}{2l+l}}$	$\sqrt{\dfrac{l+1/2-m}{2l+l}}$
$l-1/2$	$-\sqrt{\dfrac{l+1/2-m}{2l+l}}$	$\sqrt{\dfrac{l+1/2+m}{2l+l}}$

Table H.2 $(lm_l\ 1m_s|jm)$

m_s / j	1	0	-1
$l+1$	$\sqrt{\dfrac{(l+m)(l+m+1)}{(2l+1)(2l+2)}}$	$\sqrt{\dfrac{(l-m+1)(l+m+1)}{(2l+1)(2l+2)}}$	$\sqrt{\dfrac{(l-m)(l-m+1)}{(2l+1)(2l+2)}}$
l	$-\sqrt{\dfrac{(l+m)(l-m+1)}{2l(l+1)}}$	$\dfrac{m}{\sqrt{l(l+1)}}$	$\sqrt{\dfrac{(l-m)(l+m+1)}{2l(l+1)}}$
$l-1$	$\sqrt{\dfrac{(l-m)(l-m+1)}{2l(2l+1)}}$	$-\sqrt{\dfrac{(l-m)(l+m)}{l(2l+1)}}$	$\sqrt{\dfrac{(l+m+1)(l+m)}{2l(2l+1)}}$

Symmetries:

$$\begin{Bmatrix} j_1 & j_2 & j_3 \\ l_1 & l_2 & l_3 \end{Bmatrix} = \begin{Bmatrix} j_2 & j_3 & j_1 \\ l_2 & l_3 & l_1 \end{Bmatrix} = \begin{Bmatrix} j_2 & j_1 & j_3 \\ l_2 & l_1 & l_3 \end{Bmatrix} \tag{H.56}$$

$$= \ = \begin{Bmatrix} l_1 & l_2 & j_3 \\ j_1 & j_2 & l_3 \end{Bmatrix} \tag{H.57}$$

Special values:

$$\begin{aligned} j_{\text{tot}} &= j_1 + j_2 + j_3 \,, \\ j_{1+2-3} &= j_1 + j_2 - j_3 \,, \\ j_{2+3-1} &= j_2 + j_3 - j_1 \,, \\ j_{3+1-2} &= j_3 + j_1 - j_2 \,, \end{aligned}$$

$$\begin{Bmatrix} j_1 & j_2 & j_3 \\ j_2 & j_1 & 0 \end{Bmatrix} = (-1)^{j_{\text{tot}}} \frac{1}{\hat{j}_1 \hat{j}_2} \tag{H.58}$$

$$\begin{aligned} \begin{Bmatrix} j_1 & j_2 & j_3 \\ j_2+\frac{1}{2} & j_1+\frac{1}{2} & \frac{1}{2} \end{Bmatrix} &= (-1)^{j_{\text{tot}}+1} \\ &\times \left[\frac{(j_{\text{tot}}+2)(j_{1+2-3}+1)}{(2j_1+1)(2j_1+2)(2j_2+1)(2j_2+2)}\right]^{1/2} \end{aligned} \tag{H.59}$$

$$\begin{aligned} \begin{Bmatrix} j_1 & j_2 & j_3 \\ j_2+\frac{1}{2} & j_1-\frac{1}{2} & \frac{1}{2} \end{Bmatrix} &= (-1)^{j_{\text{tot}}} \\ &\times \left[\frac{j_{3+1-2}(j_{2+3-1}+1)}{2j_1(2j_1+1)(2j_2+1)(2j_2+2)}\right]^{1/2} \end{aligned} \tag{H.60}$$

$$\begin{aligned} \begin{Bmatrix} j_1 & j_2 & j_3 \\ j_2-1 & j_1-1 & 1 \end{Bmatrix} &= (-1)^{j_{\text{tot}}} \\ &\times \left[\frac{j_{\text{tot}}(j_{\text{tot}}+1)j_{1+2-3}(j_{1+2-3}-1)}{(2j_1-1)2j_1(2j_1+1)(2j_2-1)2j_2(2j_2+1)}\right]^{1/2} \end{aligned} \tag{H.61}$$

$$\begin{aligned} \begin{Bmatrix} j_1 & j_2 & j_3 \\ j_2-1 & j_1 & 1 \end{Bmatrix} &= (-1)^{j_{\text{tot}}} \\ &\times \left[\frac{2(j_{\text{tot}}+1)j_{1+2-3}j_{2+3-1}(j_{3+1-2}+1)}{2j_1(2j_1+1)(2j_1+2)(2j_2-1)2j_2(2j_2+1)}\right]^{1/2} \end{aligned} \tag{H.62}$$

$$\begin{aligned} \begin{Bmatrix} j_1 & j_2 & j_3 \\ j_2+1 & j_1-1 & 1 \end{Bmatrix} &= (-1)^{j_{\text{tot}}} \\ &\times \left[\frac{(j_{3+1-2}-1)j_{3+1-2}(j_{2+3-1}+1)(j_{2+3-1}+2)}{(2j_2+1)(2j_2+2)(2j_2+3)(2j_1-1)2j_1(2j_1+1)}\right]^{1/2} \end{aligned} \tag{H.63}$$

$$\begin{aligned} \begin{Bmatrix} j_1 & j_2 & j_3 \\ j_2 & j_1 & 1 \end{Bmatrix} &= \frac{1}{2}(-1)^{j_{\text{tot}}+1} \\ &\times \frac{j_1(j_1+1)+j_2(j_2+1)-j_3(j_3+1)}{[j_1(j_1+1)(2j_1+1)j_2(j_2+1)(2j_2+1)]^{1/2}} \end{aligned} \tag{H.64}$$

Index

Bibliography

K. Abe et al., Phys. Rev. Lett., **75**:25, 1995.
E. M. Abramowitz and I. Stegun, *Handbook of Mathematical Functions*, Dover publ. inc., New York, 1970.
M. Ademollo and R. Gatto, Phys. Rev. Lett., **13**:264, 1964.
B. Adeva et al., Phys. Lett., **B302**:533, 1993.
G. S. Adkins and C. R. Nappi, Phys. Lett., **B137**:251, 1984.
G. S. Adkins, C. R. Nappi, and E. Witten, Nucl. Phys., **B228**:552, 1983.
S. L. Adler and R. Dashen, *Current Algebra and Applications to Particle Physic*, Benjamin, 1968.
G. Altarelli and G. G. Ross, Phys. Lett., **B212**:391, 1988.
R. D. Amado, R. Bijker, and M. Oka, Phys. Rev. Lett., **58**:654, 1987.
P. Anthony et al., Phys. Rev. Lett., **17**:959, 1993.
A. Arima and F. Iachello, Advances in Nuclear Physics, **13**:139, 1984.
M. Arima, K. Shimizu, and K. Yazaki, Nucl. Phys., **A543**:613, 1992.
J. Ashman et al., Phys. Lett., **B206**:364, 1988.
J. Ashman et al., Nucl. Phys., **B328**:1, 1989.
A. P. Balachandran, V. P. Nair, S. G. Rajeev, and A. Stern, Phys. Rev., **D27**:1153, 1983.
M. Bando, T. Kugo, S. Uehara, K. Yamawaki, and T. Yanagida, Phys. Rev. Lett., **54**:1215, 1985.
M. Bando, T. Kugo, and K. Yamawaki, Prog. Theor. Phys., **73**:1541, 1985.
M. Bando, T. Kugo, and K. Yamawaki, Phys. Rept., **164**:217, 1988.
C. M. Bender and P. Hays, Phys. Rev., **D14**:2622, 1976.
R. K. Bhaduri, B. K. Jennings, and J. C. Waddington, Phys. Rev., **D29**:2051, 1984.
R. K. Bhaduri, *Models of the Nucleon*, Addison-Wesley, 1988.
J. Bjorken and S. Drell, *Relativistic Quantum Fields*, McGraw-Hill, New York, 1965.
P. Bogoliubov, Ann. Inst. Henri Poincaré, **8**:163, 1967.

A. Bohr and B. Mottelson, *Nuclear Structure*, Benjamin Inc., 1975.

K. C. Bowler, P. J. Corvi, A. J. G. Hey, and P. D. Jarvis, Phys. Rev. Lett., **45**:97, 1980.

E. Braaten, S. Townsend, and L. Carson, Phys. Lett., **B235**:147, 1990.

J. D. Breit and C. Nappi, Phys. Rev. Lett., **53**:889, 1984.

G. E. Brown and M. Rho, Phys. Lett., **B82**:177, 1979.

G. E. Brown and M. Rho, Comments Nucl. Part. Phys., **18**:1, 1988.

G. E. Brown and W. Weise, Phys. Rept., **22**:279, 1975.

G. E. Brown, M. Rho, and V. Vento, Phys. Lett., **B84**:383, 1979.

G. Brown, J. Durso, and M. Johnson, Nucl. Phys., **A397**:447, 1983.

G. E. Brown, A. D. Jackson, M. Rho, and V. Vento, Phys. Lett., **B140**:285, 1984.

G. E. Brown, M. Rho, and W. Weise, Nucl. Phys., **A454**:669, 1986.

E. by M. Rho and D. Wilkinson, *Mesons in Nuclei*, North-Holland, 1979.

J. C. G. Callan, S. Coleman, J. Wess, and B. Zumino, Phys. Rev., **177**:2247, 1969.

N. Cabbibo, Phys. Rev. Lett., **10**:531, 1963.

S. Capstick, Phys. Rev., **D46**:2864, 1992.

T. Cheng and L. Li, *Gauge Theory of Elementary Particles*, Oxford, 1984.

H.-Y. Cheng, Int. J. Mod. Phys., **A11**:5109, 1996.

A. Chodos and C. B. Thorn, Phys. Rev., **D12**:2733, 1975.

A. Chodos, R. Jaffe, K. Johnson, C. Thorn, and V. Wesskopf, Phys. Rev., **D9**:3471, 1974.

A. Chodos, R. L. Jaffe, K. Johnson, and C. B. Thorn, Phys. Rev., **D10**:2599, 1974.

C. V. Christov, K. Goeke, and P. V. Pobylitsa, Phys. Rev., **C52**:425, 1995.

F. E. Close and Z.-P. Li, Phys. Rev., **D42**:2194, 1990.

T. D. Cohen and M. K. Banerjee, Phys. Lett., **B230**:129, 1989.

S. Coleman, J. Wess, and B. Zumino, Phys. Rev., **177**:2239, 1969.

S. Coleman, Phys. Rev., **D11**:2088, 1975.

S. Coleman, *Aspects of Symmetry*, Cambridge University Press, Cambridge, 1985.

R. E. Cutkosky, R. E. Hendrick, and R. L. Kelly, Phys. Rev. Lett., **37**:645, 1976.

R. Dashen and A. V. Manohar, Phys. Lett., **B315**:438, 1993.

R. Dashen and A. V. Manohar, Phys. Lett., **B315**:425, 1993.

R. Dashen, E. Jenkins, and A. V. Manohar, Phys. Rev., **D49**:4713, 1994.

R. F. Dashen, E. Jenkins, and A. V. Manohar, Phys. Rev., **D51**:3697, 1995.

T. DeGrand, R. L. Jaffe, K. Johnson, and J. Kiskis, Phys. Rev., **D12**:2060, 1975.

C. DeTar and T. Kunihiro, Phys. Rev., **D39**:2805, 1989.

A. Dhar, R. Shankar, and S. R. Wadia, Phys. Rev., **D31**:3256, 1985.

D. Diakonov, V. Y. Petrov, and P. V. Pobylitsa, Nucl. Phys., **B306**:809, 1988.

D. Diakonov, V. Y. Petrov, and M. Praszalowicz, Nucl. Phys., **B323**:53, 1989.

Y. B. Dong, K. Shimizu, A. Faessler, and A. J. Buchmann, Phys. Rev., **C60**:035203, 1999.

N. Dorey and M. P. Mattis, Phys. Rev., **D52**:2891, 1995.

N. Dorey, J. Hughes, and M. P. Mattis, Phys. Rev. Lett., **73**:1211, 1994.
H. Dreiner, J. Ellis, and R. A. Flores, Phys. Lett., **B221**:167, 1989.
D. Ebert and H. Reinhardt, Nucl. Phys., **B271**:188, 1986.
G. Ecker, Prog. Part. Nucl. Phys., **35**:1, 1995.
T. Eguchi, Phys. Rev., p. 2755, 1976.
Z. F. Ezawa and T. Yanagida, Phys. Rev., **D33**:247, 1986.
R. P. Feynman, M. Kislinger, and F. Ravndal, Phys. Rev., **D3**:2706, 1971.
K. Fujimura, T. Kobayashi, and M. Namiki, Prog. Theor. Phys., **43**:73, 1970.
Y. Fujiwara, Prog. Theor. Phys., **90**:105, 1993.
Y. Fukuda et al., Phys. Rev. Lett., **81**:1562, 1998.
Y. Fukuda et al., Phys. Rev. Lett., **82**:2430, 1999.
S. Gasiorowicz and D. A. Geffen, Rev. Mod. Phys., **41**:531, 1969.
J. Gasser and H. Leutwyler, Phys. Rept., **87**:77, 1982.
J. Gasser and H. Leutwyler, Ann. Phys., **158**:142, 1984.
J. Gasser and H. Leutwyler, Nucl. Phys., **B250**:465, 1985.
M. Gell-Mann and M. Levy, Nuovo Cim., **16**:705, 1960.
M. Gell-Mann and Y. Ne'eman, *The Eightfold Way*, Benjamin, New York, 1964.
M. Gell-Mann, Phys. Rev., **92**:833, 1953.
M. Gell-Mann, Phys. Rev., **125**:1067, 1962.
H. Georgi, *Lie Algebras in Particle Physics*, Benjamin, 1982.
J. L. Gervais and B. Sakita, Phys. Rev., **D30**:1795, 1984.
J. L. Gervais and B. Sakita, Phys. Rev. Lett., **52**:87, 1984.
S. L. Glashow, J. Iliopoulos, and L. Maiani, Phys. Rev., **D2**:1285, 1970.
L. Y. Glozman and D. O. Riska, Phys. Rept., **268**:263, 1996.
M. L. Goldberger and S. B. Treiman, Phys. Rev., **110**:1178, 1958.
M. L. Goldberger and K. M. Watson, *Collision Theory*, John Wiley & Sons, 1965.
J. Goldstone and R. L. Jaffe, Phys. Rev. Lett., **51**:1518, 1983.
J. Goldstone and F. Wilczek, Phys. Rev. Lett., **47**:986, 1981.
J. Goldstone, Proc. Roy. Soc. London, **A239**:267, 1957.
J. Goldstone, Nuovo Cim., **19**:154, 1961.
T. Goto, Prog. Theor. Phys., **46**:1560, 1971.
M. Gourdin and P. Salin, Nuovo Cim., **27**:193, 1963.
D. J. Gross and F. Wilczek, Phys. Rev. Lett., **30**:1343, 1973.
M. Y. Han, *Quarks and Gluons*, World Scientific, 1999.
T. H. Hansson and R. L. Jaffe, Phys. Rev., **D28**:882, 1983.
T. Hatsuda, Nucl. Phys., **A458**:583, 1986.
A. Hayashi and G. Holzwarth, Phys. Lett., **B140**:175, 1984.
A. Hayashi, G. Eckart, G. Holzwarth, and H. Walliser, Phys. Lett., **B147**:5, 1984.
A. J. G. Hey and R. L. Kelly, Phys. Rept., **96**:71, 1983.
P. W. Higgs, Phys. Rev. Lett., **12**:132, 1964.
P. W. Higgs, Phys. Rev., **145**:1156, 1966.
A. Hosaka and H. Toki, Phys. Lett., **B167**:153, 1986.
A. Hosaka and H. Toki, Z. Phys., **A330**:111, 1988.

A. Hosaka and H. Toki, Int. J. Mod. Phys., **A3**:1413, 1988.
A. Hosaka and H. Toki, Prog. Theor. Phys. Suppl., **109**:137, 1992.
A. Hosaka and H. Toki, Phys. Lett., **B322**:1, 1994.
A. Hosaka and H. Toki, Phys. Lett., **B343**:1, 1995.
A. Hosaka and H. Toki, Phys. Rept., **277**:65, 1996.
A. Hosaka and N. R. Walet, Phys. Lett., **B358**:184, 1995.
A. Hosaka and W. Weise, Phys. Lett., **B232**:442, 1989.
A. Hosaka, K. Kusaka, H. Takashita, and H. Toki, Prog. Theor. Phys., **76**:315, 1986.
A. Hosaka, H. Toki, and W. Weise, Z. Phys., **A332**:97, 1989.
A. Hosaka, H. Toki, and W. Weise, Nucl. Phys., **A506**:501, 1990.
A. Hosaka, H. Toki, and M. Takayama, Mod. Phys. Lett., **A13**:1699, 1998.
A. Hosaka, M. Takayama, and H. Toki, Nucl. Phys., **A678**:147, 2000.
A. Hosaka, Phys. Lett., **B207**:249, 1988.
A. Hosaka, Nucl. Phys., **A546**:493, 1992.
I. Hulthage, F. Myhrer, and Z. Xu, Nucl. Phys., **A364**:322, 1981.
D. R. Inglis, Phys. Rev., **96**:1059, 1954.
T. Inoue and T. Maskawa, Prog. Theor. Phys., **54**:1833, 1975.
N. Isgur and G. Karl, Phys. Lett., **B72**:109, 1977.
N. Isgur and G. Karl, Phys. Rev., **D18**:4187, 1978.
N. Isgur and G. Karl, Phys. Lett., **B74**:353, 1978.
N. Isgur and G. Karl, Phys. Rev., **D20**:1191, 1979.
N. Isgur and G. Karl, Phys. Rev., **D19**:2653, 1979.
S. Ishida, Prog. Theor. Phys., **46**:1570, 1971.
S. Ishida, Prog. Theor. Phys., **46**:1905, 1971.
C. Itzykson and J.-B. Zuber, *Quantum Field Theory*, MacGraw Hill, 1980.
R. Jackiw and C. Rebbi, Phys. Rev., **D13**:3398, 1976.
A. D. Jackson, D. E. Kahana, L. Vepstas, H. Verschelde, and E. Wuest, Nucl. Phys., **A462**:661, 1987.
W. D. Jackson, *Classical Electrodynamics*, John Wiley & Sons, 1974.
M. Jacob and G. C. Wick, Ann. Phys., **7**:404, 1959.
R. L. Jaffe, 1979, Lectures presented at Erice School of Subnuclear Physics, Erice, Sicily, Jul 31 - Aug 10, 1979.
E. Jenkins, Phys. Lett., **B315**:441, 1993.
E. Jenkins, Phys. Lett., **B315**:431, 1993.
E. G. J.F. Donoghue and B. Holstine, *Dynamics of the Standard Model*, Cambridge monographs on particle physics, 1992.
X. Ji, hep-ph//9510362, 1995.
D. Jido, T. Hatsuda, and T. Kunihiro, Phys. Rev. Lett., **84**:3252, 2000.
D. Jido, Y. Nemoto, M. Oka, and A. Hosaka, Nucl. Phys., **A671**:471, 2000.
R. Johnson, N. W. Park, J. Schechter, V. Soni, and H. Weigel, Phys. Rev., **D42**:2998, 1990.
D. E. Kahana and J. Milana, Nucl. Phys., **A468**:493, 1987.
S. Kahana and G. Ripka, Nucl. Phys., **A429**:462, 1984.

D. E. Kahana, A. D. Jackson, and G. Ripka, Nucl. Phys., **A459**:663, 1986.
A. Kamlah, Z. Phys., **216**:52, 1968.
K. Kawarabayashi and M. Suzuki, Phys. Rev. Lett., **16**:255, 1966.
O. Kaymakcalan, S. Rajeev, and J. Schechter, Phys. Rev., **D30**:594, 1984.
M. Kimura et al., Phys. Rev., **C62**:015206, 2000.
D. Klabucar and G. E. Brown, Nucl. Phys., **A454**:589, 1986.
M. Kobayashi and T. Maskawa, Prog. Theor. Phys., **49**:652, 1973.
R. Koniuk and N. Isgur, Phys. Rev., **D21**:1868, 1980.
O. Krehl, C. Hanhart, S. Krewald, and J. Speth, Phys. Rev., **C62**:025207, 2000.
N. Kroll and M. Ruderman, Phys. Rev., **93**:233, 1954.
T. Kugo, *Quantum Theory of Gauge Field (in Japanese)*, Baifukan, 1989.
K. S. Kumar and Y. Nogami, Phys. Rev., **D21**:1834, 1980.
K. Kusaka and H. Toki, Nucl. Phys., **A477**:596, 1988.
B. W. Lee, *Chiral Dynamics*, Academic Press, 1970.
N. Levinson, Danske Videnskab. Selskab, Mat.-fys. Medd., **25, No.9**, 1949.
E. H. Lifschitz and L. P. Pitaevskii, *Realtivistic Quantum Theory*, Pergamon Press, 1971.
M. A. Luty and J. March-Russell, Nucl. Phys., **B426**:71, 1994.
R. Machleidt, K. Holinde, and C. Elster, Phys. Rept., **149**:1, 1987.
S. Mandelstam, Phys. Rev., **D11**:3026, 1975.
A. V. Manohar, Nucl. Phys., **B248**:19, 1984.
N. S. Manton, Commun. Math. Phys., **111**:469, 1987.
M. P. Mattis and M. Karliner, Phys. Rev., **D31**:2833, 1985.
M. P. Mattis and M. E. Peskin, Phys. Rev., **D32**:58, 1985.
T. Meissner and K. Goeke, Nucl. Phys., **A524**:719, 1991.
U. G. Meissner, Phys. Rept., **161**:213, 1988.
U. Meyer, A. J. Buchmann, and A. Faessler, Phys. Lett., **B408**:19, 1997.
K. A. Milton, J. Lester L. DeRaad, and J. Schwinger, Ann. Phys., **115**:388, 1978.
K. A. Milton, Phys. Rev., **D22**:1441, 1980.
K. A. Milton, Phys. Rev., **D22**:1444, 1980.
K. A. Milton, Ann. Phys., **150**:432, 1983.
V. Miransky, *Dynmaical Symmetry Breaking*, World Scientific, 1995.
R. G. Moorhouse and N. H. Parsons, Nucl. Phys, p. 109, 1973.
R. G. Moorhouse, Electromagnetic interactions of hadrons, p. 83, 1978.
P. J. Mulders, Phys. Rev., **D30**:1073, 1984.
M. V. N. Murthy and R. K. Bhaduri, Phys. Rev. Lett., **54**:745, 1985.
M. V. N. Murthy, M. Dey, J. Dey, and R. K. Bhaduri, Phys. Rev., **D30**:152, 1984.
T. Muta, *Foundations of Qunatum Chromodynamics*, World Scientific, 1997.
S. Nadkarni, H. B. Nielsen, and I. Zahed, Nucl. Phys., **B253**:308, 1985.
Y. Nambu and G. Jona-Lasinio, Phys. Rev., **122**:345, 1961.
Y. Nambu and G. Jona-Lasinio, Phys. Rev., **124**:246, 1961.
Y. Nambu and D. Lurie, Phys. Rev., **125**:1429, 1962.
Y. Nambu, Phys. Rev. Lett., **4**:380, 1960.

Y. Nambu, in *Proceeding of Int. Conf. on Symmetries and Quark Models*, 1969.
Y. Nambu, Phys. Rev., **D10**:4262, 1974.
Y. Nambu, *Quarks:frontiers in elementary particle physics*, World Scientific, 1985.
B. M. K. Nefkens et al., Phys. Rev., **D18**:3911, 1978.
A. J. Niemi and G. W. Semenoff, Phys. Rept., **135**:99, 1986.
K. Nishijima and T. Nakano, Prog. Theor. Phys., **10**:581, 1953.
E. Noether, Nachr. Kgl. Geo. Wiss. Gottinger, **235**, 1918.
M. Oka, R. Bijker, A. Bulgac, and R. D. Amado, Phys. Rev., **C36**:1727, 1987.
S. Okubo, Prog. Theor. Phys., **27**:949, 1962.
J. A. Oller and E. Oset, Nucl. Phys., **A629**:739, 1998.
B.-Y. Park and M. Rho, Z. Phys., **A331**:151, 1988.
Particle Data Group, Rev. Mod. Phys., **48, Part II**:1, 1976.
Particle Data Group, Phys. Rev., **D54**:1, 1996.
Particle Data Group, Euro. Phys. J., **C15**:1, 2000.
A. Pich, Rept. Prog. Phys., **58**:563, 1995.
H. D. Politzer, Phys. Rev. Lett., **30**:1346, 1973.
H. D. Politzer, Phys. Rept., **14**:129, 1974.
R. Rajaraman, *Solitons and Instantons*, North-Holland, Amsterdam, 1982.
T. Regge, Nuovo Cim., **14**:951, 1959.
T. Regge, Nuovo Cim., **18**:4171, 1960.
M. Rho, G. E. Brown, and B. Y. Park, Phys. Rev., **C39**:1173, 1989.
M. Rho, Ann. Rev. Nucl. Part. Sci., **34**:531, 1984.
Riazuddin and Fayyazuddin, Phys. Rev., **147**:1071, 1966.
P. Ring and P. Shuck, *The Nuclear Many Body Problem*, Springer, New York, 1980.
M. E. Rose, *Elementary Theory of Angular Momentum*, Wiley, 1957.
J. Sakurai, *Currents and Mesons*, Chicago Univ. Press, Chicago, 1969.
A. Salam, 1968, Originally printed in *Svartholm: Elementary Particle Theory, Proceedings Of The Nobel Symposium Held 1968 At Lerum, Sweden*, Stockholm 1968, 367-377.
V. I. Sanyuk, Int. J. Mod. Phys., **A7**:1, 1992.
T. Sato and T. S. H. Lee, Phys. Rev., **C54**:2660, 1996.
J. Schechter and H. Weigel, Mod. Phys. Lett., **A10**:885, 1995.
J. Schechter and H. Weigel, Phys. Rev., **D51**:6296, 1995.
E. V. Shuryak, Phys. Rept., **61**:71, 1980.
T. H. R. Skyrme, Proc. Roy. Soc. Lond., **A247**:260, 1958.
T. H. R. Skyrme, Proc. Roy. Soc. Lond., **A262**:237, 1961.
T. H. R. Skyrme, Nucl. Phys., **31**:556, 1962.
E. C. G. Stueckelberg, Helv. Phys. Acta, **14**:588, 1941.
H. Suganuma, S. Sasaki, and H. Toki, Nucl. Phys., **B435**:207, 1995.
L. Susskind, Nuovo Cim., **69**:210, 1970.
J. T. deForest and J. D. Walecka, Adv. in Phys., **15**:1, 1966.
G. 't Hooft, Nucl. Phys., **B72**:461, 1974.

G. 't Hooft, Phys. Rept., **142**:357, 1986.
H. Takashita, H. Toki, and S. Yoro, Nucl. Phys., **A485**:589, 1988.
M. Takayama, H. Toki, and A. Hosaka, Prog. Theor. Phys., **101**:1271, 1999.
M. Takayama, A. Hosaka, and H. Toki, Nucl. Phys., **A663-664**:695, 2000.
R. Tegen, R. Brockmann, and W. Weise, Z. Phys., **A307**:339, 1982.
A. W. Thomas, Adv. Nucl. Phys., **13**:1, 1984.
H. Toki, J. Dey, and M. Dey, Phys. Lett., **B133**:20, 1983.
J. C. U. Ascher and R. Russell, ACM Trans. Math. Software, **7**:209, 1981.
D. A. Varshalovich, A. N. Moskalev, and V. K. Khersonskii, *Quantum Theory of Angular Momentum*, World Scientific, Singapore, 1988.
V. Vento, M. Rho, E. M. Nyman, J. H. Jun, and G. E. Brown, Nucl. Phys., **A345**:413, 1980.
V. Vento, Phys. Lett., **B107**:5, 1981.
L. Vepstas and A. D. Jackson, Phys. Rept., **187**:109, 1990.
L. Vepstas, A. D. Jackson, and A. S. Goldhaber, Phys. Lett., **B140**:280, 1984.
M. Wakamatsu and T. Watabe, Phys. Lett., **B312**:184, 1993.
M. Wakamatsu and W. Weise, Z. Phys., **A331**:173, 1988.
M. Wakamatsu and H. Yoshiki, Nucl. Phys., **A524**:561, 1991.
M. Wakamatsu, Phys. Lett., **B232**:251, 1989.
M. Wakamatsu, Phys. Lett., **B331**:425, 1994.
M. Wakamatsu, Phys. Lett., **B349**:204, 1995.
R. L. Walker, Phys. Rev., **182**:1729, 1969.
T. Watabe and H. Toki, Prog. Theor. Phys., **87**:651, 1992.
H. Weigel, R. Alkofer, and H. Reinhardt, Nucl. Phys., **A582**:484, 1995.
S. Weinberg, Phys. Rev. Lett., **19**:1264, 1967.
S. Weinberg, Phys. Rev. Lett., **18**:507, 1967.
S. Weinberg, Phys. Rev., **166**:1568, 1968.
S. Weinberg, *Quantum Theory of Fields*, Cambrige Univ. Press, Cambrige, 1985.
E. P. Wigner, *Group Theory and its Application to the Quantum Mechanics of Atomic Spectra*, Academic Press, 1959.
E. Witten, Nucl. Phys., **B160**:57, 1979.
E. Wuest, L. Vepstas, and A. D. Jackson, Phys. Lett., **B173**:217, 1986.
K. Yamawaki, Phys. Rev., **D35**:412, 1987.
S. I. Yoro and T. Tatsumi, Prog. Theor. Phys., **78**:1392, 1987.
I. Zahed and G. E. Brown, Phys. Rept., **142**:1, 1986.
I. Zahed, U.-G. Meissner, and U. Kaulfuss, Nucl. Phys., **A426**:525, 1984.
I. Zahed, A. Wirzba, and U. G. Meissner, Annals Phys., **165**:406, 1985.
G. Zweig, CERN Report, **8419/TH**:412, 1963.

www.ingramcontent.com/pod-product-compliance
Ingram Content Group UK Ltd.
Pitfield, Milton Keynes, MK11 3LW, UK
UKHW020401150926
8624IPUK00004B/274

9 789810 246297